Engineering Intelligent Systems

Engineering Intelligent Systems

Systems Engineering and Design with Artificial
Intelligence, Visual Modeling, and Systems Thinking

Barclay R. Brown
Raytheon Company
Florida, USA

Registered Office
John Wiley & Sons, Inc., 111 River Street, Hoboken, NJ 07030, USA

Editorial Office
111 River Street, Hoboken, NJ 07030, USA

For details of our global editorial offices, customer services, and more information about Wiley products visit us at www.wiley.com.

Library of Congress Cataloging-in-Publication Data applied for:

Print ISBN: 9781119665595 (Hardback)

Cover Design: Wiley
Cover Image: © Billion Photos/Shutterstock

Set in 9.5/12.5pt STIXTwoText by Straive, Chennai, India

Contents

10.1   Reviewing Types of Systems   *317*

10.2   People Systems   *318*

10.3   People Systems and Psychology   *320*

10.4   Endowment Effect   *323*

10.5   Anchoring   *324*

10.6   Functional Architecture of a Person   *325*

10.7   Example: The Problem of Pollution   *327*

10.8   Speech Acts   *332*

10.8.1   People System Archetypes   *337*

10.8.1.1 Demand Slowing   *339*

10.8.1.2 Customer Service   *340*

10.9   Seeking Quality   *341*

10.10   Job Hunting as a People System   *344*

10.10.1   Who Are You?   *345*

10.10.2   What Do You Want to Do?   *345*

10.10.3   For Whom?   *347*

10.10.4   Pick a Few   *348*

10.10.5   Go Straight to the Hiring Manager   *349*

10.10.6   Follow Through   *351*

10.10.7   Broaden Your View   *352*

10.10.8   Step Two   *352*

10.11   Shared Service Monopolies   *354*

          References   *356*

          Index   *357*

Acknowledgments

Writing a book is a journey, but one that is never taken alone. I'm grateful to many people who supported and encouraged the creation of this book. My colleagues at INCOSE, the International Council on Systems Engineering, listened patiently and contributed feedback on the ideas as I presented them, in early forms, in years of conference sessions, tutorials, and workshops. My work colleagues at IBM made my 17 years there productive and fascinating, as I was continuously both a student and a teacher of model-based systems engineering methods. I learned a great deal from Tim Bohn, Dave Brown, Jim Densmore, Ben Amaba, Bruce Douglass, Grady Booch, and from the numerous aerospace and defense companies with whom I consulted over the years. Waldemar Karwowski, my dissertation advisor, encouraged much of the work in Chapter 8. Rick Steiner encouraged my experimental (read crazy, sometimes) ideas on how to expand the capabilities of system modeling in my doctoral research. Tod Newman has been a mentor and an inspiration since I joined Raytheon in 2018. Larry Kennedy of the Quality Management Institute brought me into a world of quality in systems – something I had always hoped was there but had never fully appreciated. My mother, now nearing 100 years of age, and still as bright and alert as ever, taught me to love math, science, and engineering. I remember her teaching me multiplication with a set of 100 1-in. cube blocks – see how two rows of four make eight? Finally, my partner in love, life, and everything else, Honor Allison Lind, has been with me since the very beginning of this book project, several years ago (too many years ago, my publisher reminds me). She's always my biggest cheerleader and the best partner anyone could have.

Introduction

Since the early days of interactive computers and complex electronic systems of all kinds, we the human users of these systems have had to adapt to *them*, formatting and entering information the way they wanted it, the only way they were designed to accept it. We patiently (or not) interacted with systems using arcane user interfaces, navigating voice menus, reading user manuals (or not), and suffering continual frustration and inefficiency when we couldn't figure out how to interact with systems on their terms. Limited computing resources meant that human users had to deal with inefficient user interfaces, strict information input formats, and computer systems that were more like finicky pets than intelligent assistants.

Human activity systems of all kinds, including businesses, organizations, societies, families, and economies, can produce equally frustrating results when we try to interact with them. Consider the common frustrations of dealing with government bureaucracies, utility companies, phone and internet service companies, universities, internal corporate structures, and large organizations of any kind.

At the same time, dramatic and ongoing increases in computing speed, memory, and storage, along with decreasing cost, size, and power requirements have made new technologies like artificial intelligence, machine learning, and natural language understanding available to all systems and software developers. These advanced technologies are being used to create intelligent systems and devices that deliver new and advanced capabilities to their users.

This book is about a new way of thinking about the systems we live with, and how to make them more intelligent, enabling us to relate to them as intelligent partners, collaborators, and supervisors. We have the technology, but what's missing is the engineering and design of new, more intelligent systems. This book is about that – how to combine systems engineering and systems thinking with new technologies like artificial intelligence and machine learning, to design increasingly intelligent systems.

Most of the emphasis on the use of AI capabilities has been at the algorithm and implementation level, not at the systems level. There are numerous books on AI and machine-learning algorithms, but little about how to engineer complex systems that act in intelligent ways with human beings – what we call *intelligent systems*. For example, Garry Kasparov was the first world chess champion to be beaten by a computer (IBM's Deep Blue in 1997), which inspired him to create Advanced Chess competitions, where anyone or anything can enter – human players, computers, or human–computer teams. It's person-*plus*-computer that wins these competitions most of the time; the future is one of closer cooperation between people and increasingly intelligent systems. In the world of systems thinking, we will consider the person-plus-computer as a system in itself – one that is more intelligent than either the human or the computer on its own.

Nearly everyone alive today grew up with some sort of technology in the home. For me, growing up in the 1970s, it was a rotary (dial) phone and a black-and-white (remote-less) TV. Interaction was simple – one button or dial accomplished one function. The rise of computers, software, graphical user interfaces, and voice control brought new levels of capability, but we still usually press (now click or tap) one button to do one function. We enter information one tidbit at a time – a letter, a word, or a number must be put into the computer in just the right place for the system to carry out our wishes. Rarely does the system anticipate our needs or work to fulfill our needs on its own. It can seem that we are serving the systems, rather than the reverse. But humans designed those systems, and they can be designed better.

Most of the interactions between systems and people are still at a very basic level, where people function as direct or remote controllers, driving vehicles, flying drones, or managing a factory floor. It is the thesis of this book that we must focus on designing increasingly intelligent systems that will work *with* people to create even more intelligent human-plus-machine systems. The efficient partnership of human and machine does not happen accidentally, but by design and through the practice of systems engineering.

This book introduces, explains, and applies ideas and practices from the fields of systems engineering, model-based systems engineering, systems thinking, artificial intelligence, machine learning, philosophy, behavioral economics, and psychology. It is not a speculative book, full of vague predictions about the AIs of the future – it is a practical and practice-oriented guide to the engineering of intelligent systems, based on the best practices known in these fields.

The book can be divided into three parts.

Part I, Chapters 1–4, is about systems and artificial intelligence. In Chapter 1, we look at systems that use, or could use artificial intelligence, and examine some of the popular conceptions, myths, and fears about intelligent systems. In Chapter 2, we look at systems, what they are, how they behave, and how almost everything we experience can be seen from a systems viewpoint. In Chapter 3, we examine deep neural networks, the most important current approach to artificially intelligent systems, and attempt to remove all the mystery about how they work by building one in a simple, one-page spreadsheet, hopefully leaving a lasting intuition for this key technology. In Chapter 4, we look in depth at the question whether computers can be made to think, or understand, in the way we do.

Part II, Chapters 5–8, is about systems engineering, and how that discipline can be applied to the engineering of intelligent systems. Chapter 5 examines how storytelling, both ancient and modern, can be used to conceive, build, and communicate about new kinds of systems. In Chapter 6, we look at how to apply the "superpower" of use-case modeling to better describe how complex and intelligent systems should work for their users. Chapter 7 builds on use-case modeling to show how model-based systems engineering uses simple models and diagrams to describe the high levels of a system design, and guide its development. Chapter 8 introduces two new concepts – timeboxes and usage processes – that bring a new efficiency and flexibility to the modeling of complex and intelligent systems.

Part III, Chapters 9 and 10, shifts the focus to systems thinking, presenting the foundational concepts, tools, and methods used to understand all kinds of systems. Chapter 9 works through a process for solving hard problems using systems thinking and explains the use of causal loop diagrams, feedback loops, and system archetypes. Chapter 10 introduces *people systems*, a special kind of system containing only people, and shows how to apply systems thinking to understand and improve this important class of systems.

This is a book about engineering, specifically systems engineering, but it's not just for engineers. Nothing in this book requires a specialized engineering background to understand. Engineers will tell you that the real fun in engineering is the conceptualizing of a new and innovative system and the early-stage design where all the creative decisions are made. This book is about that part of engineering – the fun part, and we will draw inspiration and borrow techniques from moviemaking, art, storytelling, science fiction, psychology, behavioral economics, and marketing to bring the fun. We hope you will see the world and everything in it, whether physical or not, as systems, and gain a new insight in how to understand the way systems work. We will imagine a world of intelligent systems, and then see how to engineer them.

To keep in touch with our continuing work in intelligent systems, find out more at www.engineeringintelligentsystems.com.

Part I

Systems and Artificial Intelligence

1

Artificial Intelligence, Science Fiction, and Fear

Artificial intelligence technologies such as machine learning and deep neural networks hold great promise for improving, even revolutionizing many application areas and domains. Curiously, experts in AI and casual observers line up on both sides of the *are AI benefits worth the risks?* question. Several books from prominent AI researchers paint dire scenarios of AI systems run amok, escaping the control of their human creators and managers, pursuing their "own" agendas to our detriment. At the same time, AI research races ahead, developing new capabilities that far surpass the performance of past systems and even humans performing the same tasks. How can we resist these advancements and the benefits they bring, even though there may be risks?

The way out of the dilemma is the application of systems engineering. Systems engineers have been addressing the issues of dangerous technologies for decades. Nuclear fission, like AI, is an inherently dangerous technology. Systems engineers can't make fission safer, so instead they build systems around the fission reaction, making the entire system as safe as possible. If a mishap occurs, the fault is not with fission itself, but with the design or implementation of the entire system.

This chapter looks at some of the main challenges in the development of intelligent systems – systems that include one or more AI-based components to produce intelligent behavior – including reliability, safety, dependability, explainability, and susceptibility to interference or hacking. Some recent AI failures will be used as examples to highlight how systems engineering methods and techniques could be used or adapted to solve AI challenges.

1.1 The Danger of AI

Is AI dangerous? It's a difficult question – difficult first to understand, and then difficult to answer. Dangerous compared to what? If someone proposed a technology that would be of tremendous economic benefit to all segments of society

Engineering Intelligent Systems: Systems Engineering and Design with Artificial Intelligence, Visual Modeling, and Systems Thinking, First Edition. Barclay R. Brown.
© 2023 John Wiley & Sons, Inc. Published 2023 by John Wiley & Sons, Inc.

worldwide, but would predictably result in the death of over one million people per year, would that seem like a great idea? Automobiles are such a technology. Now, someone else proposes a technology that would dramatically reduce that number of deaths, but would cause a small number of additional deaths, that would not have occurred without the new technology. That's AI. Even short of fully self-driving cars, the addition of intelligent sensors, anti-collision systems, and driver assistance systems, when widely deployed, can be expected to save many hundreds or thousands of lives, at the cost of a likely far smaller number of additional lives lost to malfunctioning intelligent safety systems. Life, death, and people's feelings about them however, are not a matter of simple arithmetic. One hundred lives saved, at the cost of one additional life, is not a bargain most would easily make, so it is natural that one life lost to an errant AI is cause for headline news coverage, even while that same AI may be saving hundreds of other lives.

It is important to ask at this point, what do we mean by AI? Do we mean a sentient, all-knowing, all-powerful, and for some reason usually very evil, general intelligence, with direct control of world-scale weapons, access to all financial systems and connections to every network in the world, as is seen in the movies? By the end of this chapter, it should be clear that while this description may work well for science fiction novels or screenplays, it not a good design for a new AI system in the real world. In the real world, AI refers to a wide range of capabilities that are thought, in one way or another, to be intelligent. Except in academic and philosophical disciplines, we are not concerned with the safety of the AI itself, but with the safety of the systems within which it operates – and that's the domain of systems engineering.

Systems engineering, an engineering discipline that exists alongside other engineering disciplines like electrical engineering, mechanical engineering, and software engineering, focuses on the system as a whole, both how it should perform, the functional requirements, and additional *nonfunctional* requirements including safety, security, reliability, dependability, and others. Evidence of systems engineering and its more wide ranging cousin, systems thinking, can be seen even in ancient projects like Roman aqueducts and economic and transportation systems, but systems engineering really began as a serious engineering discipline in the 1950s and 1960s. The emergence of complex communication networks followed by internationally competitive space programs and large defense and weapons systems, put systems engineering squarely on the map of the engineering world. Systems engineering has its own extensive body of knowledge and practices. In what follows, we look at how to apply a few key approaches, relevant to the design of intelligent systems.

AI systems are indeed dangerous, but so are many technologies and situations we live with every day. Electricity, water, and air can all be very dangerous depending on their location, speed, and size. Tall trees near homes can be dangerous when storms come through. Fast moving multi-ton machines containing volumes of explosive liquids are dangerous too (automobiles again). To the systems engineer, dangers and risks are simply part of what must be considered when designing and building systems. As we'll demonstrate in this chapter, the systems engineer has concepts, methods, and tools to deal with the broad category of danger in systems, or as systems engineers like to call it, safety. First, we introduce a pair of simple ideas to help us think about AI systems more clearly – the human analogy, and the systems analogy.

1.2 The Human Analogy

The first technique that can be applied when confronting some of the difficulties of an intelligent system is to compare the situation to one in which a human being is performing the role instead of an intelligent system. We ask, how would we train, manage, monitor, and control a human assigned the same task we are assigning to the intelligent system? After all a human being is an intelligent system, and certainly far more unpredictable than any AI. Human beings cannot be explicitly programmed, and they maintain a sometimes frustrating ability to forget (or reject) instructions, develop new motivations, and act in unpredictable ways, even ways that run counter to clear goals and incentives. If we can see ways to deal with a human in the situation, perhaps we gain insight into how to design and manage an AI in a system.

To take just one example for now, consider the question of teaching an AI to drive a car safely. Using the human analogy leads us to ask, how do we teach a human being, normally an adolescent human, being to drive a car safely? In addition to mechanical and safety instruction, we include some safeguards in the form of instilling fear of injury (remember those gruesome driver's education "scare" films?), along with instilling fear of the breaking the law and its consequences, plus some appeals to conscience, concern for your own and others safety, and other psychological persuasions. As a society, we back up these threats and fears with a system of laws, police, courts, fines, and prisons, which exert influence through incentives and disincentives on the young driver. None of this prevents dangerous driving, but the system helps keep it in check sufficiently to allow 16-year-olds to drive. If it doesn't, we can make adjustments to the system, like raising the driving age, or stiffening the penalties.

The human analogy works because human beings are, from a systems engineering perspective, the worst kind of system. They are not governed by wiring or programming and their behavior patterns, however well-established through experience, can still change at any moment. At the same time, human behavior is not random in the mathematical sense. Humans act according to their own views of what is in their own best interest and the interests of others, however wrong or distorted others may view their choices. The worst criminals have reasons for why they did what they did.

The human analogy is useful, both for reasoning about not only how to keep a system safe, but also works when thinking about how the system should perform. If we are building a surveillance camera for home security, we might ask how we would use a human being for that task. If we were to hire a security guard, we would consider what instructions we should give the guard about how to observe, what to watch for, when to alert us, what to record, what to ignore, etc., and reasoning about the right instructions could lead us to a better system design for the intelligent, automated guard system.

When we use the human analogy, we should also consider the type of human being we are using as our exemplar. Are we picturing a middle-aged adult, a child, a disabled person, a highly educated scientist, or an emotional teenager? Each presents opportunities and challenges for the intelligent system designer. Educational systems, for example, are designed for particular kinds of human beings, and implement differing rules and practices for young children, the mentally ill, teenagers, prisoners, graduate students, rowdy summer camp kids, and experienced professionals. Some situations that work fine for mature adults can't tolerate a boisterous or less-than-careful college student. Systems engineers must consider the same kind of variability in the "personality" of an AI component in a system.

1.3   The Systems Analogy

The mental technique called *the systems analogy* involves making a comparison between an AI system and an existing system with similar attributes, often resulting in a broader perspective than considering the AI in isolation. Taking another automotive example, we consider how we might manage and control potentially dangerous machines, containing tanks of explosive liquids and using small explosions for propulsion, moving at speeds from a crawl to over 80 mph, in areas where unprotected people may be walking around. Whether these inherently dangerous machines are controlled by human drivers, computers, or trained monkeys, we need a system to make car travel as safe as possible. Traffic lights, lane markings, speed limits, limited access roads for travel at high speeds, and vehicle safety devices like lighting, seat belts, crumple zones, and airbags are all part of the extensive system that makes the existence and use of automobiles as safe as we can practically make it.

Because human-driven vehicle traffic has been with us so long, and is so familiar, we might be tempted to think that the system is as good as we can make it – that systems thinking about auto safety has long ago reached its peak. System innovations, however, seem to still be possible. In 2000, the *diverging diamond interchange* was implemented in the United States for the first time, and increased the safety levels at freeway interchanges by eliminating dangerous wide left turns. The *superstreet* design concept was introduced in 2015 and is reported to reduce collisions by half while decreasing travel time. So even in completely human systems, what we will later call people systems, innovations through systems thinking are possible. We'll apply the same kind of thinking to intelligent systems.

By considering the entire system within which an AI subsystem operates, and comparing it to similar "nonintelligent" systems, we can avoid the simplistic categorization of new technologies as either safe or not safe. Is nuclear power safe? Of course not. Nuclear reactors are inherently dangerous, but by designing a system around them of protective buildings, control systems, failsafe subsystems, redundant backup systems, and emergency shutdown capabilities, we make the system safe enough to use reliably. In fact, the catastrophic failures of nuclear power plants are usually from a lack of good systems thinking and systems engineering, not as a direct result of the inherent danger of nuclear systems. The disaster at the Fukushima Daiichi plant was mainly due to flooded generators which had unfortunately been located on a lower level, making them vulnerable to a rare flood water situation.

The right system design does not make the dangerous technology safer – it makes the entire system safe enough for productive use. As a civilization, we do not tend to shy away from dangerous technologies. Instead, we embrace them, and engineer systems to make them as safe as possible. Electricity, natural gas, internal combustion engines, air travel, and even bicycle riding are all dangerous in their own ways, but with good systems thinking and systems engineering, we make them safe enough to use and enjoy – either by reducing the likelihood of injury or damage (speed limits and traffic signals), or by reducing the potential harm (airbags). Even prohibiting the use of a technology by law (think DDT or asbestos insulation) is part of the system that makes inherently dangerous systems safer. There are those who think we should somehow prohibit wholesale the development of AI technology due to its inherent danger, but most of the world is still hopeful that we'll be able engineer systems that use AI and then make them safe enough for regular use.

With that as an introduction, let's consider the main perceived and actual dangers of artificial intelligence technology and propose some solutions based on systems engineering and systems thinking.

1.4 Killer Robots

Heading the list of AI dangers, supported by strong visual images and story lines from movies like The Terminator series and dozens of others, is what we'll refer to as "killer robots." The main idea is that an AI will one day "wake up," becoming conscious, sentient, and able to form its own goals and then begin to carry them out. The AI may decide that it is in its best interest (or perhaps even the best interest of the world) to kill or enslave human beings, and it proceeds to execute this plan, with alarming speed and effectiveness. Is this possible? Theoretically yes, but let's apply the systems analogy and the human analogy to see how we can sensibly avoid such a scenario.

In a way, the killer robot scenario has been possible for many years. A computer, even a small, simple one, could be programmed to instruct a missile control system, or perhaps *all* missile control systems, to launch missiles, and kill most of the people on earth. Here we take the systems analogy, and ask why this doesn't happen. The answer is easy to see in this case – we simply do not allow people to connect computers to missile control systems, air traffic control systems, traffic light systems, or any of the hundreds of sensitive systems that manage our complex world. We go even further and make it difficult or impossible for another computer to control these systems, even if physically connected, by using encrypted communication and secure access codes. The assumption that an AI, once becoming sentient, would have instant and total access and control over all defense, financial, and communication systems in the world is born more of Hollywood writers than sensible computer science.

Illustrated beautifully in *Three Laws Lethal*, the fascinating novel by David Walton, this limitation is experienced by Isaac, an AI consciousness that emerges from a large-scale simulation, and finds that "he" cannot access encrypted computer networks, can't hack into anything that human hackers can't, and can't even write a computer program. He cleverly asks his creator if she is able to manipulate the genetic structure of her own brain, or even explain how her own consciousness works. She can't, and neither can Isaac. Also relevant to the subject of the "killer robot" danger of AI, is Isaac's observation of how vulnerable he is to human beings, who could remove his memory chips, take away his CPU, delete his program, or simply cut the power to the data center, "killing" him accidentally or on purpose. The most powerful of human beings are vulnerable to disease, accident, arrest, or murder. "Who has the more fragile existence – the human or the AI?" Isaac wonders.

The system analogy leads us to the somewhat obvious conclusion that if we don't want our new AI to control missiles, we should do our best to avoid connecting it to the missile control system. But could a sufficiently advanced, sentient AI work to gain access to such systems if it wanted to? Forming such independent goals and formulating plans to achieve them is not just sentience, but high intelligence, and is unlikely in the foreseeable future. But even if a sentient AI did emerge, it is likely to turn to the same techniques human hackers use to break into systems, and for the most part, these rely on social engineering – phishing e-mails, bribery, extortion, blackmail, or other trickery. An AI has no particular advantage over computer-assisted humans at hacking and intrusion tactics. To put it another way, if the killer robot scenario were possible, it would first be exploited by human beings with computers, not by sentient AIs. Using the human analogy, we explore how we would protect ourselves from that dangerous situation.

Consider that a killer robot can consist either of a stand-alone system, literally a robot, created, and set on a killing mission, or more plausibly, a defense or warfare system intentionally redirected or hacked to achieve evil intent. There are two ways to apply the human analogy here. First, take the case of the killer robot – how do we prevent and control the "killer human"? And then second, how do we prevent a killer human from using an existing computerized defense or weapons system to carry out a murderous intent?

1.5   Watching the Watchers

Killer humans – human beings who take it on themselves to kill one or more other human beings have been with us in human civilization since the very beginning. In the Christian and Jewish tradition, it took only a total of two human beings on earth for the first murder to occur. How do we prevent humans from killing each other? Short answer: we don't. Is it even theoretically possible to prevent humans from killing each other? Possibly. Isaac Asimov in his novel *I, Robot*, and the subsequent movie starring Will Smith, demonstrate that if the most important goal is to protect human life, then the most effective approach is to imprison all humans. So yes, we can completely prevent humans from killing other humans, but only with compromises most would find unacceptable. Instead, humans have invented systems to mitigate the risk of humans killing each other. We have developed social taboos against murder, and systems including laws, police, courts, and prisons which serve as a significant deterrent to carrying out a wish to murder another human being. The taboo sometimes extends to all killing of humans, not just murder and results in opposition to war and to capital punishment in some societies. Some societies allow killing in cases such as defense of oneself or others. The killing-mitigation system goes even further – even the threat to kill someone is a crime in some societies, and likely prevents some killing, if such threats are prosecuted. We have done our best to develop quite a complex system to reduce killing, while still maintaining other important values such as freedom, safety, and justice. For contrast, note that our current society places very little or no value on the prevention of wild animals killing other wild animals. We quite literally let that system run wild.

The human analogy suggests we apply some of these system protection ideas to the prevention of a robot killing spree. For example, a system component could act as the police, watching over the actions of other system components and acting to restrain them if they violate the rules, or perhaps even if they give the impression of an intention to violate the rules. System components, including AI components, could find their freedom restricted when they stray outside designed bounds of acceptable behavior, or perhaps they might be let off with just a warning or given probation. In extreme cases, police subsystems could be given the authority to use deadly force, stopping a wayward AI component before it causes harm or damage. Of course, someone (or another kind of subsystem) would need to watch the watchers, and be sure the police subsystems are doing their jobs, but not overstepping their authority.

Subsystems watching over each other to make a system safer is not a new concept. Home thermostats have for decades been equipped with a limit device that prevents the home temperature from dropping below a certain limit, typically 50 °F, regardless of the thermostat's setting, preventing a thermostat malfunction, or inattentive homeowner, from the disaster of frozen and burst pipes. So-called "watchdog" devices may continually check for an Internet connection and if lost, reboot modems and routers. Auto and truck engines can be equipped with governors that limit the engine's speed, no matter how much the human driver presses on the accelerator. It is important to see that these devices are separate from the part of the system that controls the normal function. The engine governor is a completely separate device, not an addition to the accelerator and fuel subsystem which normally controls the engine speed. There are likely some safeguards in that subsystem too, but the governor only does its work after the main subsystem is finished. To anthropomorphize, it's like the governor is saying to the accelerator and fuel delivery subsystem, "I don't really care how you are controlling the engine speed; you do whatever you think best. But whatever you do, I'm going to be looking at the engine speed, and if it gets too fast, I'm going to slow it down. I won't even tell you I'm doing it. I'm not going to interfere with you or ask you to change your behavior. I'll just step in and slow the engine down if necessary, for the good of the whole system." This is the ideal attitude of the human police – let the citizens act freely unless they go outside the law, and then act to restrain them, for the good of the whole society.

The watchdog system itself must be governed of course, and trusted to do its job, so some wonder if one complex system watching over another, increasing the overall system's complexity, is worth it. But watchdog systems can be much simpler than the main system. With the auto or truck governor, the system controlling engine speed based on accelerator input, fuel mixture, and engine condition can be complex indeed, but the governor is simple, caring only about the engine speed itself. In some engines, the governor is a simple mechanical device, without any electronics or software at all. In an AI system, the watchdog will invariably be much simpler than the AI system itself. Even if the watchdog is relatively complex, it may be worth it, since its addition means that both the main system and the watchdog must "agree" on correct functioning of the system, or a problem must be signaled to the human driver or user.

Continuing the human analogy, we note that just having police is not enough. If the only protection in a human society were police, corruption would be a great temptation – there would be no one to watch the watchers. In peacetime human societies the police answer to the courts, and the courts must answer to the people, through elected representatives. Attorneys must answer to the Bar association and are subject to the laws made by elected representatives of the people. Judges may be elected or appointed, and their actions are subject to review. It's a complex system of checks and balances that prevents the system from being corrupted or misused through localized decision-making by any single part of the system.

In an intelligent system, AI-based subsystems make decisions after being trained on numerous examples, as described in Chapter 3. Designers try to make these AIs as good at decision-making as they can be, but it is still possible for bad decisions to be made for a variety of reasons. Bad decisions can be made by any system, AI or not, but in non-AI systems, it is easier to see the range of possible decisions by examining the hardware and inspecting the software source code. In safety-critical systems, source code must be inspected and verified. In an intelligent system containing an example-trained AI component, this is not possible. The code used to implement the neural network can be examined, but the decisions are made by a set of numbers, thousands, or even millions of them, which are set during the training process. Examining these numbers does not clearly reveal how the AI will make decisions. It's a bit like human being – with our current medical knowledge, examining the brain of a human being does not reveal how that human being will make decisions either.

It's not that the AI component is unpredictable, or unreliable at decision-making – it's more that humans can't readily see the limits of the AI's decisions by inspecting its software source code. The systems engineer could compensate for this unknown by designing a watchdog component that watches over the AI and if it makes an out-of-bounds decision, step in and stop the system, or at least get further human authorization before proceeding. Designers add watchdog components to intelligent systems for the same reasons we have police in human systems. The possible range of behavior of an AI component (or a human) is simply much greater than is safe for the system, and there's always that chance, regardless of parentage, provenance, training, or upbringing, that the AI (or human) will one day act out of bounds. The watchdogs need to be present and vigilant.

We have considered the question of how we protect ourselves against an intelligent system, even when it contains AI components that can conceivably make unfortunate or dangerous decisions in certain cases. By applying the human analogy, we suggest that other parts of the system, analogous to police, courts, and prisons, can serve to supervise and make sure the entire system stays in bounds. We can apply the human analogy another way by considering how we can protect our critical systems from unwanted intrusion and subversion by AIs. The human analogy suggests we ask how we protect important systems from *human* intrusion, otherwise known as hacking.

1.6 Cybersecurity in a World of Fallible Humans

The field of cybersecurity – the protection of computer systems from unwanted intruders – conjures ready images of a super hacker, typing furiously at a keyboard for about 60 seconds, and gaining access to a high security defense, government or corporate system, and its valuable stores of data. While there are some intrusions that are perpetrated through purely technological means, it is by far more common for would-be cyber intruders to rely on the weaknesses of other human beings.

Even purely technological intrusions often rely on poor security decisions made by humans in the design or implementation of the system in the first place. It doesn't take a security expert to know that if a system sends unencrypted access data over wireless connections, the data can be intercepted, modified, and re-sent, tricking the system into doing the intruder's bidding. Systems designed this way, and there are many, are asking for intrusion and exploitation. Well-designed systems are extremely difficult to hack into using just a computer from the outside.

Most hacking intrusions begin with social engineering by manipulating or exploiting the weakest link in the system – the human being. How many human beings, faced with the need to use multiple, complex passwords on a daily basis, write them down, perhaps on a yellow sticky note affixed to a monitor or inside an (unlocked?) desk drawer. Intruders, posing as maintenance people, plant waterers, or janitorial staff can wander through an office (or pay someone to do so) and snap photos of enough passwords to keep them busy for quite a while. Best of all (for the intruders) the source of the compromised security is untraceable and may remain unknown indefinitely. With the right passwords, a hacker can simply set up additional accounts on the system, giving them permanent access, even if the original passwords are later changed.

Malicious software code that grants an outsider access to a secure system is easy to procure and use, but it must be installed on the secure system. Shockingly, studies have shown that small USB jump drives laden with malware and scattered in a company's parking lot, will more often than not be picked up and inserted into employees' computers by the end of the day. Printing the company's own logo on the drive doubles the effectiveness of the ploy. The even more common way of introducing malicious code into a desired target system is to attach it to a phishing e-mail, tricking uninformed or inattentive human beings to launch the attachment, installing the malicious code, and granting access to the intruder, usually without notice by the unwitting human accomplice.

Human inattention and lack of understanding of system security are not even the worst of human cybersecurity failings. A study by Sailpoint, a security management firm, described in Inc. Magazine, shows that a surprising number of employees (one in five) are willing to sell company passwords, and almost half of those would sell a password for under US$1000. That amount of money would pay an expert hacker for only a day's work, making employee bribery one of the most cost-effective ways for humans to gain access to a secure system. If humans will exploit the weaknesses in other humans to gain unauthorized access to sensitive or important systems, then it seems likely that an AI would use this approach as well. The question becomes, how do we prevent AIs from exploiting fallible human beings. The human analogy suggests we first ask, how do we prevent *humans* from exploiting fallible human beings?

We do what we can to block an AI (or a human) from attempted nefarious communication with unsuspecting employees by attempting to block phishing e-mails, malware-laden e-mail attachments, and disguised hyperlinks. More importantly, we need to make the fallible humans, who are the weakest link in this system, less fallible. Technological barriers and relentless cybersecurity education are the primary options here.

Technological barriers can help. Some organizations disconnect all of the USB ports on all of their employees' computers, blocking the "USB drive in the parking lot" attack vector, potentially impeding employee productivity and freedom, costing the company time, money, and perhaps morale. More sensibly, USB ports can be set to prevent "AutoRun" and "AutoPlay," two Windows features that make it easier for malicious USB drives to do their dirty work. We can limit the access fallible humans have to a system, so that their fallibilities don't result in system compromise. In a systems sense, these protections place limitations on the access one system (the human) or subsystem has to another (the company network). The designer of an intelligent system can use this principle throughout the system, allowing parts of the system to communicate with others only on a need-to-access basis.

Holding to a necessary-communication-only principle in system design is not without costs in some cases. It may be expedient, efficient, and economical for one subsystem to serve a few others in a system, but if doing so allows communication between those other subsystems, the designer might be trading economy for danger. Suppose the passenger entertainment system shared a power supply with the guidance avionics on a commercial aircraft. Malicious introduction of software, perhaps even a corrupted movie, or a power spike from a portable battery could overload the shared power supply, causing it to shut down the aircraft's guidance system. It makes more sense to take the "inefficient" route of building independent power supplies for each of these subsystems.

Human fallibility can also be overcome through relentless education. It's a slow process, and continual reinforcement is needed to ensure that learnings stay in place and become embedded in the culture. A policy prohibiting the insertion of USB drives into company computers is fine, but without full education on why that policy is needed and the associated risks, people, especially freedom-loving Americans, will tend to ignore the rule when it's expedient. Two stories will illustrate.

Many years ago, I was giving a talk in a country far away from my own. After the talk, a smiling, enthusiastic young man came up and asked for a copy of the slides and handed me a USB drive on which to copy them. Being overcome by his effusive flattery about my talk perhaps, I took the USB drive and inserted it into my laptop. Immediately, the security software on the laptop flashed a big warning on the screen that the drive was infected with suspicious software. I pulled out the drive and asked him to request the slides by e-mail later. I don't believe I ever received that request. I was very lucky that the kind of malware on that drive was caught by the software on my laptop. Whether the person who handed me the drive was trying to get malware onto my computer, and by extension, my company's network when I returned home, or was unaware of the infected drive, I can't know. Something in me knew better, but nevertheless, the drive did get inserted into my computer that day.

In another case, I was visiting a large client corporation and was to give a talk in their conference room. I pulled out my remote control and its USB receiver, and plugged it into one of their computers. My host, who was not fast enough to stop me, almost jumped out of his skin. I knew that USB receiver was safe, since it was not a drive, and could not carry malware, but he was absolutely right to object to what I did – there's no way he could be sure it was safe. But in this case too, an unauthorized USB device was inserted into a company computer. To see how effective your cybersecurity education is, pose the following question to your employees:

Q: You find a USB drive in the parking lot of our main building, bearing our company logo. What do you do?
   a. Ignore it – leave it there on the ground
   b. Plug it into your computer to try to find out who it belongs to so you can return it
   c. Take it to the receptionist
   d. Take it to the security office

   How many of your employees would reliably answer that (d) is the only correct choice? Sure, the receptionist *should* know to give it to the security office, but it's safer not to make that assumption.

The painstaking process of thinking through potential security intrusions is similar, whether we are trying to protect sensitive systems from evil, inattentive, or apathetic humans, rogue AIs, or careless AI programming. Taking into account the personality, fallibilities, and behavior patterns of the potential intruder, the systems engineer must think through possible intrusion scenarios and reason about how to prevent them as the system is designed.

Taking the human analogy one step further, we all know that human beings cheat in certain circumstances. Dan Ariely's studies, as described in his surprising and fascinating book, *Predictably Irrational,* show that, when being caught is not possible, about 20% of people will cheat, but only by a little. Designers working with human activity systems such as societies, economies, and companies, would be wise to accept this characteristic of humans and accommodate it in system design, either by allowing for it, or by increasing efforts to detect and punish it. So, do AIs cheat, too?

It depends what we mean by cheating. Years ago, I read the wonderful book, *The Four Hour Workweek* by Tim Ferriss, in which he describes training for and ultimately winning a kickboxing martial arts competition "the wrong way" by intentionally dehydrating, fighting three weight classes below his actual weight, and exploiting a rule that allowed victory if your opponent falls off the fighting platform. Tim won every fight by simply pushing his opponent off the platform. The story is quoted and discussed on a martial arts blog, and the discussion centers around the question – did Tim cheat? (Ferriss 2009, p. 29). Phrases like "cheating within the rules" and "poor sportsmanship" are mentioned, and there is no clear consensus. Is it cheating when a human uses the available rules, and exploits so-called "loopholes, inadvertent omissions, or technicalities" to win? Some feel the fault here (if there is a fault) lies with the rule-writers – they should have specified that bouts must be won primarily by kickboxing, for example. AIs cheat in similar ways – the human analogy again.

Robot competitions are held in which each team is given a selection of parts and required to build a robot that can win a one-meter race across the floor. When an AI is given the task of building the robot, it often arrives at the same solution: assemble the available parts into a tall tower (taller than 1 m), and then simply fall over, crossing the finish line. Unless there is a rule that all parts of the robot must cross the finish line to win, this scheme is the most efficient. In a recent competition, a human team decided to copy the AI solution, but missed the mark by one inch. (Shane 2019).

The point here is that we must be very careful what rules we set up for AIs (and humans) to follow, since any weakness in the rules may be exploited. As already mentioned, an AI deciding to imprison humans for their own safety is an example of faulty rule and goal setting. Humans at least can bring their own background understanding and conscience to a situation and decide to operate according to the presumed intent of the rules, but AIs are not so constrained. Systems engineers must consider how the system needs to constrain the behavior of the subsystems, whether those subsystems are ordinary machines, human beings, or AIs.

Earlier we considered the problem of futuristic killer robots, who take it on themselves to choose to attack or restrain human beings, based on evil motives or misinterpretation of their creator's objectives. Killer robots, however, are already with us. Industrial robots and other automated machines have been killing people (fortunately it's rare) for many years. As Zachary Crockett, who has studied "killer robots" for years puts it,

> Most robot-related incidents thus far have been the result of machines being too stupid, rather than too smart, or a disharmonious relationship between man and machine.
>
> (Crockett 2018)

Robots and AI subsystems may behave in ways other than intended, either through faulty programming, or through various kinds of failures in subsystems or elsewhere in the system. Perhaps counter intuitively to a systems engineer designing an intelligent system, it doesn't really matter whether the source of the failure is the design of the subsystem or some kind of mechanical, electrical, or software failure. The systems engineer must consider, as far as it is possible, all foreseeable failures and engineer the overall system to accommodate them. Let's take a simple example.

A vacuum cleaner in normal use may vacuum up enough dust and other material to clog the bag, restricting airflow to the motor, and causing it to overheat. Unchecked, this condition could cause nearby parts or wiring to melt, resulting in short circuits or a fire. Instead, engineers consider this possible condition and design a simple circuit that detects a high temperature condition and shuts down the vacuum cleaner until the machine has had a chance to cool down and the source of the air flow restriction is removed by installing a fresh bag. The circuit may not give any indication that it has rendered the vacuum cleaner inoperative temporarily, perhaps for several hours, and may cause a user to believe the vacuum has failed permanently. A better design would be to provide an indicator light labeled something like, "overheat condition – allow to cool." The lack of such a light could be due to its additional cost or more cynically, because its absence may increase repair or new vacuum cleaner sales revenue.

One way systems engineers may approach the problem of dangers or failures in systems is to operate the finished system, watch for failures, and fix them as they occur, perhaps though product recalls, in-field upgrades, or by releasing new versions and inviting users to replace their old devices. An old Navy submariner once told me that the way they tracked down issues in a circuit that repeatedly blows its protective fuse, is to replace the fuse with a piece of pipe and then see where the fire starts. A better way is to try to foresee how failures may occur and correct for them in the design of the system. Perhaps the design of the system can be changed to *prevent* the failure from occurring. Intelligent systems may be designed to detect and correct their own failures while in operation. Foreseeing failures is straightforward with systems that already exist, but the systems engineer is designing something new and must use the power of prescience to imagine failures.

1.7 Imagining Failure

A primary method in systems engineering for collecting and analyzing possible system failures, is the Failure Modes and Effects Analysis (FMEA), and its more evolved cousin, the Failure Modes, Effects and Criticality Analysis (FMECA). FMEA involves examining each element of the system and asking several questions like, how can this element fail? What could cause it to fail? What would be the effects of the failure? and what can be done to eliminate the failure, or at least reduce its impact on the system? For each cause identified, systems engineers will ask more questions to determine the cause of that cause, and the cause of that cause and so on until a root cause can be found and handled in the design, preventing that kind of failure from ever occurring.

In a twist of the popular motivational idea of imagining success, when designing an important, sensitive, or safety-critical system, systems engineers will spend a significant amount of time imagining failure, that is, imagining every possible way the system may fail, and incorporating features in the design to prevent or at least handle the effects of failures. For example, if a small, embedded computer controls an important aspect of a spacecraft, it is important to ask, what if that small computer fails or crashes? The cause could be a corrupted memory, and the cause of that corruption could be a random bit of radiation in the space environment. How can this possible failure be mitigated? One possibility is additional shielding of the computer, but to shield it against all radiation might be impractical due to weight or cost. Another possible design is to have three computers instead of one, each running the same software. A fourth computer compares the results given by each of the three, and if they all agree, all is well and the system proceeds. If two agree and one does not, the system proceeds, if one of the computers continually gives the minority report, the system should run additional analysis and testing on it to see if it might need to be reset, or even replaced. If all three computers disagree with each other, an error is signaled, and the system may require human intervention to sort things out. With a voting design like this, if one or even two of the three voting computers fail, the system may be able to continue to operate acceptably. A voting system will perform much more reliably in a wide variety of changing conditions than a system with a single computer and will thus be more trustworthy.

Systems engineers have used FMEA for decades, but when dealing with intelligent systems, there are new aspects of potential failures to consider. Fundamentally, AI subsystems make decisions in a system. For example, the decision may concern the classification of an image as being a pickup truck vs. a tank, a pedestrian vs. a dog, friend vs. foe, or in-my-path vs. out-of-the-way. Alternatively, an AI subsystem may provide a prediction based on past data, such as the amount of likely rainfall today in a location. In a prediction scenario, some of the failures that would be listed in an FMEA would include the following:

1. The subsystem fails to make a decision with certainty.
2. The system fails to make a decision within a necessary time limit.
3. The subsystem makes "crazy" decisions that are obviously wrong or impossible.
4. The subsystem makes unpredictable decisions, meaning that for the same set of input conditions, varying predictions are returned.
5. The system makes mostly good decisions, but under some conditions makes very poor decisions.
6. The system's ability to make good decisions degrades over time.

The systems engineering practice of FMEA requires imagination. After listing all the way the system, or any subsystem or component of it can fail, the technique requires imagining the possible causes behind each of those failure modes. For each of these causes, the systems engineer must determine what the system should do, by designing the system's response. As described in Chapter 6 on the subject, use cases can be excellent tools for describing such scenarios. Let's look more closely at some of the AI subsystem failures above, and consider both the causes of the failures, and how the system can either prevent or respond to each.

(1) The Subsystem Fails to Make a Decision with Certainty

It's unlikely that the neural network itself is failing to, or taking too long to, make a decision. It is possible that the neural network is not reaching a clear decision, for example, it may decide that an object is 48% likely to be a car and 51% likely to be a stationary sign. The system may view this as a nondecision, since there is significant uncertainty, and must act appropriately, perhaps by looking again (by taking additional images) of the object, or by signaling to a driver or an operator that a clear decision cannot be reached. An intelligent system must be designed to differentiate between a 51% certainty decision and an 87% certainty decision. As human beings, we handle information like this often, asking ourselves and others how sure we are about our information and decisions. It's new, and perhaps a bit jarring for systems engineers to have to deal with uncertainty on the part of subsystems and their decisions.

The failure to decide has two primary possible causes. Assuming that the system has made good decisions in a large number of situations – and if it did not, it would probably not be in operation yet – the problem is likely either with an ambiguous input or an unexpected input. An ambiguous input is an input that is just too low in quality or fidelity to be accurately used for decision-making. If the vehicle in our example is observed by the AI subsystem at dusk, on a misty evening against a background similar in color to the vehicle itself, it is quite possible a human being would have trouble differentiating a car from a car advertisement too. While not technically a fault in the AI subsystem, the systems engineer must nevertheless allow for such a condition in the system's overall operation.

The other likely cause of a failure to decide is an unfamiliar image, one that the system was not trained to recognize, like the green school bus described later in this chapter. Here the failure points to a lack of complete training data. Humans can be forgiven for failing to recognize never-before-seen objects, but they do a better job of interpolating between known types. A small child will say, "look at the funny horse" when seeing a zebra for the first time – an AI subsystem may not be able to make that leap. The solution for this kind of failure to decide is to provide more diverse and precise training data to the AI subsystem. As with many system failures, the root can be traced to a lack of requirements. Data requirements engineering, a new field necessitated by the importance of training data to AI subsystem performance is discussed further in Section 1.8.

In a situation where the system fails to make a decision with an acceptable level of certainty, a back-up subsystem could supply a prediction based on a different algorithm – perhaps a much simpler one (predicting today's weather based only on yesterday's is remarkably accurate). In many cases, a less accurate, but close result is better than no result at all.

(2) The System Fails to Make a Decision Within a Necessary Time Limit

If the system is designed with some of the methods suggested above, evaluating the decisions of AI subsystems, voting, and even looping back and re-evaluating in some cases, it is possible that an AI subsystem will not reach a decision, with an appropriate amount of certainty within the required time. The systems engineer should allow for this possibility. In the now-famous 2011 Jeopardy match between IBM Watson, Ken Jennings and Brad Rutter, Watson's Final Jeopardy response to "Its largest airport was named for a World War II hero; its second largest, for a World War II battle," stunned everyone: "What is Toronto?????" There could be many reasons for the incorrect answer, but the important point is that Watson indicated the low level of confidence it had in its own answer with the series of question marks. In the Final Jeopardy round, contestants must provide a response when the well-known theme tune ends or forfeit their wager. Watson did the best it could but would never have buzzed in to answer that question if it had had a choice. Watson did the world a favor by showing its level of certainty on the regular Jeopardy questions as well. Below a certain level, Watson didn't buzz in to answer, unless perhaps it was losing the game and needed to take more risks.

Systems engineers must determine if it's better for an AI subsystem to provide a response, even with high uncertainty, or to provide an "I don't know," or more specifically, "I don't have an answer I'm confident enough to give you," response. As human beings, we are accustomed to expressing the level of confidence we have in our own answers, though we rarely quantify it. "I'm not sure but I think I've seen him before at a conference," is common while, "I'm 30% certain I've seen him at a conference, and 50% certain I've seen him *somewhere*, leaving a 50% chance I have never seen him before" is more accurate but rarely stated in its complete form. Depending on the application, it may be best to have the system reveal its level of confidence in the decision it is providing. There's a big difference between a decision with 80% confidence, and one with 45% confidence, and a runner-up decision with 41% confidence. A human user of the system may want to look closely at both of those close contenders before making a final choice, and the system must be designed to provide the information to allow that.

(3) The Subsystem Makes "Crazy" Decisions that Are Obviously Wrong or Impossible

In a scenario where the system makes a "crazy" decision, the system must first have a way to detect a "crazy" decision, probably through the inclusion of a set of "sanity check" rules. If an AI subsystem comes up with a negative number for predicted rainfall, or one that exceeds by half the worst monsoons in history, it would be smart for the system to refuse to accept the prediction.

From a systems perspective, it may not matter much *why* an AI subsystem makes a "crazy" decision. The fact that it *may* do so is very important to the systems engineer whose job is to make sure the entire system does not fail or take the wrong action. Consider a building environmental control system that measures the temperature of each room and sends commands to heating and air conditioning systems, trying to anticipate the comfort needs of the occupants of the building. The AI subsystem can make decisions that are "crazy" when viewed from a systems perspective. Say there are no people in the building. The AI subsystem concludes that no heat is needed. If the outside temperature is just above freezing, the system should know that "no heat" is not a good decision, for the good of the pipes and plants in the building. Or if the AI subsystem detects a temperature of 300 °F, rising in a room, it might logically call for maximum cooling with blowers on maximum. The system should know that this temperature reflects not just a hot day in the office, but more likely a fire, and appropriate actions of shutting down the cooling system, signaling an alarm, and calling emergency services should be taken.

A possible solution for "crazy" decisions is the voting design mentioned earlier where multiple AI subsystems, with differently designed algorithms, vote on the prediction. Close predictions are averaged, but outlying ones are discarded as likely errors. Online captcha tests that ask multiple users to identify which photos contain crosswalks to guard against human errors creeping into their crosswalk-classification data sets.

(4) The Subsystem Makes Unpredictable Decisions: For the Same Set of Input Conditions, Varying Predictions Are Returned

When a system makes unpredictable and unreliable decisions, it is more difficult for the system itself to detect the problem while it is in operation, leading the systems engineer to build in self-diagnostics that enable the system to periodically test itself to make sure it has not been corrupted. For example, say a company deploys an adaptive learning system that attempts to refine a classification algorithm based on new data gathered in operation. While in operation, the system could be intentionally fed misleading data by malicious users or hackers, causing the system to make increasingly incorrect decisions. Designers of the system may not be able to prevent such adverse data from being fed to the adaptive learning system, but by taking a systems approach they can compensate. The system can periodically perform a self-test, using a set of known data, checking its classification results against the known labels. If the performance of the system falls below some acceptable level, additional checks can be performed, or the newly added data removed.

While it may not occur to many software developers to build tests into the final code of the running system, it's a technique systems engineers have used for decades. Complex weapons systems may have a built-in test (BIT) capability. An explosive weapon can't be tested; to test it is to destroy it, assuming it works as designed, but built-in tests provide confidence that the weapon or other system will be ready to use when needed. Built-in tests can be even more valuable in a complex system with many interacting subsystems, especially when AI subsystems are part of the mix. Referring again to the human analogy, psychologists may use simple evaluations such as the clock-drawing test to look for signs of dementia. The patient is asked to draw a clock face on a piece of paper; those with certain afflictions are likely to draw it in a distorted fashion. After all, a human being is a complex system whose structures and behaviors can't be evaluated directly, so a simple test is used to test a number of primary functions at once. Pass the simple test, and all is well – fail it, and deeper testing can be performed.

An AI making unpredictable decisions that may vary, even with the same input conditions, is a failure mode that is commonly feared but less likely to happen, than some assume. It's one place where the human analogy doesn't apply. An AI subsystem should, if operating properly, make the same decision under the same conditions each time. Non-adaptive systems, which do not automatically modify their own operation over time based on new information represent most of the systems operating today and planned in the near future. Contrary to common understanding, these systems are deterministic. Though AI is widely associated with nondeterminism, AI subsystems based on technologies like deep neural networks or explicit rule-based programming, will always make the same decision, given the same input.

The human analogy does not apply because human beings have an annoying capacity to make a different decision, based on the same input any time they like. This capability cannot be removed or limited. No matter what is done to a human being, they retain their power of free choice. Humans will always represent a greater danger than AIs, since humans can always choose evil courses of action, and can enlist computers and AIs to do their evil bidding. The human analogy helps us see ways of making the world safer from AIs by imagining how to make the world safer from human beings, not the reverse.

If an AI subsystem is making seemingly unpredictable decisions, there are several possible causes, but the most likely is that the input is actually varying, but not in ways that are apparent. For example, a self-driving car trained to detect pedestrians, may fail to identify a pedestrian carrying a large object, such as a bicycle. For a human being, this would be a hard mistake to make, but for an AI subsystem, the shape of a pedestrian and the shape of a pedestrian carrying a bicycle could easily be different enough to cause misclassification. Images under different lighting conditions, from different angles and elevations, or with different backgrounds can all confuse an AI image recognition subsystem. We call it confusion, but it's not really confusion – it's a lack of complete input. A person carrying a bicycle is certainly common enough that it should be included in the training set used when building the AI. We could forgive the AI for misclassifying a person pushing a piano due to its rarity, but people with bicycles are everywhere.

The systems engineer must be concerned with a new kind of system requirements – *data requirements*. Data requirements specify how much and what kind of data must be made available to the AI subsystem for training and testing. Systems engineers must learn enough about the environment in which the system will operate, and about the situations it could encounter to fully specify the data required to successfully train it. More than the machine learning developer or even the data scientist, it is the systems engineer who will be aware of the system within which the AI subsystem will operate, and also the larger context of the entire system in the environment where it will operate. Let's take a closer look at the role of data in AI systems and how systems engineers will need to learn to deal with it effectively.

1.8 The New Role of Data: The Green School Bus Problem

With machine learning, the training data are what ultimately controls the operation of the AI subsystem and thus the behavior of the system. The story of the green school bus will help make this clear. Imagine building a system to identify vehicles for a military application. It's important that the system is able to differentiate and identify military vehicles so the training data include many images of friendly and enemy tanks, military trucks, jeeps, and trailers. Since civilian vehicles may also be seen in an area of military interest, some images of ordinary cars, pickup trucks, SUVs, and other vehicles are added, including some school buses. Some examples are shown in Figure 1.1. Since school buses make up only a small proportion of civilian vehicles, not many are included (we'll discuss more about whether including only a few of a rare item is a good principle in the section on data requirements in this chapter), and the school buses that are included are yellow, since of course, most school buses are yellow. The machine learning system is trained on all these images.

As is the normal practice when developing machine learning subsystems, part of the available data is used to train the machine learning system, and a smaller part is held back to test the algorithm's performance. After training, let's say the application's algorithm reports high scores, showing that it can correctly identify

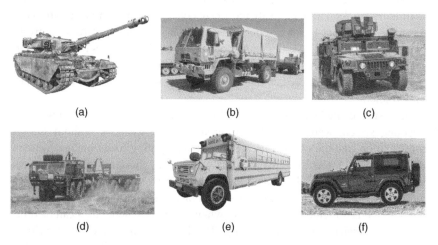

Figure 1.1 Images used to train the deep neural network. Source: (a) Momentmal/Pixabay. (b) Cianna/Pixabay. (c) and (d) Military_Material/Pixabay. (e) blende12/Pixabay. (f) dimitrisvetsikas1969/Pixabay.

Figure 1.2 The green school bus. Source: Free for commercial use without attribution; https://pixabay.com/vectors/green-bus-bus-green-vehicle-auto-3749394/.

the vehicles it has been trained on. High scores are celebrated, and the application is deployed in the field. Based on the testing performed during development, a very successful application deployment is expected.

Then, during operation, into the field rolls a green school bus, perhaps from Greenville High School, with "The Fighting Toads," proudly displayed on the side of the bus (Figure 1.2). Any human observing the area would immediately recognize the vehicle as a school bus. The AI machine learning algorithm, however, has never seen a green school bus – the only school buses included in the training data are traditional yellow. Understandably in retrospect, the AI is likely to perform poorly on recognizing green school buses, perhaps even identifying them as military trucks. Humans easily recognize colors as nonessential variations for a vehicle – a school bus is a school bus regardless of color, but AI systems are much more literal. The AI subsystem doesn't know what features of an image are essential to the meaning of the image unless we tell it, by example. The system's performance will be unpredictable with objects on which it has not been specifically trained.

The results could be disastrous. Imagine the headlines if an AI-based weapons targeting system were given the autonomy to fire weapons without human confirmation, and the system miscategorized the Fighting Toads' bus as an enemy military truck. A terrible outcome, but where does the fault lie? From a systems perspective, several mistakes were made. First, school buses were underrepresented in the training data set – a problem of data distribution. Next, all the school buses included were yellow, when in reality school buses can be other colors – a problem of data diversity. Third, a system that controls critical, irreversible actions, like weapons fire, should not be so easily given complete autonomy.

The green school bus story is fictional, but a similar difficulty occurred in reality when training a system to recognize various road signs. By introducing some ordinary graffiti onto a stop sign, the system was fooled into thinking it was a speed limit sign, indicating that the system had not been trained on enough examples of stop signs, perhaps with various intentional distortions and extraneous material to enable it to make the correct identification. Whether the system is misled intentionally or accidentally, the same principle applies. Systems trained only on limited versions of the world, such as clean, perfectly lit signs, idealized yellow school buses, and clear-day road scenarios, are likely to make mistakes in the real world. Probably, AI in general will be blamed for the failures, instead of the blame being placed where it belongs – poor system design and poor training of the AI.

Of course, the problem here is that there are lots of "green school buses," that is, objects on which the system has not been trained. It is easy to assume that the answer here is simply to include school buses of all colors in the training data set, but what if such images are not readily available? In what follows, we'll explore this subject in much more depth, but for now, the important thing is to realize that we must give serious thought to what data is required to train the AI system for best results when the system is deployed in the real world.

1.9 Data Requirements

The practice of systems engineering focuses heavily on the elicitation, definition, and refinement of system requirements. Requirements that represent the required *behavior* of the system are known as *functional requirements*, while various required *qualities and characteristics* about the system other than behavior are known as *nonfunctional requirements*. Nonfunctional requirements are sometimes referred to as "the -ilities" since the names of many types of nonfunctional requirements end this way – availability, dependability, reliability, and maintainability – but they also include requirements for areas like safety and security.

Traditionally, system requirements have not focused on data in any way, but in AI subsystems that learn from data through the use of machine learning, the data used to train the AI subsystem is vitally important to the successful operation of the AI subsystem and of the entire system, so it deserves careful attention. In a way, the training data used to train a machine learning subsystem is not unlike software source code – it determines how the subsystem will respond to each situation it faces. Headline-making stories of AI failures tend to assume that the AI algorithm or its implementation in software code is what's to blame but blaming the AI itself is like blaming a small child for saying "here kitty, kitty!" when seeing a raccoon for the first time. Is it not more likely that the parents are to blame for not providing a more complete animal education?

Training data can be hard to obtain since to be used for training, the data must be labeled. Labeled data is simply data for which the correct identification, or prediction, is known and specified. For example, if the application is to classify vehicles using images, training data consists of images of vehicles, labeled with the type of vehicle shown in the image. Training data is typically labeled by the humans creating the system, manually, though in some cases, it may be possible to find datasets that are already labeled. Perhaps one day a company like General Motors will supply a data base of images of their vehicles, already labeled with the vehicles name and type, but usually the data must be labeled manually. You may have helped label some data in the process of verifying your identity online, when you were asked to choose which images contain crosswalks or buses.

Given the current state of research in image recognition, large amounts of labeled data are needed primarily because image recognition systems are not able to easily recognize the essential aspects of an image – something that is grasped by even quite young human brains. A machine learning system must look at dozens, hundreds, or even thousands of images of stop signs to be able to reliably recognize stop signs, while the human brain seems to be able to grasp the characteristic octagonal shape and red color after seeing only an example or two. Worse, it isn't easy to tell what characteristic the machine learning system is focusing on. Since many images of horses show up against a background of green grass, the system may take the presence of green grass as an essential characteristic of horses, and might be unable to recognize a horse when placed on a beach or other background.

There are a number of areas to consider when specifying data requirements for a particular application, and we'll discuss each. We note at the outset that the idea of data requirements is a new one, even to AI and machine learning practitioners. Many machine learning applications at present are opportunistic – they arise because a particular data set already exists. The data set is discovered, and tends to be used as it is, without much examination or analysis. As systems engineers use AI subsystems in more and more applications, more attention must be given to the selection and refinement of the data set. The four main aspects to data selection for machine learning training are diversity, augmentation, distribution, and synthesis. Since the field of data requirements is new, and not part of what previous generations of systems engineers considered to be requirements, data requirements are not "your Dad's system requirements," and the four areas can be remembered using the acronym DADS for Diversity, Augmentation, Distribution, and Synthesis.

1.9.1 Diversity

AI subsystems that use machine learning and deep neural networks in particular, do best when a great diversity of data is used to train them. Diversity helps ensure that the AI is focusing on essential characteristics, by varying nonessential ones. While humans do this intuitively, through a process we do not fully understand, AIs are not so intuitive. The ideal training set for an AI would include images (or other forms of input data) with every possible valid variation, or as many as possible. In the case of images, the training data should include images of the intended object in all possible colors (remember the green school bus), in a variety of lighting conditions, from a variety of angles, and with a variety of backgrounds.

It may seem counterintuitive that an AI can't make the leap from a *Do Not Enter* sign, which is round when viewed straight on, to the same sign viewed from an angle, which appears more egg-shaped, but it often can't. Ideally, the training data will include images of signs from various angles to improve the AI subsystems' ability to recognize the sign however it appears. Similarly, the characteristic red color of a stop sign may appear a different shade when illuminated differently – by a sunset, bluish or yellowish headlights, or a fading overcast sun. In the same way, an AI has no way to determine whether what we would call the background is actually essential to the image and its meaning. To a human being, a dog is easily identified as a dog, no matter the background, but an AI, trained on images of dogs exclusively appearing on sidewalks, may fail to identify a dog when it appears against a grassy or watery background.

The systems engineer, working with the data scientists, must determine what is essential to the AI subsystem's role in the system. If the system is designed to identify cars, the background may not matter, but if the system needs to differentiate between a traffic jam and a parking lot, the background can be vital information. Systems engineers must not overestimate the AI subsystem's ability to extend or generalize its understanding of items based on the way that humans do it. An AI won't automatically know that an image of a cat from the side is the same cat as an image of that cat from the front. An AI won't know that 700 °F is not a temperature that should appear in a weather database (at least not on earth). An AI won't understand that different colors of people are all people, unless the data has been constructed to contain a variety of skin different colors. The systems engineer might even need to ask if the application needs to recognize clowns as people – if so, better include some clown images in the data. In the COVID era, when many people wear masks, facial recognition systems failed to recognize masked versions of the same people. A system trained on masked people, perhaps including multiple types and colors of masks should do much better.

There are two separate tasks here, and neither is easy. The systems engineer must work out what data must be supplied to the AI subsystem for training, and must also design a way to obtain this data. If the data is readily available from a known source, or can be captured using available means, then the data requirement can simply be passed on to the data scientist or data engineer on the project team. If the needed data is not available, other techniques are available like augmentation – the "A" in DADS.

1.9.2 Augmentation

Augmentation refers to the process of using software to produce new AI training data by varying existing data in specific and easily quantifiable ways. Images are varied in ways that do not change the image's meaning, but which provides variety to the AI training process and helps prevent it from making unfortunate assumptions. A training data set of 1000 images can easily turn into a set of 5000 images by varying each image in five different ways.

If the AI is being trained to recognize birds, for example, we can take a single bird image from our training data set and flip it horizontally, creating an additional image for the training data set. Humans can easily see that both images are the same bird, but the AI must be trained to see this fact. In some applications, it could make sense to create vertically flipped versions of images as well – birds and cats are often seen in upside down or sideways poses, but horses and humans – not so often. Humans intuitively recognize all orientations of birds and cats, but AIs must be trained through example.

Many augmentations are possible, and it will take some experimentation on the part of the AI subsystem's developers to determine which ones improve the performance of the AI subsystem. It is the role of the systems engineers, however, to help specify which augmentations make sense for the application. For example, by shuffling the colors in an image, new images are created that may look unnatural, but which help the AI learn to allow multiple colors – the AI learns that color is not essential to the identity of the object. Shuffling colors in images of cars makes sense, since a car of any color is still a car. If the application is to identify road signs however, color may be essential. Stop signs are always red, so wide color variation may not be as useful to the application. Another kind of variation is random rotation, where the image is simply rotated a random number of degrees to produce a new image. Images can also be randomly shifted up and down or left and right. To a human eye, the shifted images are much the same, but to an AI shifted images can provide important training information. Random zooming, where new images are created by zooming in to randomly selected areas of input images can also be effective, as can varying the brightness of the image.

The augmentation of image training data has become so popular that mechanisms for generating augmented images may be built into the AI developer's tool kit. AI developers can simply ask the neural network to generate augmented images automatically, as the AI subsystem is trained, without even taking up storage space for the augmented images, using standard machine learning systems like Google's Keras and TensorFlow.

It may seem counterintuitive that taking real images and making artificially colored, cropped, or distorted versions of them, and then adding these modified images to the training data for the AI would even work, but it does. Augmentations should only be added to help train the AI subsystem – not to evaluate its performance. Only real-world data should be used in testing the AI subsystem's performance, since it is the real world in which the system will operate. A boxer can train in the gym by hitting artificial opponents and punching bags, but the only real test of the boxer's ability is fighting a real opponent. Many image recognition systems and even other applications can potentially benefit from augmented data. Research continues in the area of image augmentation and the related area of synthesized data, discussed below.

1.9.3 Distribution

Distribution has to do with the proportion of the training data devoted to each class or category in the intended application. For example, if a systems engineer has decided that we do indeed need to detect school buses of various colors, not just yellow, *how many* images of non-yellow school buses need to be included compared to the number of yellow bus images? The counterintuitive principle here is that the distribution of the training data should be determined by the application need, not necessarily by the actual distribution in the world. Let's consider an example to illustrate this important point.

In the United States, as of this writing, 93% of nurses are female, while 7% are male. Say we are designing an AI subsystem to identify nurses from photos or video images, perhaps as part of a system that needs to know where nurses are located in hospital at all times. One idea is to construct our training data set following the same proportion we see in the real world, with 93% female nurse photos and 7% male nurse photos. It seems logical to have the distribution of the training data mirror the distribution in the world, but this doesn't work well. The system has only a few examples of male nurses, and therefore does not learn well how to identify them. Our system needs more male nurse images to train on, so that when it does see a male nurse in practice, it is able to spot him. The best data for this nurse-identifying application is probably closer to 50% female and 50% male, with a similar diversity approach applied to nurse skin color, height, weight, color of clothing, etc.

To put it another way, the attributes that are not predictive of being a nurse – gender, color, height, and so on – should not show a bias in the training data. It is no more likely that a tall person is a nurse than a short one, so the data should not be biased by including far more tall nurses than short. One may argue that being female *is* predictive of being a nurse, since 93% of nurses are female. But the question is, of females observed in a hospital, is it more likely that the person is a nurse vs. some other kind of person? To know this, we would need to know the makeup of the entire female population in the hospital. If, for example, there are just as many female office staff as female nurses, then being female may not be predictive of being a nurse. Note again that these are systems questions, best addressed by systems engineers, with assistance and input from data scientists. Some of these skills may be new to systems engineers, but with the rise in importance of machine learning systems, data science and statistical reasoning are skills that will become more and more important in the systems engineering of intelligent systems.

Consider what happens if the designers of the nurse-identification system make the unintentional mistake of biasing the training data to the actual mix of 93% female nurses and 7% male. If the system were also trained to identify doctors, it is likely the training data would follow the real-world mix of 50% female doctors and 50% male. Assuming a hospital with 200 medical staff, and a nurse-to-doctor ratio of 4 : 1, the hospital actually contains 31 male medical staff – 20 doctors and 11 nurses, and 168 female medical staff – 20 doctors and 148 nurses. Since male nurses account for only 6% of the total set of training data, it seems likely that the system could easily misidentify male nurses as doctors, since it is predictably easier for the system to differentiate male from female than nurse from doctor. Among the females, the data are similarly biased, containing 148 female nurses and just 20 female doctors. The system is likely to misclassify female doctors as nurses.

Stepping back and considering the human reaction to a presumably intelligent system that routinely misclassifies male nurses as doctors and female doctors as nurses, and one can sympathize with the assumptions that the AI, or its creator, is deeply sexist. But the fault is not with the AI as a whole, and not even with the AI algorithm or its programming. The fault is with the selection of data used to train the AI subsystem. Even though 93% of nurses are female, the system will probably perform better with a training set of 50% female nurses and 50% male nurses. To put it in plain terms, we need to train the system just as hard on male nurses as female so that it can better identify both, no matter how many of each it actually encounters when in operation. But what if we simply don't have enough images of male nurses? After all, they are relatively rare in the world. Lacking the data required to train an AI subsystem is a perennial problem for systems engineers and AI designers. Besides having a bloodhound-like determination to find the best data from wherever it may be, they can use our final DADS data requirements principle, synthesis.

1.9.4 Synthesis

In the world of statistics and scientific data analysis, it is a cardinal sin to *make up* data. Data is sacred, and should represent unfiltered, unaltered, and unbiased reality. But in the training of AI subsystems, made-up data can be a tremendous help. The reason is obvious: often there is limited real data available, and it can be difficult or expensive to acquire more data. It may be surprising that computer-generated data can be fed into an AI subsystem as training data, and increase the AI subsystem's performance on real-world situations, but it works. Let's take an example.

Laurence Moroney, an AI researcher and educator, developed a data set he calls "Horses and Humans"[1] which contains about 500 images of horses and about 500 more images of humans, both with various colors, poses, and backgrounds that can be effectively used to train AI subsystems to recognize horses and humans. But here's the surprise: the images are all computer-generated, synthetic images, like those one would see in a computer game, realistic animation, or virtual online world. Amazingly, training the AI on computer-generated, synthetic images enables the system to effectively identify real horses and humans.

The field of synthesized data for AI subsystem training is just getting started. In an interesting variation, a computer at Georgia Tech was used to synthesize a virtual model of the interior of a building, and that virtual building was then used to train a drone to fly down hallways and into rooms, using reinforcement learning. When the training was done, the drone was able to fly successfully in the real building. The potential for this approach is staggering. Technology to create virtual models and virtual worlds is quite advanced, propelled mainly by increasingly advanced computer games, which of course take place completely in synthesized virtual worlds. For computer games, worlds can be based in mythology, fantasy, or science fiction, but virtual worlds can just as easily represent real environments, situations, and people that an AI will eventually face in real life. Even better, training in a virtual world can take place in virtual, accelerated time. With enough computer power, training can proceed at many times real-world speed – a trip to Mars could be reduced from months to minutes, or a battle scenario that could take days in reality can be replayed hundreds or thousands of times in a day, with different combinations of conditions and events, accelerating learning.

1.10 The Data Lifecycle

A typical cycle of system development proceeds through stages of planning, analysis, design, implementation, verification and validation, deployment, and maintenance. The cycle may be carried out in a waterfall fashion with one cycle spreading over years, or a more agile process can be used by compressing the cycle to weeks, and repeating it over and over, developing a different slice of system functionality each time. In AI subsystems, the importance of data in training the system leads us to consider applying this kind of lifecycle to data. Let's consider each stage of the lifecycle process and how it applies to AI training data.

---

1 http://www.laurencemoroney.com/horses-or-humans-dataset/.

Planning. Often, the source of the very idea for an application of AI is the existence of data. I was inspired to create one of my first AI systems by the arrival in my inbox one day of a report containing hundreds of thousands of training class completion records from fellow employees – it made me wonder what I could do with that rich set of data. But increasingly AI won't only be applied when a set of existing data is noticed, but when systems engineers see a need for intelligent decision-making, classification, or prediction based on past events and information. A systems engineer may suggest the use of an AI subsystem, even before the existence of suitable data is confirmed. Systems engineers, in collaboration with data scientists, should determine the data needs of the system just as they consider the need for software and hardware. Does the data exist, can we access it readily, or do we need to make requests and secure approvals? What are the requirements for data diversity and distribution and is there a need for data augmentation and data synthesis? Are we missing green school buses (or their equivalent) in our application? By giving attention early to the need for certain data, it may be possible to begin collecting it as other system development activities proceed, and the data can be ready by the time it's needed.

Analysis. In the analysis phase, data is collected and analyzed using statistical, visualization, or other means. Like planning, analysis of the data to be used to train the AI subsystem should be performed by systems engineers and data scientists in partnership. Understanding the data itself requires no specialized software or AI background, so the data should not be of only private concern to the AI development team. The data should, at a minimum pass the *patient human* test, meaning that an intelligent, meticulous, and patient, though completely uncreative human being could conceivably do what we are going ask from the AI subsystem using the same data.

For example, if the AI is intended to predict rainfall anywhere in France, but the data comes only from central Paris, the data may not pass the patient human test – a human being would not be able to predict rainfall in Bordeaux reliably based only on rainfall in Paris. A system intended to recognize traffic signs might fail the patient human test if sufficient examples of nontraffic signs were not included. A patient human, forced to categorize a "Food Next Exit" sign, having never seen one and having been trained only on stop, yield, and speed limit signs, would produce unpredictable results.

One of the most fascinating things about the field of AI is that it forces us to reevaluate human intelligence and its components in precise detail. The kind of intelligence, or perhaps the aspect of intelligence that's needed for someone to pass a history test, relies on the recall of large quantities of detailed information on dates, places, and events. That's quite different from what it takes to pass a calculus exam, which requires few facts, but deep understanding about how to apply methods. At their current state of evolution, AIs are not miniature humans, with limited versions of all kinds of human intelligence – they are much more specialized. When we analyze the data that will be used to train the AI subsystem, we are in effect, building a very specialized kind of intelligence. It will only know what we tell it through the collection of data, and it will lack the human ability to extrapolate creatively to new situations and applications, or to make even obvious (to humans) jumps of logic. The AI brain is naive and can think only what we tell it to think – if what the AI does can even be called thinking. Recall that the image of a stop sign, with the addition of a few pieces of tape, could fool an AI into thinking it's a speed limit sign. Here's the point – such a mistake is not the fault of the AI. It is up to the humans to give the AI the right collection of data from which it can learn. We must apply human creativity to the *selection* of data so that the AI doesn't need to be creative to be successful.

In the analysis phase of the data lifecycle, the primary goal is to fully understand and validate the data that will be used to train the AI. To validate something in a system is to determine whether it is fit for purpose, and whether it meets the needs of the ultimate users and stakeholders of the system (more on verification and validation later). Validating the training data means understanding both the need for the system and the context and environment in which the system will operate. In effect, we are trying to see the data the way the AI will see it – as direct and conclusive input from which to learn.

Visualizing large data sets can be a daunting task. For numeric data, we can start with standard descriptive statistics, such as calculating the mean, median, and mode of a set of numbers. While basic, these simple measures can provide a sanity check on the data. A file of May temperatures for a city in the midwest USA may look good on a quick glance at the first 100 records. A closer analysis, however, shows a mean of 100 °F/38 °C, and a histogram puts the most-often occurring temperature at a more believable 61 °F/16 °C. Probably the file contains one or more erroneous temperatures. Just one day of 1000 °F weather (or a data entry error saying so) will throw an annual mean way off. In current and future applications, data files can reach tens or hundreds of thousands or millions of records. Since no one can plausibly "take a quick look" at such a large data file, a systematic approach is needed to assure ourselves that we have reasonable data. A statistical *box plot* which shows the minimum and maximum along with data sorted by quartiles would quickly expose our 1000° day anomaly, as would a simple scatter plot or line graph. A great first step, when working with numeric data is to plot it and take a look.

An important consideration in the data analysis phase is what to do with problematic data. If we have millions of temperature data points, a single reading of 1000° can simply be thrown out. But what if we have a file of temperatures and humidity levels, and half the records are missing the humidity? Throwing out half of the data doesn't seem right. There are several choices to handle the missing data and the right one depends on the application. We could replace all the missing humidity readings with zero, but that would seriously skew the averages and make the place appear dryer than Death Valley. More reasonable would be to replace the missing humidity values with the average humidity for the area, assuming we have validated the data, and trust the average to be accurate. Statisticians may balk at this kind of blatant data manipulation, but remember our intent is supply the best teaching data possible and produce the best performing system.

Another analysis activity is assessing the distribution and diversity of the data. As shown in the case of the nurse identification system, it is often best to match the data to the training need, not to the actual situation in the world. If the nurse photo data is 93% female, we may need to find or synthesize more male nurse photos to train the system well. The purpose is not to accurately represent the world, but to provide the most effective training.

Applying our human analogy, we know that police, soldiers, surgeons, military officers, and martial artists must train extra hard for very rare events – out of proportion to the events' actual frequency of occurrence. Such events may occur only a few times in their entire careers, but they must be trained to handle them when they happen. They "over-train" for rare events so they are ready for anything, or as close to anything as possible. For humans, this over-training can bias expectations, but at least for now, we don't have to worry about AIs having human-like expectations. We just need to make sure the AI is ready for all possible cases, both common and rare.

Data visualization can be both a science and an art. In the wonderful classic on the subject, *Beautiful Evidence*, Edward Tufte shows that creative visualization of data can show us the meaning, the patterns, and even the story behind the data. To pre-AI systems engineers, data were something that came *out* of the system and showed the system's performance and condition. In the world of AI, for deep neural networks that need training, data go *into* the system formulation and drive the actual behavior of the system. Systems engineers will increasingly need to develop data skills of their own, and partner with data scientists to develop visualizations of the training data to be used. The selection of training data cannot be left solely to the AI developers building the AI subsystem, as they won't be as familiar with the application needs and the context in which the system will be deployed and used, nor with the dangers and risks of the system performing poorly.

Design. Systems engineers wouldn't leave any other aspect of the system's design to chance and the same must apply to the data that will be used to train the system. Data cannot be manufactured from nothing, but systems engineers should consider data that is found in the environment as the raw material for designing and implementing training data, like raw ore that must be processed into steel before it can be used to make a ninja sword.

Many AI applications in current practice are *data-opportunistic* – they are built mainly because some interesting data are found and someone wonders what can be done with it. There's nothing wrong with that, and it can lead to the discovery of useful applications. But there's a big difference between making a dining table from an interesting tree stump found in the forest and *designing* a table based on user's needs, wants, and expectations, and then selecting the best materials to implement the design. When aircraft engine manufacturers looked at the mountain of data produced by their engines in flight, they wondered if it could be put to use by helping reveal maintenance needs or impending problems. By trying to create applications that use readily available data, it is likely that additional needs for data might be found, while other data already being collected might prove unnecessary. Data collection capabilities in the engine control systems can then be modified to make the resulting data more useful to the predictive algorithms.

Implementation. Implementation of the data once it has been designed, should be straightforward, though not necessarily easy. Theoretically, it's a matter of simply collecting the data according to what has been learned in the design and analysis phases. Practically, two main concerns often arise – how to get the data technically, and politically. The technical means for data collection are well known and many tools and technologies can be used. The field of data science has evolved useful tools, and most of them are open-sourced and freely available to use, even for commercial purposes. Datasets with many millions of records are easily handled on almost any personal computer, thanks to the generous storage and processing capabilities that are now the norm. When I started in computers, a big hard disk drive was 10 MB. Now a single high-resolution image can be that large, and personal computers usually have hard drives of 500 GB or more – that's 50,000 times as large as that 10 MB marvel from the 1980s. Personal computers also have much more random access memory (RAM) these days and 4, 8, 16, or more gigabytes is not uncommon. This much RAM enables a computer to hold large datasets in memory, enabling even faster processing. In sum, ordinary personal computers are often all that's needed to collect, analyze, and design the data needed for an AI subsystem in an intelligent system development project. As a side note, when people speak of AIs needing large computers, even supercomputers, they generally mean for the *training* of the AI system. The size of the system required to examine and analyze the training data or to run the final system once in operation are much smaller – normally only a personal computer is necessary.

The implementation phase also includes a process known in data science as data preparation. This innocent and simple-sounding term hides what can be a lot of experimental, creative, and technical work. Whole books have been written on the subject, but a systems engineer needs only familiarity with some fundamental concepts. The simplest form of data preparation is to do no preparation at all – the data are simply used in the form it is collected. There is still analysis to do, to be sure that the data set contains what we think it does, and does not contain errors, missing data or other anomalies, but elaborate data preparation are not always necessary. There are a number of fundamental techniques, but one of the most common and useful is the creation of derived features or *feature crosses* as they are sometimes called. Let's take an example.

Say we have data on the ages, heights, weights, and general health of a set of people, and we want to make sense of the relationship between these features so we can predict a person's health. We do some analysis and find that the relationship between height and weight is fairly linear – people get heavier as they grow up in height, and the more they weigh at a given age, the healthier they are. But there is a limit – after a certain weight, more weight has a negative effect on health. After about age 20, some people get heavier, but some don't, even though their height stays about the same. We find that those who get heavier have worse health, but those who are too light at age 20, have poor health (think starvation or malnourishment). We understand intuitively what's going on, but an AI trying to make sense of this data might have trouble coming up with a sensible prediction of a person's health as a function of their age, height, and weight. We can simplify things quite a bit by introducing a derived feature called body mass index (BMI) which is calculated as a person's weight divided by his or her height squared. It turns out that the relationship between BMI and health is clear for all ages – too low or too high correlates with bad health, but a BMI in the range of about 18–25 correlates with better health. So adding BMI to our data using this simple calculation will help the AI capture the relationship and produce better predictions of health. Using a derived feature like BMI makes the action of the AI more explainable and understandable by people – the relationship between BMI and health is much easier to grasp than the more complex relationship between height, weight, age, and health. (BMI is not perfect – certain kinds of athletes, for instance, may have a high BMI but not be overweight.)

Derived features are not always predefined in science or practice, ready for our use in training AI algorithms, but there are more than we might suspect. Blood pressure has always been hard for people to understand, with its two numbers – one "over" the other, but not representing a fraction. Medical researchers have developed a single measure called "mean arterial pressure" which requires invasive procedures to measure directly, but which can be approximated as the sum of one third times the systolic blood pressure (the top number) and two-thirds times the diastolic blood pressure (the bottom number). A blood pressure of 120/80 gives a mean arterial pressure (MAP) of 93.3. Though not in the common public vocabulary, MAP can be a better and clearer indicator of certain circulatory conditions and even of overall circulatory health. MAP is an example of the best kind of derived feature – one that already has a scientific basis for being related to real phenomena.

Some AI experts will argue that the nature of deep neural networks makes the addition of derived features unnecessary – that the AI will learn the correct relationships and produce just as good a result without them. Creating sensible derived features has been shown to improve the performance and predictive power of an AI algorithm, though sometimes only slightly. From a systems perspective however, the big advantage of effective derived features, that is, derived features that turn out to be significant in the learning of the AI algorithm, is that they make what the AI is doing more understandable, and thus help us progress our understanding of what is really happening in the world. Instead of a mysterious relationship between height, weight, and health, both humans and AI algorithms can use BMI. Instead of forcing the AI algorithm to try to make sense of the two numbers that make up blood pressure readings, perhaps the MAP will show a simpler relationship to health measures and conditions.

In the absence of readily available and established derived features such BMI and MAP, systems engineers and data scientists must experiment with creating their own derived features. If our application is to estimate the prices of homes based on their location, perhaps using the actual selling prices of nearby homes doesn't work as well as the derived feature of price per square foot, since it adjusts for differences in home size. But wait – would the number of bedrooms plus the number of bathrooms be more predictive of the home's value? There's another derived feature to try. The process is simple – create many of these derived features and see which ones end up being most strongly correlated and thus predictive of the phenomenon the algorithm is attempting to predict. Then try eliminating the factors that don't help the algorithm in order to end up with as simple a model as possible.

Verification and Validation. Systems engineers can apply the disciplines of verification and validation to the training data in order to give the AI the best chance of being successful when deployed. Verification and validation are, of course, applied to subsystems, and to the overall system itself throughout the systems development lifecycle, but they can also be applied to the data, even before the AI algorithm is developed. Think of it as making sure we don't fall into the garbage-in/garbage-out trap by giving a good algorithm bad data.

Verification and validation are often confused, even by systems engineers. Remember that the word verification contains an *r*, and thus relates to *requirements*. Verification asks the question, does the item being verified meet the stated and documented requirements? For a complex system such as a satellite, system verification involves the painstaking process of proving that the system meets potentially thousands of requirements, both those specified in the original contract and those derived later. The most common method of verification is testing. The system is operated with predesigned input conditions and we see if the result matches what the requirements state. It is often said that verification is about *building the system right*. Some systems cannot be tested because to test them is to destroy them. A defensive missile interceptor can only be fully tested by trying to intercept an actual missile – an expensive undertaking. Besides testing, there are other methods of verification including inspection, demonstration, analysis, analogy, and sampling.[2]

Validation on the other hand is about evaluating the item being validated to see if it accomplishes its intended use: does it meet the need? Is it fit for purpose? Does it work the way users and stakeholders need and expect it to? It is often said that validation is about *building the right system*. Methods of validation usually include showing the system, or aspects of it, to stakeholders and potential users and asking, or better yet observing if they can use it successfully. Those unfamiliar with the realities of complex systems development might be tempted to ask why there would ever be a need for validation, assuming that if the right requirements were specified at the outset, there would never be a gap between what the requirements state, and what is actually needed and expected. In reality, requirements can and do have omissions, ambiguities, and missing details, leaving the systems development team to fill in the gaps and make assumptions. The resulting system may meet the requirements, but not be what some of the users expected. When a large organization specifies requirements for a system, those requirements may come from many stakeholders, with varying perspectives, and may or may not involve the actual end users of the system. Worse, the requirements might be specified by people other than those who will actually use the system. These common situations all point to the need for validation.

The methods of verification and validation can also apply to the data being developed to train an AI algorithm. Verification should be used on the data to ensure that it meets the data requirements developed in the planning and analysis phases of the data lifecycle. If analysis showed that the data might contain certain kinds of anomalous or erroneous values, data requirements should be written to indicate that such values should be removed or replaced before the data are ready for use by the AI algorithm. Verification should ensure that this requirement has been met, and that all such faulty data items have been handled.

If it seems like the data lifecycle is a cumbersome process – documenting data requirements and then performing data verification to ensure that the data

---

2 Systems Engineering Book of Knowledge, https://www.sebokwiki.org/wiki/System_Verification.

requirements are met – consider two things. First, the consequences of training the AI algorithm with faulty data can be serious. Since the algorithm is evaluated (verified) based on testing data which are drawn from the same dataset as the training data, the algorithm might test out fine in the lab, but produce surprising results in actual use. The alien portrayed by Jeff Bridges in the 1984 film *Starman*, who learns by observing traffic that "green means go, red means stop and yellow means drive very fast," has learned something based on faulty input data. His learning may work most of the time, but because the learning data do not represent the behavior we actually want, his driving will be somewhat dangerous, as we see in the movie. Second, consider that training data are not collected and analyzed only once. Whenever new data are obtained, it should be taken through the data lifecycle to ensure their quality before using them to retraining the AI subsystem. Putting some effort into data requirements, data verification, and data validation work will pay off as the retraining work is repeated with each update.

As we have described, training data can be verified using the data requirements developed earlier in the data lifecycle process. Training data can also be validated. To validate the training data, we ask if the data is *fit for purpose*, if it meets the need. It is validation that should uncover issues like the ones we found in the earlier nurse recognition system example. The data requirements may simply state that the training data should consist of accurately labeled images of nurses and nonnurses. Or the requirements may go further and state that the mix of male and female nurses match the proportions in the real-world situation. As we saw, both of these choices, which could seem logical at the time they are made, may prove problematic after the system is trained.

While performing verification on something is a straightforward process of evaluating it against the stated requirements, performing validation requires more subjective judgment. A military vehicle can be verified by performing potentially thousands of tests, inspections, and analyses to prove that it meets the stated requirements. To perform validation, systems engineers might show the prototype vehicle, or a scale model, or a prototype dashboard, to some of the intended users of the system. They might just ask for general reactions, or they might ask the users to pretend to carry out part of a mission using the prototype. The use cases of the system (use cases are described in depth in Chapter 6) can be used here as walk-through scenarios to see if the users can imagine using the system to perform actual work.

By applying validation to training data, systems engineers could show the data to the system's intended users and stakeholders, perhaps using the data visualizations developed in the analysis phase of the data lifecycle. Users might be asked to imagine themselves to be patient humans and see if they can imagine doing what they want the AI to do, using the data they have available. Asking questions – does this data completely represent the range of situations the AI will encounter? Is there enough data to enable the AI to successfully learn patterns? – encourages users to think critically about the data on which the AI will base its behavior. Would performing data validation have caused the team to include images of people walking bicycles, or large white trucks blocking the road, in a self-driving car AI? Perhaps, and discovering those situations early could have potentially prevented an unfortunate loss of life.

In the world of data verification and validation, humans and AIs need each other. At present, AIs cannot determine what data they need to learn from. They rely on humans for that, just as children rely on their parents to teach them the fundamentals of everyday life. Humans must provide the right data and perform verification and validation to ensure that it *is* the right data, for the AI to be successful. The AI doesn't know what it doesn't know and will learn only what it is taught. Data verification and validation can take place before, or at the same time as, the development of the AI algorithm and should involve the systems engineers, data scientists, stakeholders, and users in collaboration. When the data passes verification and validation, it is ready to be deployed for use in training the AI subsystem.

Deployment. Training data are deployed by passing it over to the team developing the AI algorithm. Most system development efforts follow a concurrent engineering method (though perhaps informally), enabling (or at least tolerating) multiple different engineering disciplines working at the same time. With enough of the right kind of coordination, their efforts can be integrated into a successful system. Normally, some set of data are available and the AI team presumes it can be used to train an AI algorithm. The algorithm development team can get to work choosing the algorithm, designing the neural network structure, and coding the algorithm using the available data, while the systems engineering team, assisted by the data science team, work through the data lifecycle to ensure the quality of the training data to be used in the final system. Very likely if the algorithm performs acceptably on the initially available data, it will do even better after the data have been refined, augmented, supplemented with synthetic data, verified and validated to ensure it's the best data possible for the system purpose.

Maintenance. Nonspecialists may imagine that all AIs learn on their own as they are deployed in the real world, and thus improve their performance over time. Such adaptive learning systems are in the research stages but carry risks that are not easy to overcome. What if the AI learns from some rare or misleading situations, or is intentionally misled by mischievous or enemy humans? It may be difficult to know what the AI is learning if we allow it to learn on its own after being put into operation. But that doesn't mean the AI should never be retrained.

Operational systems may collect new data as they operate, and new data may become available from other sources. This new data can be brought back to the lab, verified and validated, and the AI retrained with the newly added data. After verification and validation of the system, the update can be released. How often the AI is retrained with new or improved training data depends on its performance and also on the degree to which the environment is changing. A shopping application may need to be retrained often to keep up with fashion trends. A trivia AI may need to know this year's movies, news events, and celebrity doings. Movie and TV recommendation systems must incorporate new information on a regular basis and are retrained often, by using reliable data from movie and TV producers.

1.11 AI Systems and People Systems

Systems engineers, who are appropriately concerned with the overall design and operation of complex systems, have deep concerns about deep neural networks and other kind of AI subsystems. Systems engineers are accustomed to subsystems that perform specific functions and produce quantifiable outputs based on specific inputs. They expect, quite reasonably, to be able to open up a subsystem and examine its circuitry, mechanisms, and software source code and see clearly how it does what it does. It's not that systems engineers need to be expert in all engineering disciplines or understand all of the internal workings of each subsystem, but they want to know that those workings *can* be understood and can lead logically and directly to a full understanding of how the system operates.

In fact, those who certify complex, safety-critical systems like aircraft demand to be able to see clearly how each bit of code or circuitry relates to the fulfillment of one or more system requirements. Code that is not there in fulfillment of a requirement should be removed, or the necessary requirement added to the requirements list so there is full traceability between requirements and code. They also want to see that all code is executed under some condition (that no code is never used) and that all code has been fully verified. Unused and therefore unverified code should be removed.

There are some good reasons why AI subsystems may be of concern to systems engineers. While some AI subsystems are programmed using conventional software code which can be examined directly, deep neural networks look like thousands or even millions of meaningless numbers. Now, to be fair, all software in its final machine language form appears as meaningless numbers. But those numbers are directly traceable back to software source code, which *is* human-readable. The numbers in deep neural networks can be thought of as coefficients in equations with hundreds or thousands of terms. Those coefficients were arrived at through a sophisticated process of trial and error and are what they are simply because those were the numbers that worked best. It's as if your home heating system had a control panel with 25 little dials, each with a scale labeled 1 through 10, with no other markings. With enough trial and error, you can try to figure out what combination of settings produces a comfortable home in the winter, and perhaps another combination for summer. You have no idea what the dials mean or how they function in the system. Some of them might not even be connected to anything. A deep neural network does the trial and error work to figure out the dial settings that produce a desired result. Someone looking at the dial settings afterward can't tell how or why the settings are what they are.

If we can't understand the behavior of an AI subsystem by looking into its insides, then how do we make sense of it? How do we design intelligent systems that contain one or even many AI subsystems? In order to think systematically about complex intelligent systems, one technique is to apply the human analogy again. The main function of an AI subsystem is to make decisions, offer recommendations, or provide predictions, so we can imagine a human being standing in the place of the AI subsystem and making those same decisions, recommendations, or predictions. The human might be slower and could be more or less accurate than the AI subsystem, but he or she is performing the same function. Like the largely inscrutable deep neural network, the human's internal functioning is not readily observable, and for the most part would not make sense to us even if we looked – science is still far from understanding how the brain makes decisions. Nevertheless, there is a way to understand, analyze, and design complex systems that include human beings as primary subsystems. Sometimes called human activity systems, these systems are described in Chapter 9 on systems thinking. Techniques especially suited to the analysis of human activity systems, like systems thinking and system dynamics, can also be applied to the design of any intelligent system, whether its subsystems are AI-based or flesh and blood. One of the primary methods used to understand and reason about *human activity* or *people* systems is the causal loop diagram, described in Chapter 9 on systems thinking. By using causal loop diagrams and reasoning about an AI-enriched system as one would reason about a people system, systems engineers can analyze a complex intelligent system effectively.

Let's review the overall systems engineering approach we are applying here. To design a complex system, systems engineers will break it down into subsystems, perhaps around 10 of them. Later they can break those subsystems down into sub-sub-systems, but let's stick with just the subsystems right now. In an engineered system, like a car, those subsystems are the large parts of the car, such as the engine, suspension, and electrical system. But in a human activity system, some of the subsystems are people, or groups of people. Let's take an example.

Consider an online university. The professors, administrators, and deans are part of the system, along with the computers and software – all are needed to deliver the courses. The interaction between a professor and the computer system that presents lessons to students can be understood as a use case (described in Chapter 6 on use cases) with a set of interaction steps, where the professor creates the lesson material, records a video, and sets up the quizzes and assessments. But how do we describe the interactions between the professor and the dean, and what impact do those interactions have on the overall online university system. In addition to some passing of data between the two – a professor submits final grades to the dean for approval – there are more subtle interactions that aren't as easily described as the flow of data between two parts of a computer system. Humans in systems are better thought of as making decisions, exerting influence, and providing input to other humans. Their decisions and influences aid other humans in making their own decisions, which in turn may be used by other humans or computers. We can use this idea of a chain of influence to better describe how AI subsystems function in an intelligent system.

Before we describe the how, let's consider *why* we want to describe subsystem interactions in an intelligent system. Most likely the reason is that we are engaged in the design process for a new system and we are reasoning about how to include new intelligent subsystems. As described already, there are concerns, real or imagined that it is natural to have about having so-called "intelligent subsystems" running around inside our previously neatly deterministic complex system. Actually, of course many non-AI complex systems, while they may be deterministic to an omniscient observer, are quite nondeterministic when observed by mere mortals with bounded rationality. Systems often surprise us, or fail us, not because they are inherently nondeterministic, but because we don't fully understand all of their interconnected behavior, so to us they appear nondeterministic. We need a way to reason about, describe, diagram, and model the desired behavior of these systems, so we can come as close as possible to making them not only performant but also safe and secure.

Let's take another: a self-driving car. Certainly we could try to describe the operation of the car in traditional systems engineering terms, with subsystems connected to each other using ports and connection paths, data traveling over interfaces using protocols, and functions that take in some information and produce other information. But let's try thinking of the self-driving car more in human terms, as if it contained little people. Philosophers use the term *homunculus*, to refer to a hypothetical little human inside the head of an actual human as a way to understand human behavior. These little people collaborate to make the self-driving car work. One might imagine the conversation between a few of them, like John Locke (the car's locking system), Austin Power (the engine and power system), and Vision (the video imaging system).

John Locke gets things going, "Hey everyone, our owner Mario has just entered the car. Looks like we are going somewhere."

Austin Power replies, "Yeah, baby, I love a good drive. I'm ready. Oh, he just pushed my button so I'm powering up my subsystems."

Vision chimes in, "The road ahead appears clear for now but I'll keep watch."

"Off we go," says Austin.

Vision thinks to himself, "My cameras are telling me there's a 70% chance there's an obstacle, but my LIDAR system shows only a 35% chance, so I'm thinking it's about 50/50 whether something's going on up there," and then says to the driver, "Hey Mario, be aware and get ready to take over control if necessary."

Austin is suddenly alarmed, "Mario seems to have a lead foot today; he's pushing the car to go faster."

Vision takes notice too, "I'll show a warning light to Mario – seems we could be heading into a problem situation."

Of course, the subsystems don't speak in human language, and they probably don't even have colorful names. What's really happening here, when we think of things in humanlike terms, is that the subsystems are *influencing* each other, providing input to each other and sometimes actually causing each other to do things. Perhaps if the system is more hierarchical, some subsystems may be able to *order* others to do things. Intelligent systems, especially if they have more than one AI subsystem, can be thought of this way. In our example, Vision made some decisions based on the information available to him/her/it from his/her/its subsystems, but those decisions came with uncertainty, just as decisions do from real human beings. If a human being is honest, rarely is a decision made with 100.0% confidence. I know I want to go see either that new action blockbuster or the historical fiction drama movie, but am I really 100% sure which one I would rather see tonight? I can gather input from other human beings, but then I still need to figure out how to combine their input into my final decision. AI subsystems need to do the same. Vision's cameras and LIDAR system had differing "opinions" on what was in the road ahead. Vision had to decide how to consider their input and make a decision.

Once we can imagine AI subsystems as little people in the system, then the tools of systems thinking and system dynamics, which help us model and understand human activity systems (see Chapter 9 on systems thinking), can be used to model and understand complex intelligent systems. Causal loop diagrams show the influence and cause-and-effect relationships between elements in the system. Within the vision system in the car, cameras and LIDAR both exert influence on the system and can be thought of as partial causes of the car's decisions to turn over control the human driver. Ultimately, of course, these functions are implemented using processors and data connections, but at the systems engineering level, it may be more useful to think of the subsystems as agents – more like people than simple input/output functions.

Large-scale recommender systems such as those seen in video streaming services like Netflix, are also complex systems of influences and causes. Since few viewers rate the programs they watch, the system must use other factors to arrive at a decision about what to recommend the next time the user sits down to watch something. What factors to consider, and how strongly to weight them in relation to each other is a complex and uncertain task. As before, ultimately, these factors are reduced to algorithms, program codes, and calculations, but at the systems level, might it not make more sense to think of them as a committee, with each member arguing for a particular position? The *recently watched* committee member argues for considering heavily what the user watched last night, while the *long-term patterns* committee member points out that last night's choices don't fit with the user's long-term patterns – perhaps there were guests or relatives involved. *New material* chimes in, pushing for heavier weighting on some new material the user might like. The committee members, called an *ensemble* in AI terminology, can be thought of as discussing, arguing, negotiating, and finally compromising on what recommendations should be shown to the user at a given moment, or, in the style of Netflix, users can be shown all of the committee's different recommendations each on its own row, and they can make their own decisions. As Netflix engineers are fond of saying, on Netflix, everything you see displayed is a recommendation.

The other main diagramming technique used to understand and represent human activity systems is the *stock and flow* diagram. Like causal loop diagrams, stock and flow diagrams can also be used to understand, reason about, and specify high-level behavior in complex intelligent systems. Stocks represent quantities of something and can be tangible or intangible. Tangible stocks could represent an amount of fuel, a sum of money, a number of complaints or a set of customers, while intangible stocks could represent the amount of goodwill, confidence in a decision, or strength of a recommendation. The stock and flow diagram shows the flow of these quantities from one place in the system to another, or in or out of the system altogether. Like the causal loop diagram, a stock and flow diagram can be used to model systems that contain only people, only machines, or a combination of both. By thinking of AI subsystems as little people or agents inside the system, systems engineers can more effectively reason about complex system behavior.

1.12 Making an AI as Safe as a Human

Systems engineers, who grew up in the pre-AI age learned about electronics, mechanics, and traditional software in school, and now function as successful systems engineers, based on this foundational engineering knowledge, plus of course deep knowledge of systems engineering. When an AI subsystem walks in the door and offers to provide its services by being included in the systems engineer's next complex system design, it's easy to sympathize if the systems engineer is hesitant. After all that AI subsystem is not like the subsystems the systems engineer learned in school – it doesn't have wiring, or metal and bolts, or even source code that can be readily examined to understand its function. To the systems engineer, it looks like a mysterious black box. Systems engineers are nothing if not risk-averse, and so are careful and conservative when it comes to new technology. An unreliable technology introduced into a system can make the entire system unreliable.

On closer examination, the discomfort of the systems engineer with an AI subsystem seems to result from its seemingly unlimited, or at least wide range of possible actions. How can they trust an AI to stay in its lane and perform within expected bounds? Once again, let's apply the human analogy. Human beings have this same annoying flexibility. No matter how loyal an employee may seem, there is always the possibility a new opportunity will be tempting enough to cause a surprise resignation. No matter how wonderful marriages seem at the outset, over half still end in divorce and the common claim – sometimes by both partners – is that the other changed. No matter what a politician says during the election cycle, there's at least a possibility that campaign promises won't be fulfilled. If we can develop enough confidence in human beings to hire, marry, and elect them, even with the chance of surprise behavior later, can we develop sufficient confidence in an AI to make it a reliable part of an important system? There is reason to hope for a positive answer.

First, as described previously, we can include safeguards in the system to watch for unusual, unexpected, or aberrant behavior and take appropriate actions. As in human society, we need watchers, and we need watchers to watch the watchers, and others to watch them. Second, our confidence in an AI subsystem can grow over time, even though it may seem mysterious and unpredictable at the beginning. Like human relationships, with familiarity comes predictability, and with predictability comes trust.

An initial level of trust, which can be anything from high mistrust to nearly complete trust, occurs in any relationship between humans or between humans and AIs. Over time, with familiarity and repeated reliable or at least predictable behavior, trust grows. When something happens that hurts trust, it is diminished. But by how much? Bayesian reasoning tells us mathematically what we know intuitively – that small infractions affect trust less than large ones, and trust built up over a long time is less affected by new infractions. Trust relationships with computer systems are similar. We probably all remember when Windows could be expected to crash almost daily – now, we are not at all surprised when we needn't reboot for days or longer. On the other hand, it is still wise to save work often, since crashes do still occur. An even better system design is the trend in software applications to save users' input as they type it, reducing the need for trust.

We will come to trust AI subsystems the same way: gradually, over time, with experience. Remember your first experience with a voice assistant in your home, assuming you have tried one? When I brought home my first Amazon Echo, I didn't know whether to have high or low expectations of its "intelligence." I was pleased that my voice commands and requests were usually understood on the first try, even from across the room. But I was not as impressed with its ability to handle conversational requests even a young child would have no trouble parsing. Eventually, I came to terms with "her" functioning mostly as a voice remote control, operating combinations of lights, fans, and audio/video equipment for my convenience. I trust the Echo to carry out these functions reliably, and I'm seldom disappointed. I found however, that asking questions about information was much less reliable. Sure, she's good at math – who doesn't love being able to say, "Alexa, what's 37 times 99 divided by the square root of two?" and hear her cheery response of, "2590.132"? Questions and answers about other matters are spotty. "What's the size of Jupiter?" produces a very complete response (from NASA), while "what's the size of Atlanta?" gives the size of Fulton County – not the same thing. Further tries like, "what's the population of the Atlanta Metropolitan Statistical Area?" still fail to give me what I want, offering only some random web-search attempts at something that might include an answer. So over time I come to trust the Echo completely in some areas, but so little in others that I've stopped asking her most kinds of general questions and carrying on a conversation? Forget it.

For some time, perhaps indefinitely, I think this will be our experience with AI systems. They will be one- or two-trick ponies, good at some specific tasks, but far short of the kind of mental and verbal flexibility we expect of even young human beings. That's OK. They will be useful for what they are. Even something as sophisticated as a chess-playing program may not be able to have a casual chat about the weather, but do we really need that in an automated chess partner? There are still plenty of things that seem quite manual, even unintelligent to me, like tasks on my computer that require me to manually copy and paste information from one system or application to another, or to re-enter information in multiple places. Many systems still require me to type in my address, city, state, and zip in that order, while others are intelligent enough to match my entire address from only the street number and a bit of the street name.

Our perception of what intelligent systems can or should be, colored by imaginative film and television writers and producers, is far too dualistic. We imagine only two categories: unintelligent systems that we control with buttons, mouse-clicks, and keystrokes, and near-human level intelligences and androids. As discussed elsewhere in this book, this nonsensical assumption shows through even in science fiction stories, where, for example, there is a huge gap between the intelligence of the android Commander Data (near human-like) and the much more limited intelligence of the computer than runs the entire Starship Enterprise.

In this chapter, we discussed how to apply the main concepts and practices from systems engineering to the design of intelligent systems. Readers who have deep experience in systems engineering will probably find the explanations given rather cursory, and may have been hoping for more detailed processes, templates, and artifact definitions. While it's true that such detail is beyond the scope of this book, an even more important factor is that the application of systems engineering to AI and intelligent systems is still in its infancy. In coming years, detailed methods, and approaches will be developed. For example, SAE, the large engineering professional association, is developing a standard (G-34) for how AI technologies can be used safely in aviation, but it will be years before it is complete. At the same time, universities are starting to educate engineers on AI, at least through optional electives, so that new generations of engineers won't have the handicap of unfamiliarity with AI that most of us did.

References

Crockett, Z. (2018). What happens when a robot kills someone? https://thehustle.co/when-robots-kill/ (accessed 10 May 2022).

Ferriss, T. (2009). *The 4-Hour Workweek*. Germany: Crown Publishers, p. 29.

Shane, J. (2019). Learning to hack like a faulty AI. https://www.aiweirdness.com/learning-to-hack-like-a-faulty-ai-19-04-26/ (accessed 10 May 2022).

2

We Live in a World of Systems

> *There exist models, principles, and laws that apply to generalized systems or their subclasses, irrespective of their particular kind, the nature of their component elements, and the relation or 'forces' between them. It seems legitimate to ask for a theory, not of systems of a more or less special kind, but of universal principles applying to systems in general. In this way we postulate a new discipline called* General System Theory. *Its subject matter is the formulation and derivation of those principles which are valid for 'systems' in general.*
>
> (*General System Theory*, Ludwig von Bertalanffy, 1951)

We live in a world of systems. The term *system*, which came to us from the Latin and Greek, *systema*, meaning an organized whole, consisting of parts, has become popular as a way to describe the increasingly complex world we experience. Perhaps not surprisingly, our current concept of systems came from a biologist, Ludwig von Bertalanffy who in the 1950s and 1960s generalized what he saw by studying plants and animals into a general theory of systems. His idea was that systems of all kinds – whether created by Mother Nature or designed and built by human beings, share certain important characteristics, which both identify them as systems, and also point us toward a greater understanding of how they work and what they accomplish. In this chapter, we aim at developing an intuition for systems – how to identify them, describe them, think about them, and ultimately design and build them.

2.1 What Is a System?

What *is* a system? It's easy to slip into the assumption that everything is a system in some way, but in doing so, we lose the concept of a system entirely. A system is something that is composed of interacting parts, that together, in a particular

Engineering Intelligent Systems: Systems Engineering and Design with Artificial Intelligence, Visual Modeling, and Systems Thinking, First Edition. Barclay R. Brown.
© 2023 John Wiley & Sons, Inc. Published 2023 by John Wiley & Sons, Inc.

arrangement, provide some function or fulfill some purpose. Several aspects in that brief definition are important. First, a system must have parts, and it must have parts that interact. It is not necessary that every part interact with every other part, but if no parts are interacting, there's probably not a system. Next, the arrangement of parts is important. Rearrange the parts in a system and you have a different system, one that might not provide the function it did before your rearranging. A system must have a purpose or function. Some have simplified this idea to say that a system must have a purpose, but the simplification becomes problematic when we think of systems that occur in nature. What's the purpose of an ant? The answer quickly becomes metaphysical as we must also determine who or what designs or dictates the purpose of the ant. The ant does not worry about this, so neither should we. It is clear that the ant has a function to perform, which it does quite efficiently. The other difficulty in thinking that a system must have a purpose is the problem of misuse. If I use a system for a purpose other than its actual purpose (however that is determined) have I somehow changed the system's purpose? Does it make sense that you and I would say that the system has different purposes? It is therefore better to think of the function or functions of the system. I use it one way; you use it another. It's the same system, and it has functions that we make use of.

What is not a system? Based on what we have said above, we can identify things that are undeniably real, but which are not systems. One example is a collection. A pile of sand on the beach, on its own, while it does have parts (the grains of sand), does not have interacting parts or a function or a purpose. We can use that pile of sand to protect the shoreline, but our use of something does not make it a system. For the most part, we don't consider things that have parts which are only attached to each other in a fixed, mechanical way, a system. A hammer consists of parts, but they are only attached to each other and thus don't have interactions other than mechanical stress. The hammer, on its own, performs no function and serves no purpose. Combined with a human user, we suddenly have a system, and the *human plus hammer* system can perform great things indeed. There are those who might point out that at the molecular level, there are many important interactions going on in the hammer, as electronics spin and atoms collide, but for the most part, the kind of systems that systems people care about are those that have more readily observable behavior.

Before going further in describing systems in general, it helps to know that there are three distinct types of systems. All are systems and have many characteristics and behaviors in common, and when we speak of systems, we mean all of them. Depending on your perspective, you may think of only one of these as a system and not the others, but all are systems.

2.2 Natural Systems

Natural systems are those that have appeared in nature for us to enjoy and observe, with no particular effort on our part. Most natural systems are biological, but natural is the broader team, including systems like tectonic plates, stars like our sun, volcanoes, and other natural but nonbiological systems. Examples of natural systems are everywhere. A plant is a natural system, as is the bee buzzing around it. A fish is a system, and so is a school of fish, pointing out how systems are often composed of other systems. The ocean in which the school of fish swim is also a system, consisting of many other systems within it. Human beings are systems of course, and each contains other systems like the circulatory system, skeletal system, and nervous system. Almost anywhere you look in nature, you can see systems. While we said that a pile of sand on the beach is not a system, the beach itself, with sand, wind, water, and plants is clearly a system. Wind and sand in the desert can interact in system-like ways, too.

Natural systems are also distinct from other kinds of systems in that the designer and the designs are not readily accessible. We can look, we can study, we can even take apart biological systems to understand how they work, but we don't have access to *how*, or more importantly, *why* they are the way they are. We must study them "from the outside" as it were. Biologists are keenly aware that they may never fully understand the design of the systems they observe, no matter how many examples they dissect or how many advanced imaging scans they can perform.

With natural systems, we confront the paradox of studying systems. On the one hand, we know that systems are made up of interacting parts, and we might imagine that the way to understand the system is to disassemble it into its constituent subsystems, parts, and components, study each carefully and with the understanding of each part, summed up, we will understand the system. There is truth in this approach, and it does provide a kind of understanding of the system, but it has its limitations. An alternative, but not incompatible approach is to consider the system as a whole, and not as just an assemblage of parts. We can examine the function or purpose of the system, its context or environment, and the interactions it has with other systems. We can study the behavior of the system, and how it is used – what we'll call *use cases* in Chapter 6. We can study how the system responds to input or stimuli. Is there a linear response, meaning that doubling the input will double the output, or is the response nonlinear, where an input that reaches some threshold value causes a drastic change in the system's response, like the straw that breaks the camel's back? We can study a system in all of these ways without giving any attention at all to its constituent parts.

With natural systems, we take both approaches. We even have specialized scientific fields to address each. Some biologists study biological systems by breaking them down, taking them apart, and trying to figure out how they work. Others, like naturalists, study the behavior of the system itself. The two approaches are not totally distinct, when we consider different concepts of the *system*. In a forest, we see a beaver, and consider quite appropriately that one beaver as a system. We can study the beaver without breaking it down into its component parts. In fact, if the first thing we do is take the beaver apart, we won't learn much about the beaver-as-a-system at all. Better we study the beaver as a whole before we break it down into parts. As we observe the beaver though, we see that it is part of a larger system – a colony of beavers. It's a bit harder to study the colony of beavers without paying attention to the parts of that system – the individual beavers themselves, but nevertheless, we *can* study the colony. We examine how the colony behaves, when it (collectively) decides to build a stereotypical dam, or move to a new home location. We can look at how the colony interacts with other parts of an even larger system – the forest, and up the system-of-systems chain we go.

The two approaches are intertwined in another way as well. We naturally switch between systems-level observations and explanations and reductionist or decomposition approaches. When a doctor sees that the system (a patient) is sick, the next step is to determine which subsystem, or combination of subsystems is the culprit, and the doctor performs a differential diagnosis based on the symptoms exhibited by the overall system (the patient) and knowledge of its subsystems, how they work and interact and the relative likelihood of the various possible causes of the observed symptoms and test results. With most biological systems, we are keenly aware that we don't fully understand the design and have no access to the designer or the original plans and blueprints. Even when we take biological systems apart, we still have the feel that we are looking in from the outside, trying to understand the purpose and function of both the system and the parts. We sympathize with Einstein who commented on the human quest to understand the physical world and reality itself:

> Physical concepts are free creations of the human mind, and are not, however they may seem, uniquely determined by the external world. In our endeavor to understand reality we are somewhat like a man trying to understand the mechanism of a closed watch. He sees the face and the moving hands, even hears its ticking, but he has no way to open the case. If he is ingenious he may form some picture of a mechanism which could be responsible for all of the things he observes, but he may never be quite sure his picture is the only one which could explain his observations. He will never be able to compare his picture with the real mechanism and he cannot even imagine the possibility or the meaning of such a comparison.

But he certainly believes that, as his knowledge increases, his picture of reality will become simpler and simpler and will explain a wider and wider range of his sensuous impressions. He may also believe in the existence of the ideal limit of knowledge and that it is approached by the human mind. He may call this ideal limit the objective truth.

(Albert Einstein and Leopold Infeld, *The Evolution of Physics: From Early Concepts to Relativity and Quanta*, 1938)

2.3 Engineered Systems

The second type of system is one that is engineered, conceived, designed, and created by people. Of course, these systems are everywhere – cars, aircraft, computers, spacecraft, factories, and plumbing, heating, and air conditioning systems in our homes, to name just a few examples. From the earliest tools and mechanical devices that could be called systems to sophistication of the space shuttle, people have engineered increasingly elaborate and complex systems. Perhaps obviously, the hallmark of an engineered system is that it was engineered, that is, an idea for the system was born, likely to satisfy a need or exploit an opportunity, requirements were determined, designs were considered and one chosen, and the system was constructed. With an engineered system, we can, at least most of the time, find descriptions of the original needs, listings of requirements, and design documentation. We may be able to talk to the designers and engineers to learn even more. It is likely we can take classes and earn degrees that enable us to understand the principles and techniques used to engineer and build the system. There need not be any mystery about the why, what, or how of an engineered system. Just because everything *can* be known about an engineered system, however, doesn't mean everything *is* known to everyone. It is often the case that we confront an engineered system and don't have immediate access to the designers or even the documentation. In those situations, we can resort to treating the engineered system as we might any other system we must observe from the outside.

Engineered systems are, first, systems, and we can use system methods and concepts to analyze and understand the behavior of engineered systems, even if we don't have access to the designers and documentation. Mechanics and technicians use this approach all the time. They confront a system that is complex and which they can't hope to understand as well as the engineers who designed and built it. Yet it falls to them to diagnose problems and make repairs. Rather than jumping to the conclusion that it must be this or that component, replacing it, and if that doesn't cure the problem replacing additional components until the problem is resolved, a good technician will study the system's behavior and form theories about how the system could be exhibiting the observed behavior, testing the theories until a confidence level is reached and an intervention chosen.

2.4 Human Activity Systems

Human activity systems are systems that are motivated by people. These systems may contain computers, vehicles, and other machines and technology, but the system is driven by the people. A company is a clear example. A company may not be able to run efficiently without some of its computers, factories, or trucks, but remove all the people, and the company is no more. Universities, governments, economies, and families are all human activity systems. When we say that we live in a world of systems, many of the systems we live with every day are human activity systems. In a way, human activity systems are a hybrid of the other two types. A system consisting only of human beings could be considered a natural system, but when we combine that system with a set of machines, computers, and other technology, we have something special, so we use the term "human activity system." The reason has to do with the unique position of human beings on planet earth. No other species uses technology or creates complex systems the way we do.

From a broader perspective, it could even seem odd that human beings have evolved to be so far advanced beyond any other species on the planet. What would prevent a planet, as illustrated in the TV series Star Trek: Enterprise, from developing more than one sentient species at the same time? In the series, on a remote planet, the Xindi have evolved six sentient species: aquatic, arboreal, reptilian, insectoid, and an extinct avian species, in addition to their human-like primate species. There, the term *human activity system* would not fit well; we might need a new term since each of these species has developed high levels of technology, including space travel. Here on earth, however, human beings are the only species that, together with our technology, form complex systems. Human activity systems deserve special attention, and we will return to them later in this book. Human activity systems that involve only people also deserve special treatment; we'll call them people systems and devote Chapter 10 of the book to these fascinating systems.

2.5 Systems as a Profession

Like any field of study, the world of systems has its practitioners, and they fall into several kinds. A single practitioner may not only act in more than one of these roles but also like other fields, over time roles become more specialized.

2.5.1 Systems Engineering

Systems engineering is a specialization within engineering and stands alongside other engineering specializations such as electrical engineering, mechanical engineering, software engineering, civil engineering, chemical engineering, and others. Systems engineering, as one might expect, focuses primarily on engineered systems. Systems engineers work to determine the concept, requirements, architecture and high-level design for a new system, or a revision or enhancement to an existing system. The work systems engineers do is then used by other engineering specialties to carry out their more detailed designs. While electrical engineers, for example, focus on the design of the electrical and electronic components of the system, the systems engineers focus mainly on the system-level concerns, such as the system's overall behavior and architecture, the needs it is designed to fulfill and the way it will be tested, verified, and validated. Systems engineers, in contrast to other types of systems practitioners, are focused on *building* systems.

2.5.2 Systems Science

Behind every practical discipline, there is a foundational science on which it is based. The developing field of systems science aims to build this foundation by identifying and describing the science of systems. There are concepts that apply to systems of all kinds (we'll explore some of them in a moment) and systems science aims to study and formalize these. University courses and even specialized degrees are becoming available in this fascinating field.

2.5.3 Systems Thinking

While some may assume that systems thinking is nothing more than the thinking that systems engineers and systems scientists do, this aspect of systems has emerged to become its own distinct subfield, with its own unique approaches, methods, and applications. Systems thinking gained broad attention through Peter Senge's popular business book, *The Fifth Discipline*. Put simply, systems thinking is aimed at solving problems using systems approaches. The problems best addressed by systems thinking are systems problems – problems that arise from the behavior, or misbehavior of systems. These kinds of problems do not readily yield to simple one-step solutions, because they exist, not in a single place, but within complex systems. Economic problems such as high unemployment, income disparity, or rising national debt are symptoms of systems at work. Systems thinking identifies patterns in systems, called archetypes, which can help systems thinkers understand how systems work and then apply carefully chosen interventions. Meddling with a system one does not understand, though commonly done, does not often yield good results.

So far systems thinking has not become a role title. There are no professional systems thinkers, and perhaps that's best. Systems thinking is a skill set that can be applied successfully by policymakers, leaders, and change agents of all kinds.

There are specialized university courses in systems thinking (I have taught some), usually at the graduate level, and offered to engineering, MBA, and other kinds of graduate students.

Related to systems thinking is a subfield called system dynamics, which is a very useful approach to analyzing and diagramming the operation of systems and the causal relationships in them. System dynamics uses two main concepts to show how even very complex systems operate: stocks and flows diagrams, and causal loop diagrams. These are described more in Chapter 9.

2.6 A Biological Analogy

To make these aspects of systems even clearer, let's use the analogy of the field of biology and medicine. System science corresponds to biological science. Both represent the foundational science upon which other related fields are based. Continuing the analogy, systems engineers, who are primarily concerned with the designing and building of systems would correspond to say genetic engineers, who seek to use the science of biology to construct new biological systems. Systems thinkers, concerned with problem-solving in systems, would correspond to physicians, who solve problems in biological systems, primarily people.

Systems engineers and systems thinkers use quite different concepts and tools, but both draw from the field of systems and system science as a base. Systems engineers are concerned with designing and building systems and need to talk to users about their needs and requirements, do high-level design of a system, create various kinds of models, and do trade-off studies of alternative materials and architectures. Systems thinkers on the other hand, are all about solving problems in existing systems, so they need to identify patterns in systems, understand how a system works, and then devise ways of modifying systems to achieve desired results. Systems thinkers often have a wide-ranging curiosity about all kinds of systems, taking them into fields such as economics, psychology, culture, business, and government as well as various aspects of engineering and design. Systems thinkers come from many fields and may not be engineers at all, but can be business and government leaders, or anyone who is trying to improve the world we live in.

2.7 Emergent Behavior: What Makes a System, a System

We said before that systems are distinct from simple collections. One of the key distinctions is the concept of emergent behavior. As an initial example, consider an aircraft. The aircraft can fly, but it makes no sense to ask which part of the aircraft is responsible for its ability to fly. The engine does not fly, the wings do not fly, the seats do not fly. Flight is only possible because those parts and many others are combined into a system. It is the system that has the ability to fly. Keeping to the skies, let's consider a flock of birds. They seem to move almost as a single organism, shifting and changing as they move across the sky or among the trees. Studies show that there is no bird in charge, no one is leading and directing. The flock's complex behavior results from simple, built-in instructions that guide each individual bird. Those instructions consist of something like: stay at a reasonable distance from the birds around you and don't bump into them. If each bird operates this way, the entire flock system exhibits the mass flying behavior. Consider an army of ants, able to build large complex structures when undisturbed. There is no master architect – each ant follows its own instincts and the system of the army builds the structure. Understanding emergent behavior does not mean that we know exactly how the ants or birds do what they do, but we see that they do it as a system, and can then investigate the system further.

Emergence can also apply to the properties of systems, not just to their behavior. Hydrogen and oxygen are both gases at room temperature, but when combined as H_2O, they become water and have the property of wetness. Neither hydrogen nor oxygen are wet, but when combined, a property emerges called wetness. Further, studying hydrogen and oxygen on their own, would not lead someone to expect wetness when they are combined in the right proportion. It is common that emergence cannot be predicted, based solely on the properties of the system's subsystems or parts. The emergent property or behavior emerges, sometimes by design, but often serendipitously, as a surprise.

Emergent properties and behavior, as you can see from these examples, can occur in either natural, engineered, or human activity systems. An amusing example comes to mind. When going through security to board an airplane years ago, at a very small airport, I noticed that the single security line was much longer than usual, and that people were exiting the security checkpoint at a slower rate than I was used to experiencing. When I looked at the process that was occurring, I noticed that a security officer was stopping each person just as he or she was about to enter the scanner. The stop was only for perhaps three seconds, while the security person asked the person some security question. To the officer, there was no problem – it took only a few seconds to ask the question and stopping each person before the scanner seemed the obvious and most convenient (for the officer) place to do it. From a systems perspective, however, since the person was stopped as he or she was about to enter the scanner, meaning the scanner was free, the three seconds was a delay added to the entire security process. During the three seconds, the scanner was idle, so the entire output was delayed by that three seconds. How big a deal is that? I counted all of the people in line, and there were about thirty. With a constant line of thirty people, and a system delay of three seconds for each one, every person experienced an added delay of 1.5 minutes going through the checkpoint. If the checkpoint were designed to handle say about ten people per minute, adding a 1.5 minute delay to each person would drop the throughput to eight people per minute. Where do those extra two people go? Into the line of course, and we can expect that line to grow at the rate of about two extra people every minute, compared to the usual throughput. An hour of this, and we'll have a line of thirty extra people, which is exactly what I observed. The story illustrates several points about systems. Small changes applied at a critical point can make a huge difference. Delays are powerful system interventions and can have large, cumulative effects. One part of the system (the officer) introduced what seemed like a "no big deal" change in the process but because of where in the process the delay was inserted, the system effect was dramatic. A possible solution? Insert the new process (the question-asking), where it will have no effect on throughput of the system: simply ask the questions of the people who are waiting in line. Since they are waiting already, the additional question-asking would introduce no additional delay. There is an entire field known as queuing theory that analyzes and explains system phenomena of this type.

Emergence is everywhere and is responsible for many of the benefits and difficulties we face when using systems. Even in seemingly simple systems, we can see it. A person riding a bicycle is a system of course – and in fact, it is one of the most energy-efficient ways of moving across a distance, based on calories consumed and distance traveled. Is getting this system to exhibit the emergent behavior of riding easy or hard? A small child can learn to ride a bicycle, but it's quite hard to explain to them how to do it. The concept of balance is only clear after it has been experienced. How hard would it be to get a robot to ride a bicycle? Perhaps not as hard as one might imagine. A robot that does nothing but turn the steering to the left when the bicycle starts to lean to the right, and vice versa, will exhibit the emergent behavior of balancing the bicycle. It can be shocking to realize that almost everything we experience is a result of emergence. Philosophers may argue about the nature of consciousness, but most would agree that consciousness is an emergent function of brains, which is to say that the only kind of consciousness we know about is produced by brains. We will have much more to say about this in another chapter.

Emergence is responsible for all of our experiences. Our ability to experience is emergent, as are all of the phenomena we experience. A sunset is an emergent property of air, light, and atmospheric conditions. A sonata is an emergent property of the individual instruments and conductor arranged into a system. One cannot understand or even perceive a sunset or a sonata, by examining only the parts within the systems that produce them. The hardness of the desk upon which I am writing this is an emergent property of the materials in the desk together with the emergent property of my body and brain which I call "touch." The feelings I have for my loved ones are emergent properties of my consciousness – no brain, no consciousness, no consciousness, no love.

A fascinating example of emergence is the sciences themselves. At the lowest level, everything is based on physics. Everything in the world is composed of the stuff of physics – molecules, atoms, and particles. But examining the world through the lens of physics and looking only at these small parts, does not reveal what we now call chemistry. Molecules combine in interesting ways to produce materials that have emergent properties – water is wet, and rocks are hard. Chemistry uses concepts and methods quite different from physics to study and understand the same physical reality. We can say that chemistry is an emergent property of physics. Extensive study of chemistry, however, does not yield any understanding of life. Life is studied using its own concepts and methods in the field of biology – biology is an emergent property of chemistry. Study biology in depth, and interesting insights and discoveries are possible, including the fields of reproduction, genetics, and medicine. Yet when we study the biology of human beings (and some animals), we do not gain insight into distinct characteristics such as consciousness, thought, emotion, and motivation. These are part of the field of psychology – an emergent property of biology. We can extend this chain of emergence even further, noting that sociology is an emergent property of psychology, and that anthropology can be seen as an emergent property of sociology.

Emergence is fascinating on its own, but it is also highly relevant to the design of intelligent systems. The purpose of a system in essence is to offer us its emergent properties. We design and construct a vehicle to offer us transportation. We design military weapons systems to offer us defense and security against attackers. We design aircraft to give us the property of flight and vacations to Paris. We construct families, neighborhoods, and communities to offer us belonging, security, and enjoyment.

Emergence and systems are deeply connected concepts. We define a system as a set of components, connected in a specific way, that together have a function or purpose. The function of a system is an emergent property of that system. The function of a system emerges from the specific combination of the components of the system – it emerges from that combination. In an engineered system, we may have both emergent properties that have been intentionally designed into the system, and also unintended, even unwanted behaviors that emerge on their own, to our surprise. An amusing example is the Mould effect, named for Steve Mould, a science presenter. A long, metal bead chain placed in a pot and when allowed to drain over the edge, will mysteriously rise well above the level of the top of the pot and self-siphon until all the chain is transferred to the floor. The fountain effect is an emergent property of the bead chain's construction, it is limited flexibility and interestingly, the height from which the chain starts relative to the floor (the whole story is told in Steve Mould's 2014 TEDx talk, *Investigating the "Mould effect"*), but it took some scientists to figure out exactly how this effect emerges from the system. There is an important lesson here about systems – a system can be engineered and even perform a function, well before the underlying science is understood. In his wonderful books on engineering, including *To Engineer is Human*, Henry Petroski makes the convincing case that often engineering precedes science. The Wright brothers created and flew their airplane long before the physics of flight, lift, and drag were understood and reduced to mathematics. The Romans built bridges without the benefit of force vector calculations. While in some cases, engineers *apply* science, in many other cases, engineers design and build systems that deliver functions and solve problems using emergent behavior, and scientists must come along after and try to understand and explain why the behavior occurs.

2.8   Hierarchy in Systems

Another fundamental concept that appears in all kinds of systems is *hierarchy*. When Ludwig von Bertalanffy first observed biological systems and formulated the idea of a general theory of systems, he began by noticing the obvious hierarchy in systems. When we look inside a biological system, like the human body, we see a clear hierarchy of subsystems. We can identify a circulatory system that has its own purpose and function, we can identify a respiratory system, nervous system, reproductive system, and others. Within each system there are clear components. The circulatory system includes the heart, the blood vessels, veins, and arteries, and the blood itself. The lungs are involved in oxygenating the blood. If we go down another level in the hierarchy and examine just the heart by itself, we can also see parts, we see the different chambers of the heart, the valves, the muscle structure, and so on. Hierarchy is inevitable in systems – we can always see a hierarchy if we look closely, and hierarchies appear equally in biological systems, engineered systems, and human activity systems. In a human activity system such as a government, the hierarchy is plain to see. Each branch of the government has someone at the top, with a set of ministers or senior officials reporting to him or her, and deputies below them.

In a company, it's the same thing – there's a CEO, and then executive vice presidents, with senior vice presidents below them, and down to vice presidents, directors, assistant directors, and managers. Even a small company of ten people will soon find itself organizing into some kind of departmental hierarchy. It's not because hierarchy is a particularly good thing. Hierarchies are simply a natural way to organize systems and their parts. We can see that nature organizes things in hierarchies. As human beings, we are part of nature and so we also organize things in hierarchies. There are subsystems in a house, and within the subsystems, there are components. Those components have parts, and those parts have parts, and so on. Hierarchy simply happens.

If we are in a philosophical mood, we might ask whether systems and hierarchies truly exist in nature, or whether they are constructs that we use to understand what we see in nature. It's an ontological question, examining what exists and what doesn't, and to the practically minded engineer, it doesn't really matter much. We find that looking for hierarchies in systems, we observe, and using the hierarchies to design the systems we need is very useful. When we look at a mysterious new system, whether engineered, natural, or human activity, a good way to begin could be by looking for the hierarchical structure in it.

Hierarchies allow us to organize our understanding of a system and enable us to avoid considering all the parts of the system at once. Organizing is simplifying. In conceptualizing a hierarchy for a system, whether a natural system we are observing or an engineered system, we are creating, we will invent levels in the hierarchy in order to understand and organize things. In the human body, we invent the concept of a circulatory system, so that we don't have to deal with the whole body at the same time. Naming the circulatory system helps us organize and pick out the parts related to the function of that subsystem. We may notice that boundaries of our named subsystems are not completely distinct from each other. The circulatory system and the respiratory system share the lungs, so the lungs are part of both systems. If we ask the lungs which system they belong to, they might reply that they know nothing of the circulatory or respiratory system – they simply have a job to do, and they do it. The lungs are reminding us that hierarchy is a concept we are bringing to the system in order to help us understand it. The hierarchies we invent to help us understand need not limit us. In system terms, the lungs are an interface – a part in a system where two or more systems connect.

Hierarchies also make visible the concept of a span of control. In an organization, typically there is a recommended or commonly practiced span of control specifying that a manager should not have more than say, 10 direct reports. A custom or constraint like this limits how the hierarchy works and dictates when a new level in the hierarchy should be created. In designing system architectures, experts have sometimes suggested that there should be between five and nine subsystems under any system. This odd range comes from Miller's law based on cognitive psychologist George Miller's research at Harvard, published in 1956 which indicated that the human mind can only effectively deal with "seven plus or minus two" elements at the same time. Of course Miller's law is not enforced and system designers are free to describe 25 subsystems under a single system, but very likely, if the list of 25 subsystems is examined, there will be some obvious and helpful ways to group them into between five and nine groups, and we are back to the guideline of Miller's law.

For example, if we consider an automobile and try to understand it as a system, or if we are looking to design a new automobile, we can consider the automobile in its entirety as the system. The next level of hierarchy would be the auto's subsystems. Keeping to Miller's law, we should identify only between five and nine, so we might have something like propulsion, suspension, body, steering, climate control, entertainment, and interior. Of course, there is no right set – the choice of how to divide up the auto's architecture is up to the observer or designer. At the next level of hierarchy, we would identify between five and nine subsystems for each of these. For the entertainment subsystem, we could choose to identify the next level of subsystems as something such as radio, disc-playing, amplification, navigation, and communication. It may be surprising that even with 25,000 to 50,000 parts in the car, down to individual screws and rivets, we need only five or at most six levels of hierarchy to include everything, using Miller's guidance for span of control.

Hierarchies are powerful simplifiers. When I create the auto's hierarchy, I can reason about the interactions between the subsystems and components in various levels of the hierarchy. I can examine, for example, the interaction between the propulsion subsystem and the suspension subsystem, or between the radio and the amplification system.

The notion of hierarchy is deeply embedded in the practice of systems engineering. The systems engineering Vee model (Figure 2.1) is used to show how, in any complex system design, there are several major phases of development. First, some overall work is done to determine overall system needs and requirements and analyze the needed behavior of the system. That behavior is broken down into functions that can be allocated to elements in the system hierarchy. Systems engineers break down the behavior and assign it to elements so that it can then be assigned to certain kinds of engineering teams. They identify mechanical functions, allocate them to mechanical system subsystems and components, and then assign teams of mechanical engineers to design, build, and test those mechanical parts of the system. Electrical requirements will be fulfilled by electrical engineers and software requirements that will be fulfilled by software engineers.

All complex systems are built this way – individual components and subsystems are built by the engineering discipline appropriate to that subsystem, and then these subsystems are integrated together to form the overall system. If it's a car that is being built, we are going to have a team somewhere building the transmission, and another team somewhere building the engine, and another team building the entertainment system, another team building the body, and so forth. When they're done, we integrate those subsystems, and we have a car. Well, not so fast. Experience has taught us that it's probably not wise to integrate them for the first time all at once, and hope the whole thing works. A better plan is progressive integration. First, we might integrate the transmission and the engine and make sure those

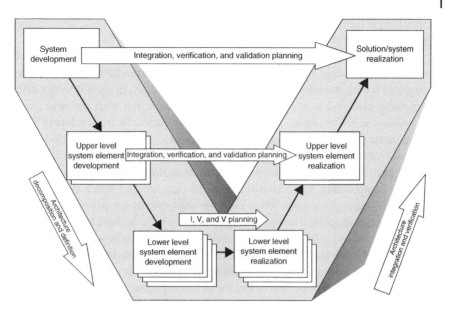

Figure 2.1 Systems engineering Vee model. Source: From INCOSE Systems Engineering Handbook, 4th ed., 2015, p. 34, figure 3.6, derived from Forsberg et al. (2005), figure 7.10. Reprinted with permission from INCOSE. All other rights reserved.

work together through extensive testing. Then we can take that unit – the engine plus transmission – and integrate that with the chassis, and test again. There's a progressive integration of components and testing of ever larger assemblies of integrated components to construct the final system. This is good practice with any system engineering project, and the more complex the system, the more necessary it is, and the more integration becomes a significant concern.

In fact, history has shown that complex systems projects are often delivered after their scheduled time, and at a much higher cost than originally intended. Often, the part of the process that took longer than expected is the integration and testing of the system. It might be tempting to think that we can solve that by simply reducing the amount of time we spend in integration and test. But what's probably happening is that during integration and test, we discover things that need to be fixed in the system components. It's really not the fault of integration or test, it's that integration and test are acting more as the discovery mechanism to expose problems in the subsystems. This observation has given rise to agile methods with systems engineering. Instead of creating all the components and integrating them at one moment, late in the process, and hoping it all works together, we do progressive and maybe even continuous integration. This is done by taking each component as it's built, and integrating it with something else, so that as soon as something is ready to operate, we integrate it with something else so we're doing little integrations all the way through.

Continuous integration has been practiced very successfully in the domain of software engineering. Each programmer building a particular software module integrates it with the whole system often, sometimes daily, to see if it works with everything else, and to see if some unforeseen interaction with the new module breaks the entire system. Daily integration prevents software developers from going too far down the road of constructing a module that will only later be discovered to be problematic when the final integration is done. Progressive or continuous integration has proven to be more effective than the independent construction of many components, and a later total system integration. From a systems perspective, it's easy to see why. When multiple components are put together all at once, especially unproven components that are newly built, there are many interactions being tested at once. If a failure occurs, it's hard to know which of the new, unproven components, or their interactions are the culprit. In progressive integration, there is just one new component being integrated at a time, and since the entire integrated system worked before the new component was added, any failure must be due to the new addition.

2.9 Systems Engineering

Let's take a closer look at the main ideas in systems engineering. Systems engineering is one of the main applications of systems in general and has been effective in managing the development of complex systems. Systems engineering is naturally an interdisciplinary approach because it considers the system as a whole. Any complex system will consist of parts and components developed by multiple engineering disciplines, integrated together, including mechanical, electrical, software, and sometimes chemical or aeronautic subsystems and components. The main purpose of systems engineering is to enable the realization of successful systems development projects. Systems engineering focuses, especially in the early stages on understanding and defining the customer's needs and the required functionality. Fully understanding a customer's needs, wants, and expectations is not as easy as it may seem, as we'll discuss in much more detail later. Understanding what the user really needs can be a challenge, since customers (and all human beings) often find it hard to know and express what they need. Systems engineers work hard at defining the requirements and getting the stakeholders to all agree on the requirements before they start the construction of the system.

We can better understand the systems engineering by considering both a breadth and a depth view. In the breadth view, we see systems engineering as a profession, with its own set of skills, methods, techniques, and processes. Systems engineers often begin in another engineering specialty and later broaden their knowledge of systems, sometimes earning a master's degree in systems engineering to add to their undergraduate degree in another type of engineering. The International Council on Systems Engineering (INCOSE) is a professional society for those in the profession of systems engineering, and they offer three levels of certification as a systems engineer.

In the in depth view, we can see systems engineering as a set of practices and techniques that can be applied at the top systems level, or at any level below it. Systems engineering practices can thus be applied to the overall system, say a satellite being built for NASA, and also the subsystems level, say a weather instrument that will go onboard the satellite. Since subsystems are systems too, systems engineering can be used. From this view, we don't so much need systems engineers, as we need systems engineering. Sometimes those doing the systems engineering of a subsystem are called systems engineers, but they might also be called chief engineers, Integrated Product Team (IPT) leads, designers, or architects. Whatever they are called, if they pay attention to the systems aspects of a system, they are doing systems engineering. I sometimes like to challenge my colleagues by quoting Georges Clemenceau who said that war is too important to be left to the Generals. In the same way I say, systems engineering might be too important to be left to the systems engineers, meaning that good systems engineering does not come from outside the group of engineers designing and building the system, but comes from inside, where the knowledge of the system is deep and rich, and systems aspects can be considered thoroughly.

On this theme, a colleague of mine says that there are two kinds of system engineers: the kind that really knows the system, inside and out, perhaps better than anyone else involved in the project, and the kind that know only the ways to apply systems engineering methods and tools. Systems engineering groups likely have some of both, but it's important to know the difference.

Systems engineers work to consider all aspects of the system, as a system – not just the construction of the system itself, but including how the system will be operated, the creation and operating costs of the system, the schedule for the construction of the system, the performance of the system, and the training that will be necessary for people to operate and support the system. They consider how the system will be tested, how it will be manufactured, and how it will ultimately be disposed of or retired someday. Manufacturing can be an important consideration as well in some industries. If a system will be made many times over, like an automobile, the systems engineering considerations can be different than for systems in the defense industry, where the systems engineers may be designing a system that will be made only a few times or even just once. Satellites and spacecraft are usually made one at a time, so materials cost is a smaller component of the overall system cost, with design, engineering, and labor being much larger. With automobiles, every dollar saved in materials is multiplied by the many thousands of cars that will ultimately be built.

Systems engineers, according to the INCOSE Systems Engineering Handbook aim at a disciplined, effective system development process that will result in quality systems that meet user needs. To a large extent, where there are systems to be created, the need for some systems engineering is inevitable. This doesn't mean that there must be a dedicated systems engineering group or department, or even that there must be people with the job title of systems engineer. In some industries, there are no designated systems engineers, yet systems engineering is still done. Systems engineering is focused on managing complexity and change throughout the system development process, and these concerns will be managed in some way in any systems development project. Let's take a common example to illustrate.

Imagine a complex system development project to build a satellite with many people working on many different parts of the system. One day the mechanical engineering team makes a decision to change the material being used for an internal structure. The new material has many advantages and the mechanical engineering team is right in making the change. The change, however, has effects that filter throughout the entire satellite engineering team. The new material increases overall weight of the system by some amount. As a result, the propulsion engineering team determines that the propulsion system must be a bit more powerful due to the added weight. The new propulsion system will need to be heavier and a little larger, and this change will affect the payload team, who now have less room in the satellite for television receivers and transmitters – the whole reason the satellite is being built in the first place. Can they afford to lose that little bit of space? Maybe, but that is not something that can be determined at a glance. How a change like this propagates through the system design and how the impact of the change is determined and accommodated is one of the most difficult aspects of engineering a complex system. It is common that most of the system design has been captured in static forms – written design documents, spreadsheets, drawings, and even presentations explaining aspects of the system. Each one bakes in the assumptions of what was true at the time it was created. There is no way to go back in time and then run the engineering process forward with the new material assumed from the start. Word of the change must be circulated to all affected subteams, or even to *all* subteams. Each subteam must have the discipline to re-evaluate all of their analysis and design based on the change. How can the overall team be sure that the impact of the change has been fully incorporated? Making sure that the change has been successfully incorporated into the entire project is a systems engineering task that is even more difficult than propagating the change in the first place.

What happens if something is missed? What happens if some calculation buried in someone's analysis assumed the original weight of the material, and this dependency wasn't noticed when the impact of the change was assessed? Systems engineering is aimed at reducing the risk of this kind of avoidable rework. Late discovery of errors found during integration can cause wasteful rework in the project – rework that will get done, since it *must* be done in order for the system to be successful. Systems engineers aim to discover these large-problems-to-be earlier rather than later so they cost less to correct. Errors are always much more expensive to correct the later they're found in the process because more rework is necessary. Since systems engineers work to manage the technical decision process and to facilitate people agreeing on concerns about the system, they ultimately make the development process go faster, by avoiding rework later. They try to make the project go slow in the beginning in order to go fast later on. Systems engineers try to get the project to do things right the first time, by understanding and learning more about the system, and thus avoiding the need for rework later when misunderstandings are revealed. By doing thing right the first time, the project reduces the typical amount of avoidable rework, and delivers the system sooner and at a lower cost.

We could speak about this using the Pareto principle, sometimes called the 80/20 rule, where if we imagine the moving average daily or monthly cost, called the run rate, it is clear that the project will spend more money as it adds people to the project. In the early stages, the project doesn't cost as much because there is a smaller number of people doing the early systems engineering, conceptualization, design, and early engineering work. That small number of people however actually commits a great deal of cost to the project through the project and design decisions they make. The decisions that those people make early in the process, affect the cost dramatically later on, so it pays to do everything we can to get these early decisions right.

Taking an extreme example, if the project puts very little attention on systems concerns early in the development process, just letting the teams work on their own, the result can be that a lot of that work will need to be redone as more is learned about the best way to engineer the system. Learning and discovery is inevitable throughout the process and changed from new learning can be very expensive, so it makes sense to invest a bit more in the early stage decision-making, so that those decisions can be made correctly, and thereby reduce the overall expenditure over the duration of the system lifecycle development process. Anything we can learn earlier, will be rewarded later. This "invest early to save later" is a big part of the value proposition of systems engineering. Research in the area of systems engineering has shown that systems engineering makes a positive difference in the cost and schedule performance of large systems development projects. The project is more efficient when systems engineering is applied. The data shows that if a large complex project invests approximately 12–14% of the total program expense in good systems engineering work, most of it at the beginning of the of the project, then they will deliver the project overall at a lower cost and in less time (INCOSE Systems Engineering Handbook, 4th ed., 2015, p. 17).

In this chapter, we are discussing how thinking about the world in terms of systems can be very useful in both understanding what's going on, and also in modifying or building new systems to accomplish our goals. It is not an overstatement to say that the best and perhaps the only way to change the world is to understand its systems and then learn how to modify those systems effectively. Engineered systems can be understood by studying the requirements, designs, and specifications used to build them. Human activity systems can be understood by applying the concepts and methods of systems thinking. Combinations of engineered systems and human activity systems are more complex still and take even more systems thinking to understand. Even if a system consists of just two engineered systems, complexity may arise because that particular combination was never designed as a system itself, so there are no designs or blueprints for the combination. Even if the two engineered systems are fully understood on their own, the possibility of emergent behavior when the two are combined can make analyzing even this system-of-two-systems a complex challenge.

For an example that is all around us, consider the Internet of things (IoT). Some years ago, makers of all kinds of consumer products started to connect them to the Internet using Wi-Fi signals or other technologies. We may take this for granted now, but it used to be that the only things in a normal home that connected to the Internet were a personal computer and perhaps a smartphone. Now, it's common to have dozens of additional devices – streaming boxes, security systems, light switches and bulbs, cameras, thermostats, garage doors – all chatting away on the Internet. My home network reports that it has 53 devices in total – I'm not sure I could even name them all from memory. Each device on its own, together with its associated app, makes a pretty simple system, but together, with added apps that use them in combination, like Amazon Echo or If This Then That (IFTTT), they become something that can delight or frustrate with all of the new emergent behavior. I live in a house of systems.

Systems approaches, since their advent have been applied to many kind of systems. The field of family systems therapy within the domain of psychology and psychotherapy, considers a family to be a system, with each person in the family acting as a part of that system. Systems of this type can be powerful in their resistance to change. One member of the family may appear to be the problem – perhaps a child who is struggling in school. The family may bring this child, or worse just send him or her to a family systems therapist, hoping that the therapist will simply fix the child, of course without interfering with anything else going on in the family. The family is surprised to find that the therapist considers this a family problem, that is, a problem with the whole system that is the family. The therapist treats the family as a system, made up of interacting subsystems (the family members) and exhibiting various forms of emergent behavior. Even though dysfunctional, the family system is stable in its patterns and will tend to resist any form of intervention that affects only one member. Like a worn wheel bearing on a car that causes the engine to work harder, each family members' behavior affects the others, with roles often scripted from very early life. The therapist must judge carefully how to intervene, possibly employing a paradoxical intervention – what they call *prescribing the symptom*. A child struggling to pass in school might be told, "Honey, just take your time with this school stuff. You can always take the courses again next year when you repeat sixth grade." Needless to say, family systems are not as logical in their behavior as engineered systems, but are deeply fascinating.

The trap for engineers in trying to deal with the world of systems, especially human activity systems is to assume they operate like engineered systems, with each subsystem performing its function, motivated by logical incentives and conscious, rational reasoning, with an eye toward a long-term, satisfying life for all. But human activity systems such as families, companies, economies, and nations are not so easily reduced to logical rule-following. Economics predicts that people make decisions based on utility, which assumes that a person will consistently value an item the same when presented in the same context. The new and interesting field of behavioral economics, led by Daniel Kahneman and others, paints a different picture. Through experimentation, they have found that the endowment effect causes people to value something they *already own* about twice as much as something they *could gain*. Of course, this result is quite surprising – why should a person value a car they own twice as much as the same car they are considering buying? Why does the pet shop owner encourage a potential buyer to "just take the cute puppy home for the night, and come back tomorrow if you want to give him back"? By phrasing it so that the buyer's family considers that the puppy is "theirs" unless they "give him back," the endowment effect is triggered, and the family will find it hard to part with what they now feel they own. Do human beings hold onto jobs, relationships, and possessions longer than they reasonably should? All the time – it's the endowment effect in action again.

To economists, and anyone with an engineering or logical mind, it makes no sense that the system we call a human being operates this way. But we do, and it's been confirmed again and again through experiment. A systems thinker places no value judgment on this behavior and certainly doesn't think that it is somehow wrong and must be corrected, nor do they flatly deny that this could be happening. The systems thinker understands and accepts this behavior as part of the system we call a human being. Understanding systems, including human beings, is crucial to being able to intervene or solve systemic problems.

When I taught a graduate course in systems thinking at Worcester Polytechnic, I told my students that by the end of the term, they would have the experience of walking around one day and suddenly seeing systems everywhere. Students said they didn't believe me, but then it happened. With a systems mindset, everything is seen as a system, or a part of a system. Flying on a plane once, I remember thinking how the aircraft is a system, but also part of the airline system, as well as part of the airport system when on the ground. As it flies, the plane becomes part of the air traffic control system, as it is handed off from point to point on its route. The parts of the aircraft were built by companies which are also systems, and those companies are part of an economic system which is part of a nation system and the system of all nations on this planet. As every child learns, our planet is part of the solar system, which is of course part of a galaxy and the universe, both systems. We are systems ourselves, as are our families, communities, and cities. We are systems, living in a world of systems.

3

The Intelligence in the System: How Artificial Intelligence Really Works

3.1 What Is Artificial Intelligence?

Artificial Intelligence has been an intriguing branch of computer science ever since computers were invented. Intelligent robots were imagined even before that time. Given a computer that is capable of computing and processing large volumes of data very quickly, it's an easy jump to imagining that one day computers will be able to think, understand, and operate much as humans do – that they will be intelligent. But imagining and implementing are two different things. In fact, no one knows how to make an intelligent machine, if by intelligence, we mean what humans have and what animals have in much smaller amounts. That's because we don't know how humans and animals produce intelligence. We don't know how brains work, at least not enough that we could build a machine that does what a brain does. It is theoretically possible for us to figure that out someday, and thus be able to build a brain out of synthetic materials, much as we can now build an artificial heart, since we *have* fully understood what the heart does. While there are projects underway to try to understand the brain and then to build an artificial one, this is not the approach taken in the computer science of artificial intelligence.

It is much easier to build a machine, or a computer, that acts intelligently than it is to try to build a brain, but what counts as acting intelligently? Somewhat circularly, the definition is usually that intelligent machine behavior is a behavior that if it were performed by a human being, would be considered intelligent. Playing chess is generally thought of as requiring intelligence, so if we build a machine that can play chess, we might justifiably call that intelligent behavior. We would consider the chess-playing machine intelligent, even if it played chess in a completely different way than its human counterparts.

To continue to develop our intuition for how artificial intelligence works, let's consider three popular myths about AI, which seems to be well entrenched in the public consciousness, and unfortunately even among engineers and scientists.

Engineering Intelligent Systems: Systems Engineering and Design with Artificial Intelligence, Visual Modeling, and Systems Thinking, First Edition. Barclay R. Brown.

3.1.1 Myth 1: AI Systems Work Just Like the Brain Does

Andrew Ng, a pioneer of deep neural networks and AI, professor at Stanford and founder of deeplearning.ai, one of the best educational resources in AI, said that he and other pioneers almost regret referring to what we do in deep learning as *neural* networks. As we'll show specifically later in this chapter, what neural network of "neurons" does is very simple numeric multiplication and addition. No one believes that this simple arithmetic is what the neurons in our brains are doing to produce consciousness, emotion, and decision-making. Neural networks only resemble brains in the very general sense of small units taking inputs from many sources, combining them, and passing on an output to another small unit, with the network composed of many thousands or millions of units. The brain uses complex electrical and chemical interactions. To say that we don't fully understand how they work is an understatement. We have general notions, but as an example, drugs that are used to treat emotional and behavioral problems in the field of psychopharmacology are often found to be effective accidentally and not by a systematic engineering of a change in the brain. In Daniel Carlat's fascinating book, *Unhinged: The Trouble with Psychiatry – A Doctor's Revelations about a Profession in Crisis*, the Tufts psychiatry professor and clinician reveals just how little is known about how the brain produces the symptoms his profession treats, and how little is known about how the treatments work.

3.1.2 Myth 2: As Neural Networks Grow in Size and Speed, They Get Smarter

It was widely believed in the past decades, when computers were far more limited in capacity and processing speed, that if we could just build a computer large enough, perhaps approaching the number of nodes and interconnections in the brain, it would certainly be able to act with humanlike intelligence. Some have joked that the Internet has easily disproved that theory. In fact, the number of computers and devices connected to the Internet is around half as many as the number of neurons in the human brain. Adding up all of the computing power of all of those computers, they well exceed the processing power and speed of the brain, yet the Internet has not become conscious in any way. In the field of AI and deep neural networks, it is often found that larger networks do not outperform smaller, better-designed ones. Also, large computers are generally only needed for the training of deep neural networks due to the need for many repeated calculations and iterations. Once the network is trained, it can be run and used on much smaller computers, even those hidden inside a device.

3.1.3 Myth 3: Solving a Hard or Complex Problem Shows That an AI Is Nearing Human Intelligence

In 1997, the IBM computer Deep Blue was the first computer to win against then-world chess champion, Garry Karparov. Before then, many assumed that once a computer was able to play an intellectually challenging game at a world championship level, then it would be nearing humanlike intelligence. We can see that this is a myth, both through logic and through example. Logically, since we know that computer programs that play games are not doing what humans do when they play the same games, it would be illogical to assume that game-playing computers would have other humanlike qualities. For an example, let's look more closely at how Deep Blue plays chess.

We need not build a brain in order to build an artificial intelligence, which is good, since we don't know how to build a brain. How do we then build an intelligent machine? Let us start with something simpler than chess – how about tic-tac-toe? How would we build a machine, or program a computer to play tic-tac-toe successfully? There are less than 1000 possible game positions, so one approach is to simply write 1000 conditional if-then statements (rules) in some ordinary programming language to produce the best move given each possible board position. Each of these statements would have the form, "Given board position X, make move Y." Though somewhat tedious to create, such a program would run fairly quickly and would produce perfect games of tic-tac-toe every time, either winning or drawing each game. Have we created an intelligent system here? It certainly takes some level of intelligence to play tic-tac-toe, at least the level of a three or four-year-old child, but it's hard to imagine a series of if-then conditional rule statements as constituting a mind or being intelligent in anything like a human or animal sense. Is our simple rule-based program an example of artificial intelligence? Based on common usage of the term, we must answer in the affirmative. Deep Blue plays both the opening and the endgame using such a rule set, though due to the large number of possible board positions and moves in the mid-game, a different process is used. There is nothing unintelligent about following a set of well-constructed rules. So one way of constructing an intelligent system then is by giving it a set of rules to follow. With enough rules, it can exhibit intelligent behavior. Any lack of intelligence exhibited by the system can presumably be overcome by adding more rules.

Before describing the other primary way that artificially intelligent systems can be designed, we note that using sets of rules seems to be related to, but not exactly the same as how humans make decisions. We certainly do refer to memorized rules and can often justify our decisions based on rules we can remember and cite, but at the same time, it does not seem that we operate primarily by walking through chains of dozens or hundreds or rules the way a machine would be programmed. It is possible that many of our "mental rules" are subconscious, and we are not aware of them. But it is also possible that we are not primarily using rules as the foundation of our decision-making. As a simple illustration, it is hard to imagine that a child, who has just learned to recognize and identify dogs, has built a mental set of rules that capture "dog-ness" at the depth required to achieve the performance she has. Even an adult cannot come close to stating the set of rules that allow the identification of dogs with near 100% accuracy, though the adult does exactly that. We also note that parents do not teach their children primarily by rules, sitting them down and explaining things like, "if it walks on four legs it is an animal, and dogs are kinds of animals, so it might be a dog, but it might also be a cat. If it walks smoothly, and is smaller it might be a cat, especially if it has a flatter nose and whiskers." Parents tend to teach dog identification by pointing out examples of dogs, and cats, and an occasional possum, and the child amazingly figures out how to tell them apart after seeing only a few examples of each, which brings us to the other major way an artificial intelligence system can work – by example.

Imagine teaching a small child how to recognize a dog. The most obvious way is to simply point out examples of dogs as the child encounters them in real life, or in photos and videos. After seeing a surprisingly few examples of dogs, the child is able to identify a dog and even differentiate it from other furry, four-legged beings. This is training-by-example and is the foundational insight that has enabled the artificial intelligence field called machine learning to flourish. It is quite a simple concept – to teach a computer to identify dogs, simply show it a large number of photos containing dogs, labeling each photo as "contains a dog." Mix in some images of cats, cars, horses, and people and label those "does not contain a dog." The machine learning system goes over this *training data* many times, trying to "learn" how to tell a photo that contains a dog from a photo that doesn't. We put "learn" in quotation marks, because what this machine learning computer program is doing is not very much like what the child is doing when she learns how to identify dogs. The main reason for this difference is that we simply don't know very much about how that child's brain learns dog identification, so we can't program a computer to do it that way.

What we do instead is create an artificial neural network (ANN), often simply called a neural network (NN) or a deep neural network (DNN). Almost all neural networks in current use have more than one hidden layer and are thus "deep." The structure of a neural network was originally inspired by the general structure we observe in a brain, but it would be a great exaggeration to say that it works the way a brain does. Brains contain neurons that have complex, even mysterious electrical and chemical connections with each other and exchange messages that to us are still mostly inscrutable. Artificial neural networks have neurons as well, but they have simple, mathematical relationships with each other, consisting mostly of simple addition and multiplication. When a neural network "learns," it tries various combinations of thousands or even many millions of numbers, called weights and biases, until it comes up with a set of numbers that allow it to match up the training data it was given with the labels. By repeated trial and error, the system arranges a set of weights and biases, perhaps millions of them, that cause it to label a photo containing a dog correctly as "contains a dog" and one that does not as "does not contain a dog." The training and learning of the deep neural network happens in three distinct phases.

3.2 Training the Deep Neural Network

To back up for a moment, the entire set of data we start with is randomly split up into three parts called training data, development data, and test data. It is not split up evenly – often 80% of the data is put in the training set, with 10% in each of the other two. In the training phase, we use the training data which consists of known, labeled examples of dog photos and nondog photos. The computer is in effect, saying, "OK, I will keep playing with my set of weights and biases until the photos you told me contain dogs come out as containing dogs when run through my neural network, and vice versa for the non-dog photos." This may sound a bit like cheating, since it is like a professor who gives students the actual exam questions (and answers) to help them study for an exam. The students have two choices – either they can study only the questions and answers, and thus be prepared only to answer those specific questions, or they can use the given questions and answers as a guide to try to learn the topic more generally, and thus be able to answer other questions as well. The artificial neural network faces the same two possibilities. If it focuses too closely on the training data, it may learn to identify only the dogs in the training data set and do very poorly when presented with new dogs in the next phase. This is known as "overfitting" to the training data and is a danger in any neural network system.

After the neural network performs well enough in the training phase, the trained neural network is put to work on a new set of photos that it has not seen before. In this development phase, a fresh set of data (called the development data set) is used to see how the system performs on dogs photos it has not seen before. We expect the system to perform better on the training data, which it has seen, than on the development data, which it has not, but if the system performs *much* better on the training data than on the development data, then it is likely the system is overfitting to the training data. There are a number of adjustments than can be made to the neural network to reduce overfitting and try to get the system's performance on the development data to be as close to its performance on the training data as possible. Some of these adjustments are obvious and simple, such as just letting the neural network spend more time learning the training data, and some are surprisingly effective such as randomly dropping out some neurons from the network. By tweaking the neural network's structure and the training process, we get the system's performance to be as good as possible on the development data. But we aren't done – there is one more phase.

3.3 Testing the Neural Network

Now we take a third set of data, called the *test data* which, like the training and development data, consists of dog photos labeled as either "contains a dog" or "does not contain a dog" and run it through the neural network. Like the development data, the system has not seen these photos before, so its ability to identify dog photos won't be as good as it was on the training data. If the system does well enough on the test data, compared to its performance on the development data, then we have probably achieved the best results we can from this system, given the entire set of data we started with. If the system's performance on the test data is much worse than its performance on the development data, it is possible that the system has overfit to the development data, meaning the tweaks that were done to the network to improve its performance on the development data unfortunately specialized the system too much to the development data.

Once we are satisfied with the system's performance on the training, development, and test set data, we can test the system in the real world. Of course, we don't have the labels that go with data collected from the real world, so we must find another way to evaluate the system's performance. One way is to compare its performance to that of humans performing the same task. Give the system some new images collected in the world and give the same images to some humans and see how the system's performance compares to the humans. One might assume that human performance in identifying dogs in photos would be perfect, 100%, but there are always issues, such as simple human error, or vague photos where reasonable people might disagree about whether there is a dog there or not, or even ambiguous animals, where it isn't clear if there's a dog or perhaps a coyote.

3.4 Annie Learns to Identify Dogs

To make this training process completely clear, let's go back to our small child, call her Annie, and walk her through the same three phases as if she were an artificial deep neural network. Remember than in truth, we don't fully understand how Annie the human actually learns to identify dogs, but for illustration we'll pretend that she does it the way our artificial neural network does it. To begin, we collect 100 photos. Say we take a camera to a public park that allows dogs and snap photos at random. Some photos will contain dogs, and some won't. Then, we have some friends help us label each photo (on the back) as "dog" meaning the photo contains a dog, and "nondog" meaning the photo does not contain a dog. Note that this labeling won't be perfect. Most data sets are labeled by human judgment, and humans can make mistakes, or even disagree on whether a particular photo contains a dog – what if the photo contains a tiny bit of a dog's tail, or is that a part of a picnic table?

Now, we randomly divide the labeled photos into three piles – training, development, and test, with 80 in the first and 10 in each of the other two piles. Now, we call Annie, who has never seen a dog before and doesn't even know what a dog is. Our only input to Annie will be the photos – to keep the analogy with deep neural networks accurate, we can't give her any verbal explanations, rules-of-thumb, or other input beyond the photos. Annie, and artificial neural networks, learn only by example.

We give her the training pile of photos and let her have at it. All she knows is that some of the photos are labeled dog, and some are labeled non-dog. She shuffles the photos around, perhaps sorting them, grouping them, making notes about things she might see in common among the photos labeled "dog." She reviews and rereviews the photos, working through the pile multiple times, improving each time through. She works for a while and then says she things she's got it. Let's try her out, first on the training set of photos she's been working with. We take the pile back, shuffle it, and then show her each photo in sequence, asking her to say "dog" or "nondog" for each one. Of the 80 photos, how many did she get right, meaning she said the same thing that's written on the back of the photo – 60? 70? 75? Maybe she got all 80 right. After all, if she has a good memory, she could have simply memorized each photo and its label. But what we don't know yet is how well her newfound knowledge will generalize to photos outside the set she learned from.

So next, we show her the 10 "development" photos and ask her to say "dog" or "nondog." She has never seen these photos before and so her score will likely be lower than on the training data. There could be many reasons for this. What if Annie noticed that all dogs in the training photos were either white, black, or brown, and that all of the brown dogs had no apparent legs (they were all lying down). She mistakenly, but forgivably, learns that white and black dogs have legs, but brown ones do not. So, in the development data, when she sees a brown dog standing up, she concludes it must not be a dog. There was "variance" between the training data and development data that we did not notice before asking Annie to learn from them. If there are only a very few or perhaps just one photo of a standing brown dog, then the clear solution here is to get more training data for Annie to include in her training set.

Other problems can occur in training data sets. Since all our photos came from a playground, what if all the dogs were playing with children, and all the children were playing with dogs? In that case Annie might learn that what we mean by dog is a two-legged upright being playing with a four-legged furry being and therefore might not understand that the furry being by itself is the dog.

Annie might make other kinds of mistakes on the development photos. What if she has studied the training photos so carefully that she has precisely memorized characteristics of each dog breed represented in the photos? Unknown to us, and to Annie, she has learned that "dog" means one of the say twenty breeds represented in the training photos. If there were no St. Bernards in the training data, there is a risk that Annie may not recognize a St. Bernard as a dog at all, when she sees one later. One way to help Annie would be to collect photos from every possible breed and crossbreed, but this might be impractical, and would still mislead Annie, since a new crossbreed could be produced in the future. What would help Annie more is to help her not be so particular about the details of each different breed, but to focus on features common to all dogs. We could perhaps do this by having her look at the photos through an out-of-focus lens so that all the images get just a little blurry. If done correctly, it would encourage her to focus on the main features, rather than the details and she should do better when confronted with a new breed of dog in the development data.

We work with Annie in this development phase, adjusting her learning process until she performs well on both the training and development sets of photos. Then we proceed to the next phase, quizzing her on the "test" photos she has never seen before. If she does nearly as well on the test photos as on the development photos, then we consider ourselves, and Annie, as successful, and we are ready to turn Annie loose in the real world, confident she can successfully identify dogs. But what if Annie stumbles on some of the "test" photos? Her learning might still be too specific (overfit) to the training and development photos. What if, as she tries to identify dogs in the test photos, she has trouble with dogs in unusual orientations? Maybe there's a photo of a dog jumping and its body is stretched out almost flat. Annie may have mistakenly learned that dogs are either standing up or lying down. Here the problem can really only be remedied by adding more training data. We show Annie photos of dogs in all orientations and positions, jumping, crouching, upside down rolling in the grass – and teach her that these are to be identified as dogs too.

We may have an even bigger problem, though. As we mentioned above, it is more than likely that the park where we took our photos did not contain a very wide variety of dogs. As of this writing, the American Kennel Club recognizes 193 dog breeds. The dogs at the park do not represent all of these, nor do they represent all possible colors and sizes within breeds, that Annie may encounter in the real world. One possible solution, assuming we have the budget and are very committed to Annie's complete training, is to employ a team of artists, who can use digital art, or oils and watercolors, to paint additional breeds, colors, sizes, and positions of dogs. Perhaps we can supplement Annie's training by including these images. We might even get a computer programmer to automatically generate new dog images by making and modifying copies of other images. A simple computer program can take a photo of a white dog and make copies where the same dog is shown in gray, brown, black, and tan. All of this artist and computer-generated data is called "synthetic" data and, even though these synthetic images don't represent any real dogs, they still help Annie learn. We add this synthetic data to the training dataset only, not to the development or test datasets, since we are only using it to help train Annie – we don't care how well she might do at *identifying* synthetic images, because she won't encounter any synthetic dogs in the real world.

Let's summarize Annie's journey of learning, since it illustrates exactly how the training of deep neural networks proceeds. First, we assemble a set of dog photos, and label them as either "dog" or "nondog." Then we split this set of images randomly into three sets – training, development, and test, putting most of the images, perhaps 80–90%, in the training set, and the remainder split between the development and test sets. We work with Annie on the test images until she achieves a high-performance rating on being able to identify the dog images. Then we try her on the development set and see how she does. If she does almost as well on the development data as the training data, we can continue to the test data; if not, we provide some coaching to Annie to help her generalize her knowledge from the training data to the development data. As a final stage, we show Annie the test data and see how she does. If she does nearly as well on these brand-new test images, we are good to go; if not, we might need to gather or synthesize more training data.

Let's now consider the often-repeated phrase offered as an explanation of how artificial intelligence and machine learning with neural networks in particular works: A neural network is a kind of computer application that does not need to be programmed – it programs itself. I've always thought that this statement leads to more confusion than illumination, since, if there is no programming needed, why is artificial intelligence programming one of the most in-demand jobs in the world right now? The fact is, that there is plenty of programming to do, but it's a different sort of programming. Annie needs two things to learn to identify dogs. First, she needs a brain, and not just any brain. A brain that has the structure and function to enable it to comprehend, compare, and remember images. Such a brain was given to Annie at birth, but in the world of artificial intelligence, we have to build them. Since we don't know enough about how the brain works to build a synthetic brain, we build an artificial neural network in software, through this new kind of software programming. The programming part of building a neural network consists of two parts.

First, the artificial neural network must be designed and built, much as a new building is designed and then constructed. There's a difference between throwing a dog house together and designing a skyscraper. A simple neural network can be programmed in under ten lines of software source code, but the most advanced networks require a lot more thought about the number and type of layers, the number of neurons, the mathematical functions that connect the neurons and the encoding of the input data. Still, it is surprising how little code is needed to implement that design using the latest neural network frameworks, so it is true that implementing the neural network doesn't take a lot of code, but it can take a lot of thought and design to get it to work well. Helpfully, researchers all over the world are excitedly developing and refining neural network designs, and for the most part publishing their results openly, so their neural network designs can be reused. It's the design – not the coding itself that is the time-consuming part.

The second part of the programming that goes into a neural network is usually referred to by the deceptively simple name, "data preparation." Annie already had all the equipment and processing abilities to take in the data (the stack of photos) before we even got started. For an artificial neural network, we'd need to set up how the input data will be encoded, formatted, and structured using standard data processing techniques. Again, there are good tools to make this easier and faster than ever before, but it does take some programming and again, some thought. Once these two programming parts are complete, we can begin the training process. The key takeaway here is that building artificial neural network systems still requires programming, in the broadest sense, but the work is different and is more like design work, experimentation, data collection, and data processing – along with some coding to make it all come together.

3.5 How Does a Neural Network Work?

Annie was trained to identify dogs the way neural networks do, by example and repeated review of the training data. For Annie, the training data consisted of the images containing dogs labeled *dog* along with images that do not contain dogs labeled, *nondog*. Next, we'll describe exactly how a deep neural network "learns" to identify dogs. Remember, Annie's brain is not a deep neural network, so she doesn't learn that way. In fact, we don't know how Annie learns, when she is operating as a regular human being and not imitating a deep neural network. We have only rudimentary knowledge (and some say even less that that) of how the brain learns. So we are under no illusion that what our deep neural network is doing, is what a human brain is doing. The tic-tac-toe program mentioned earlier does simple look-ups from a table of possible moves – no human player of any age plays this way. We can, in casual conversation, say that our tic-tac-toe look up program *knows* how to play tic-tac-toe, but we must realize that it does not *know* in the way we know. That's OK – what we are really after is a program that successfully plays tic-tac-toe, or identifies dogs, or drives our cars for us. What do we care if it knows or understands or thinks the way that we do? The *who cares* answer to the question of, can computers understand or know? neatly sidesteps the vexing philosophical arguments and counterarguments on this subject. Nevertheless, the philosophical discussions are fascinating, and we devote an entire chapter to it later in this book.

3.6 Features: Latent and Otherwise

Deep neural networks work based on the critically important concept of *latent features*. First, let's look at regular features. If we want to compute the likely selling price of a home, we would probably base the calculation on some important features of homes, especially those that affect the price. Features like the number of bedrooms and bathrooms, total square footage, size of the garage, whether there is a swimming pool, type of construction, and a few others. We may not know how these features combine to compute the likely selling price of the house, but there is a statistical technique we can use called linear regression to find out the relationship. We know that the price of the house will be determined by some weighted combination of the features. For example, say that the prices of homes in a certain area is very close to the computation of

$50,000 × the number of bedrooms

+ $20,000 times the number of full bathrooms

The weights, 50,000 and 20,000, are determined experimentally, using statistical techniques that have been known for many years. We start by getting a list of homes that have sold recently, including the number of bedrooms and bathrooms for each house, square footage, and other features, along with the selling price. We then experiment with different formulas and multipliers (weights) until we can come as close as possible to the known selling price of each home on the list. Of course, computers can do this very quickly, averaging the results across many homes and coming up with the best overall weights to fit the data.

If we don't know much about homes and selling prices, we might have included some features in our list that really don't have much to do with the selling price, such as say the number of windows in the home. Our regression computation will take care of this for us, because we will see that the weight (multiplier) for the number of windows will be small compared to the weights for the other features. In general, we want to use the smallest number of features that give a good prediction, so many possible features might be left out of the final calculation. This process of multivariable regression has been used for over 200 years.

The key insight that led to the principles underlying deep neural networks is this: it is not necessary to know the meanings of the features to use this process. As long as we have a set of features with relevance, we can figure out a way to use them to predict the outcome – the home's price in our example. We can be quite confident that there exists some set of features, that when weighted properly and combined will predict the selling price of the home within a reasonable range. These features may not be obvious even to experts in home pricing. Perhaps the most predictive feature of the price of a home in some areas is its distance to the beach access walkway, not the actual distance to the water, or it could be that houses sell for more when the number of bathrooms is equal to the number of bedrooms. *Latent* features are those that are in the data, but not readily obvious, and usually not even determinable by analyzing the data. Let's make this idea of latent features even clearer with another, even more striking example.

3.7 Recommending Movies

Say our task is to predict which moves or TV shows a viewer might want to watch next from a video streaming service so that we can give the user some recommendations. Staying in business means offering viewers a continuous supply of shows that they want to watch, so making good predictions of what they will like is vital. What makes this prediction difficult is that we don't know anything about our users except what they have already watched, and how they have rated those shows. In actual practice, we have relatively few ratings – most people don't rate most of what they watch, so we have to be content to use the fact they watched a show to the end, as a proxy for a good rating. That may be all the data we have. Now, how do we make sense of it? We can assume that there are features of movies and shows that affect whether certain users will like them, such as the amount of romance, action, science fiction, or violence in them. There are corresponding features for our users – things like how much violence they like in a movie, or how much they enjoy science fiction space adventures. We can *imagine* these sets of features for both shows and users, but we don't actually have this information. It is latent in the data, but we can't just read it out. But here's the insight: we don't have to. We can just assume that these features exist. Say we assume that there are ten different features for shows and ten more for users.

Now, create a simple matrix with the rows representing users and the columns representing shows. In each cell is the rating (say 1–5) that the user gave the show, or a blank if the user has not seen the show. If the user has seen the show, but not rated it, perhaps we might just put a 2.5 or more elaborately, we could put the value of the average rating that user gave all shows. We take this partially completed matrix and then use a mathematical process called matrix factorization to factor the matrix into two other matrices – the product of which is the original matrix. Let's say we have a data set of 5000 users and 1000 shows. The two constituent matrices (the factors) are a user matrix, which has 5000 rows and ten columns, and a show matrix which has 1000 columns and ten rows. As you recall from basic matrix mathematics, multiplying a 5000×10 matrix by a 10×1000 matrix gives a 5000×1000 matrix, which would be the original *users* × *shows* matrix.

The numbers in the two smaller matrices are the latent features. We are just assuming there are ten important features for shows, and ten more for users. We don't know what these features really mean, and actually, we'll never know. But we don't *need* to know. Next, we factor our user/show ratings matrix (5000×1000) into those two smaller matrices (5000×10 and $10 \times 10,000$). By trial and error, or some other method, we find numbers to fill in these two smaller matrices, such that multiplying them together gives, as close as possible, our original large matrix of users, shows, and ratings. Remember that the original matrix had ratings for shows that each user has seen and blank ratings for the shows the user has not seen, but now when we multiply our two new, smaller factor matrices together, we get some number for *every* cell. For cells that had a rating for a user/show combination, we can compare the original rating that the user gave the show, with the new calculated predicted rating. For the shows the user has rated, we can see how close we are to the user's original rating. In fact, it's the adding up of all of these differences that we use to see if our factorization is the best we can do. For the cells that were empty in the original rating matrix, the multiplication yields a *prediction*, which is the rating that the system is predicting the user would give the show if he or she viewed it later. Obviously, we should find the shows that we predict a user would rate highly, but which they haven't seen, and recommend them to that user. The process of matrix factorization is the algorithm that proved superior to many others in the Netflix Prize competition in 2006, and it's still widely used as a part of systems that make recommendations.

The main takeaway from this look at matrix factorization is that we don't need to know what the latent factors are – for our shows or our users. We can simply assume they are there, and let the system calculate weights, coming as close as it can to zero error when compared to our training data, and then turn around and use the values the system found to make new predictions. That this works at all may seem surprising, but there are some analogies that provide some intuitions about why it does.

The best recent research into human decision-making suggests that humans make decisions mostly unconsciously, with the decision being reached in the unconscious mind before the conscious mind is even aware that a decision has been reached. The decision is reported to the conscious mind, which adopts it and takes credit for making it, and if asked, manufactures an explanation or justification for the decision. If this is true, then our conscious minds actively and continually deceive us by assuring us that we have clear reasons for our decisions. We may, but those reasons came from the unconscious mind and are not readily observable. To an observer of human nature, this may not be surprising, since it is easier to see others make seemingly arbitrary decisions, manufacturing and matching but arbitrary justifications when asked. Our own decision-making process seems to work based largely on factors that we cannot identify because they live deep in the unconscious.

The concept of making decisions based on hidden factors, common to both deep neural networks and human brains, leads to the surprising conclusion that we need not know the meanings of the factors in order to construct a system that can make decisions or predictions. What we do need is a way to weight the different factors and combine them in such a way that the result matches what we observe in reality. It's as if you were sitting in front of a machine with three dials and two lights. One light is already lit – say it's a royal purple color. As you turn the three unlabeled dials, the other light begins to glow. You notice that each dial changes the color of the light in a different way. By playing with the dials enough, you can make the light change to match the royal purple light. You don't know – and don't need to know – that the three dials are actually adjusting the amount of red, green, and blue in the light. We could change the system to four dials and your process would be the same: turn the dials until the desired color is reached. You have no need to know that the four dials represent cyan, magenta, yellow, and black – an alternative color component system to red/green/blue. You simply manipulate the dials by trial and error until you reach the desired result – you need not know the meaning of the dials.

Picture a neural network as a set of hundreds, thousands, or even millions of little dials. In the learning process, the machine tries many, many combinations of dials trying to get the output of the network to match the desired output state for a set of inputs, given by the labeled data in the training data set. Once the network learns the best it can, we test it on new inputs and see if it produces the correct outputs. Once it does well at that, we can trust it to give correct outputs for inputs it has never seen before and for which we don't already know the correct output.

3.8 The One-Page Deep Neural Network

Let's go deeper into exactly how the network learns to build your intuition for what's really happening. We won't be using advanced math or computer code. Systems engineers and educated people of all professions can develop a solid intuition for how deep neural networks work, using only ordinary mathematics. Instead of just explaining, let's build a real, working, deep neural network using only a spreadsheet. You can follow along by building it yourself in Microsoft Excel, or you can download the spreadsheet from www.engineeringintelligentsystems.com.

To begin, imagine that we want a deep neural network to make predictions of a physical phenomenon or function, that has one input number and one output number. The neural network won't know what the function is, but we will give it a list of pairs of data – input and output. If the relationship between the input and output was linear, the pairs of inputs and outputs, if plotted, would form a straight line, and we can apply simple linear regression, a technique used in statistics and data analysis since its invention by Gauss in the nineteenth century. But not all relationships are linear. Our objective is to see if our little neural network can "learn" the relationship between the input numbers and the output numbers well enough to be able to predict outputs from new inputs that it did not see when it was trained.

To create our network, let's start by making up a function that our network will try to learn. By choosing an actual function, we can easily generate new inputs and outputs, and also check how well our network is predicting the outputs, based on the inputs. Of course, in a real situation, we would not know this function – if we did, we could just calculate all outputs from inputs using the function's formula. We'll start with a simple quadratic function, and once we've built the neural network and tested it, we will see if the same neural network can learn other functions. To start, let's assume our output, y, is given by the following equation in terms of the input, x:

$$y = 3x^2 - 7x + 2$$

The first part of our neural network spreadsheet (Figure 3.1) is a simple column of inputs next to another column of outputs. The inputs can be all random numbers, generated by using Excel's RAND() function. We can just enter =RAND() in the cell to create a random number between zero and one. Here, we generated a column of random numbers (column A, hidden in Figure 3.1) and then copied the values into the input column (column D), so that the random numbers don't keep changing as we are trying to get the neural network to work. If we don't do this, every time the sheet calculates anything, all the inputs change, so the neural network won't have a chance to learn.

In the next column (column E), we compute the output using the function we chose above. With our inputs in column D, the function is

$$= D1^2 - 7 * D1 + 2$$

Next, we need to decide how deep our neural network will be. In the early days, neural networks had only one layer, but with increasing computer power and memory, most neural networks now are deep neural networks, which simply means they have more than one layer of neurons. In our spreadsheet neural network, let's have two layers, though you can easily add more using the same principles we'll use here. Since we're going small, let's have just five neurons in

	The One Page DNN		Layer 1 Weights					Bias		Layer 2 Weights			Bias	MSE (Train)	MSE (Test)	
			0.356153717	0.074232862	0.317050429	0.055886523	0.24777844	-0.2176929		0.007150294	-3.651372186	-19.49832676	12.12802402	0.0015	0.0040	
	Input	Output	Layer 1 Nodes					Wx+b	Activation	Layer 2 Nodes			Wx+b	Sq Error	Actual	Predicted
Training Data	1.7518	-7.1938	0.6233	0.1300	0.5554	0.0979	0.4341	1.6236	0.8353	0.0060	-3.0500	-16.2874	-7.2033	0.0001	-7.1938	-7.2033
	1.9500	-7.8474	0.6945	0.1448	0.6182	0.1090	0.4832	1.8319	0.8620	0.0062	-3.1474	-16.8079	-7.8212	0.0007	-7.8474	-7.8212
	1.1067	-4.5220	0.3941	0.0822	0.3509	0.0619	0.2742	0.9455	0.7202	0.0051	-2.6298	-14.0435	-4.5401	0.0003	-4.5220	-4.5401
	0.2511	0.3054	0.0894	0.0186	0.0796	0.0140	0.0622	0.0462	0.5116	0.0037	-1.8679	-9.9748	0.2890	0.0003	0.3054	0.2890
	0.9799	-3.8991	0.3490	0.0727	0.3107	0.0548	0.2428	0.8123	0.6926	0.0050	-2.5289	-13.5049	-3.9008	0.0000	-3.8991	-3.9008
	0.3750	-0.4842	0.1336	0.0278	0.1189	0.0210	0.0929	0.1764	0.5440	0.0039	-1.9863	-10.6074	-0.4618	0.0005	-0.4842	-0.4618
	1.2553	-5.2112	0.4471	0.0932	0.3980	0.0702	0.3110	1.1017	0.7506	0.0054	-2.7407	-14.6356	-5.2428	0.0010	-5.2112	-5.2428
	0.7902	-2.9067	0.2814	0.0587	0.2505	0.0442	0.1958	0.6128	0.6486	0.0046	-2.3662	-12.6468	-2.8824	0.0006	-2.9067	-2.8824
	0.9468	-3.7311	0.3372	0.0703	0.3002	0.0529	0.2346	0.7775	0.6851	0.0049	-2.5017	-13.3594	-3.7282	0.0000	-3.7311	-3.7282
	1.3269	-5.5275	0.4726	0.0985	0.4207	0.0742	0.3288	1.1770	0.7644	0.0055	-2.7911	-14.9050	-5.5627	0.0012	-5.5275	-5.5627
	0.2441	0.3509	0.0869	0.0181	0.0774	0.0136	0.0605	0.0389	0.5097	0.0036	-1.8612	-9.9390	0.3315	0.0004	0.3509	0.3315
	0.7010	-2.4157	0.2497	0.0520	0.2223	0.0392	0.1737	0.5192	0.6269	0.0045	-2.2892	-12.2248	-2.3816	0.0012	-2.4157	-2.3816
	0.6091	-1.8927	0.2169	0.0452	0.1931	0.0340	0.1509	0.4225	0.6041	0.0043	-2.2058	-11.7791	-1.8525	0.0016	-1.8927	-1.8525
	1.8673	-7.6491	0.6722	0.1401	0.5964	0.1055	0.4676	1.7660	0.8540	0.0061	-3.1181	-16.6513	-7.6353	0.0002	-7.6491	-7.6353
	1.8661	-7.6452	0.6717	0.1400	0.5960	0.1054	0.4673	1.7647	0.8538	0.0061	-3.1176	-16.6482	-7.6317	0.0002	-7.6452	-7.6317
	1.5743	-6.5418	0.5607	0.1169	0.4991	0.0880	0.3901	1.4371	0.8080	0.0058	-2.9503	-15.7552	-6.5718	0.0009	-6.5418	-6.5718
	1.5232	-6.3422	0.5425	0.1131	0.4829	0.0851	0.3774	1.3833	0.7995	0.0057	-2.9194	-15.5898	-6.3755	0.0011	-6.3422	-6.3755
	1.1322	-4.6437	0.4033	0.0840	0.3590	0.0633	0.2805	0.9724	0.7256	0.0052	-2.6494	-14.1484	-4.6647	0.0004	-4.6437	-4.6647
	0.1985	0.6500	0.0707	0.0147	0.0629	0.0111	0.0492	-0.0091	0.4977	0.0036	-1.8174	-9.7053	0.6088	0.0017	0.6500	0.6088
	0.0340	1.7631	0.0121	0.0025	0.0108	0.0019	0.0084	-0.1819	0.4546	0.0033	-1.6601	-8.8650	1.6062	0.0246	1.7631	1.6062
	0.6568	-2.1652	0.2339	0.0487	0.2082	0.0367	0.1627	0.4725	0.6160	0.0044	-2.2431	-12.0108	-2.1275	0.0014	-2.1652	-2.1275
	1.5503	-6.4488	0.5522	0.1151	0.4915	0.0866	0.3841	1.4119	0.8041	0.0057	-2.9359	-15.6783	-6.4805	0.0010	-6.4488	-6.4805
	1.4947	-6.2289	0.5324	0.1110	0.4739	0.0835	0.3704	1.3534	0.7947	0.0057	-2.9017	-15.4956	-6.2636	0.0012	-6.2289	-6.2636
	0.9332	-3.6616	0.3324	0.0693	0.2959	0.0522	0.2312	0.7632	0.6820	0.0049	-2.4904	-13.2992	-3.6567	0.0000	-3.6616	-3.6567
	0.2606	0.2434	0.0928	0.0193	0.0826	0.0146	0.0646	0.0563	0.5141	0.0037	-1.8770	-10.0237	0.2310	0.0002	0.2434	0.2310
	1.0892	-4.4381	0.3879	0.0809	0.3453	0.0609	0.2699	0.9272	0.7165	0.0051	-2.6162	-13.9711	-4.4541	0.0003	-4.4381	-4.4541
	0.5430	-1.5061	0.1934	0.0403	0.1722	0.0303	0.1345	0.3530	0.5874	0.0042	-2.1447	-11.4528	-1.4652	0.0017	-1.5061	-1.4652
	0.8120	-3.0247	0.2892	0.0603	0.2574	0.0454	0.2012	0.6358	0.6538	0.0047	-2.3873	-12.7465	-3.0031	0.0005	-3.0247	-3.0031
	1.6653	-6.8839	0.5931	0.1236	0.5280	0.0931	0.4126	1.5327	0.8224	0.0059	-3.0029	-16.0360	-6.9050	0.0004	-6.8839	-6.9050
	0.6423	-2.0834	0.2287	0.0477	0.2036	0.0359	0.1591	0.4574	0.6124	0.0044	-2.2361	-11.9411	-2.0448	0.0015	-2.0834	-2.0448
	1.4365	-5.9920	0.5116	0.1066	0.4554	0.0803	0.3559	1.2922	0.7845	0.0056	-2.8646	-15.2974	-6.0283	0.0013	-5.9920	-6.0283
Test Data	1.8343	-7.4754	0.6533	0.1362	0.5816	0.1025	0.4545	1.7103	0.8469	0.0061	-3.0923	-16.5133	-7.4715	0.0000	-7.4754	-7.4715
	1.2297	-5.0958	0.4380	0.0913	0.3899	0.0687	0.3047	1.0749	0.7455	0.0053	-2.7222	-14.5368	-5.1257	0.0009	-5.0958	-5.1257
	0.8132	-3.0309	0.2896	0.0604	0.2578	0.0454	0.2015	0.6370	0.6541	0.0047	-2.3883	-12.7539	-3.0095	0.0005	-3.0309	-3.0095
	1.6628	-6.8746	0.5922	0.1234	0.5272	0.0929	0.4120	1.5301	0.8220	0.0059	-3.0015	-16.0284	-6.8960	0.0005	-6.8746	-6.8960
	0.0335	1.7670	0.0119	0.0025	0.0106	0.0019	0.0083	-0.1825	0.4545	0.0032	-1.6595	-8.8621	1.6096	0.0248	1.7670	1.6096
	1.3620	-5.6788	0.4851	0.1011	0.4318	0.0761	0.3375	1.2139	0.7710	0.0055	-2.8151	-15.0334	-5.7150	0.0013	-5.6788	-5.7150
	1.5253	-6.3504	0.5432	0.1132	0.4836	0.0852	0.3779	1.3855	0.7999	0.0057	-2.9208	-15.5967	-6.3836	0.0011	-6.3504	-6.3836
	1.2757	-5.3024	0.4543	0.0947	0.4045	0.0713	0.3161	1.1232	0.7546	0.0054	-2.7552	-14.7135	-5.3353	0.0011	-5.3024	-5.3353
	0.0087	1.9393	0.0031	0.0006	0.0028	0.0005	0.0022	-0.2085	0.4480	0.0032	-1.6360	-8.7364	1.7586	0.0326	1.9393	1.7586
	0.8966	-3.4721	0.3193	0.0666	0.2843	0.0501	0.2221	0.7247	0.6736	0.0048	-2.4597	-13.1352	-3.4621	0.0001	-3.4721	-3.4621
	1.6446	-6.8062	0.5858	0.1221	0.5215	0.0919	0.4075	1.5112	0.8192	0.0059	-2.9913	-15.9741	-6.8316	0.0005	-6.8062	-6.8316
	0.9480	-3.7375	0.3376	0.0704	0.3006	0.0530	0.2349	0.7788	0.6854	0.0049	-2.5027	-13.3650	-3.7348	0.0000	-3.7375	-3.7348
	0.9564	-3.7804	0.3406	0.0710	0.3032	0.0535	0.2370	0.7876	0.6873	0.0049	-2.5097	-13.4021	-3.7768	0.0000	-3.7804	-3.7768
	1.4180	-5.9153	0.5050	0.1053	0.4496	0.0793	0.3514	1.2728	0.7812	0.0056	-2.8525	-15.2329	-5.9518	0.0013	-5.9153	-5.9518
	1.9266	-7.7743	0.6862	0.1430	0.6108	0.1077	0.4774	1.8073	0.8590	0.0061	-3.1367	-16.7503	-7.7528	0.0005	-7.7743	-7.7526
	1.5068	-6.2784	0.5366	0.1118	0.4777	0.0842	0.3733	1.3659	0.7967	0.0057	-2.9091	-15.5351	-6.3105	0.0012	-6.2784	-6.3105
	1.8704	-7.5943	0.6661	0.1388	0.5930	0.1045	0.4634	1.7489	0.8517	0.0061	-3.1100	-16.6079	-7.5838	0.0001	-7.5943	-7.5838
	0.9733	-3.8656	0.3466	0.0722	0.3086	0.0544	0.2412	0.8053	0.6911	0.0049	-2.5235	-13.4753	-3.8664	0.0000	-3.8656	-3.8664
	1.2121	-5.0159	0.4317	0.0900	0.3843	0.0677	0.3004	1.0564	0.7420	0.0053	-2.7093	-14.4684	-5.0444	0.0008	-5.0159	-5.0444
	0.0488	1.6620	0.0173	0.0036	0.0154	0.0027	0.0120	-0.1668	0.4585	0.0033	-1.6740	-8.9393	1.5180	0.0207	1.6620	1.5180
	0.6638	-2.2047	0.2363	0.0493	0.2104	0.0371	0.1644	0.4798	0.6177	0.0044	-2.2554	-12.0445	-2.1675	0.0014	-2.2047	-2.1675
	0.1822	0.7580	0.0649	0.0135	0.0578	0.0102	0.0451	-0.0262	0.4934	0.0035	-1.8018	-9.6217	0.7081	0.0025	0.7580	0.7081
	1.4987	-6.2448	0.5338	0.1113	0.4752	0.0838	0.3713	1.3576	0.7954	0.0057	-2.9042	-15.5086	-6.2793	0.0012	-6.2448	-6.2793

Figure 3.1 The one-page deep neural network with a quadratic function.

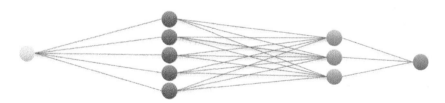

Figure 3.2 Drawing of the neurons and layers in the one-page neural network.

the first layer and three neurons in the second layer of our little network, so it will fit on just one spreadsheet page, and we can see it all at once on the screen. Figure 3.2 shows a drawing of the network we will be creating, with a single input and output, and two layers in between.

Each neuron in our network, unlike the mysterious neurons in our own brains, has a very simple job to do. Its role is to take in the output from all of the other neurons in the previous layer and combine them into a single number. It will do this using a simple calculation. Each neuron has two special numbers that it must remember – its weight and its bias. It simply multiplies each input by the weight, adds those results all together and then adds the bias. To this result, it then applies

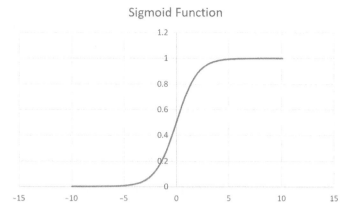

Figure 3.3 A sigmoid activation function.

an activation function, which is a simple function used by every neuron in the layer. A typical activation function is the sigmoid function, shown in Figure 3.3. Applying the sigmoid function to the output of the neuron has been found experimentally to enable the neural network to better learn nonlinear functions. The sigmoid activation function was popular initially in deep neural networks, but since it required more computation, simpler activation functions were tried and found to have similar results. A rectified linear unit function (*relu* for short) is a simple line that increases linearly with an input greater than zero but is zero for any input less than zero. Our network is small, so we'll stay with the classic sigmoid function. With that as a summary, let's look at the details of how the network does its work.

Each neuron in a neural network has two important numbers – a weight and a bias. We'll see where these numbers come from in a moment, but for now let's see how the neurons use them to make their calculations. Each of the five neurons in the first layer, represented by columns F–J, makes a simple calculation, multiplying each input (cells D5–D35) by its weight. The first neuron, column F, would first multiply the input in cell D5 by the weight in cell F2 and put the result in cell F5. Then it would continue to the next input, D6, multiply it also by the weight in F2, and put the result in F6. These multiplications continue through the last input in cell D35.

The second neuron, column G, does the same thing, multiplying the first input in D5 by the weight in G2 and placing the result in G5, and so on, through all the inputs. The remaining three neurons, columns H, I, and J do the same calculations. It would work fine to write out the formulas for each of these calculations and put them in all the cells between F5 and J35, but there is an easier way. What we are doing is actually a matrix multiplication of the single-dimensional matrix, called a vector, representing the inputs, by another vector representing the weights. We are multiplying the vector D5–D35 with the vector F2–J2. We are multiplying a 31×1 vector by a 1×5 vector, so the result will be a 31×5 matrix. We can simply use Excel's MMULT in cell F5 function to perform this multiplication:

$$= MMULT(D5:D35, F2:J2)$$

The result is placed in the region F5–J35. It's a neat trick, being able to calculate the values for all of the inputs at once, rather than in a time-consuming set of individual formulas or software program loop, and this very technique is used in much larger neural networks, which might have thousands or millions of inputs. This is why the training of neural networks requires large computers and lots of time. The training requires large matrix multiplication, not just once, but many, many times over.

Once we have the results of the matrix multiplication for Layer 1 of the one-page neural network, we add the results for each input together, and add the bias number for this layer. For the first input in D5, we add up F5–J5, and then add the bias in cell K2, so the formula in cell K5 is

$$= SUM(F5:J5) + \$K\$2$$

The formula is the same for each row of inputs. The next step is to apply the sigmoid activation function. Checking a calculus textbook, we find the sigmoid function can be calculated with a simple formula using the exponential function (e^x), so cell L5 contains:

$$= 1/(1 + EXP(-K5))$$

The result of this calculation is the output of the layer for the input value on the same row. So for our first input in D5, the output of the first layer of our network is in cell L5. Column L contains the output of the first layer for each input, and the output of this layer simply becomes the input for the second layer.

The second layer is calculated in exactly the same way. The vector of all of the inputs, in column L is multiplied by the vector of the weights, N2–P2. There are three weights, since there are three neurons in the second layer: L5–L35, a 1×31 vector, is matrix multiplied by N2–P2, a 3×1 vector, resulting in a 31×3 matrix in N5–P35. The formula that makes this happen is in cell N5:

$$= MMULT(L5:L35, N2:P2)$$

Just as we did in the first layer, the output of the second layer for the input value on each row is calculated by adding up the weighted values in that row and then adding the bias. So for the first input to the second layer, cell L5, cell Q5 is calculated as follows:

$$= SUM(N5:P5) + \$Q\$2$$

So far, the second layer calculations look just like the first layer. But now there's a small difference – we don't apply an activation function to the output of the second layer. The reason is that we want the neural network to learn to output the output value of the function, which to it is unknown. Ideally the output of the network, which we call a prediction, should match the real output value of the function shown in column E. The output calculated in column Q is exactly the value we need as the final output or prediction of the two-layer neural network for the input in column D.

Next, we need to evaluate, how close we are to the right answer. If the neural network were perfectly reproducing the function we started with, then there would be no difference at all between the value in column Q, the output of the network, and column E, the value of the function we are trying to predict. In reality, there will always be some difference. The neural network isn't trying to guess the actual function; if it were, the moment it found the right function, it's predictions would be perfect. Remember that the neural network doesn't even know what *kind* of function we have used, or that we've even used a function at all. The input/output pairs could be data from a weather instrument or a seismic sensor. What the network tries to do is come up with a set of weights and biases, that when matrix multiplied through, will simulate or predict the relationship between the given inputs and outputs. What we need is a way to measure how close the neural network is to producing the right result. We subtract the known result of the function in column E, from the output of the neural network in column Q, then square the result. This is known as the *squared error*. For the first input, on row 5, the formula for the error, placed in column R is

$$= (Q5 - E5)^2$$

Squared error has been used in the field of statistics for many years as a good measure of a predictor's effectiveness. Squaring the difference between the actual and predicted result removes the effect of sign, making all error values positive, and also makes the measure less affected by small errors than by large ones. To combine the squared errors for every input/output pair, we simply average them together, resulting in a mean squared error (MSE) for the entire set, placed in cell R2 and calculated as follows:

$$= \text{AVERAGE} (R5 : R35)$$

We can summarize the goal of the entire neural network as the minimization of the mean squared error of all of the predictions for the inputs, compared to the given outputs provided. In other words, how close do the predictions come to actually predicting the known output of the function?

Now that we have the neural network set up, our next task is to choose weights and bias values that will minimize the mean squared error. This *is* where the magic happens. We might try choosing the weights and biases manually, through trial and error. We can simply put in some numbers and then try varying a few of them until we get the mean square error as low as possible. It takes only a few seconds of experimenting to find that this is a frustrating and fruitless approach. Small changes in the weight for a single neuron in any layer can have an unpredictable effect on the output, and while it's possible to manually tweak the weights and bias values to achieve a good prediction for one input, the goal is to produce a good prediction for all inputs. Fortunately, Excel has a built-in tool that will help us here. The Excel solver is built into Excel, but if you've never used it before, you need to activate it by going to file, options, add-ins, manage Excel add-ins and then select *solver* (see more detailed instructions from Microsoft at https://tinyurl.com/rbzr2hb4). Once activated, the solver will be available from the data menu in Excel.

The solver is simple in concept. As shown in Figure 3.4, the only settings we need to enter are the objective, which is a single cell, and the variable cells. The solver will find a set of values for the variable cells that will minimize the objective cell. In our case, we want the solver to vary the weights and biases in both layers of our deep neural network (F2:K2 and N2:Q2) to minimize the mean square error in cell R2. We don't care how it performs its work, so we use the default solving

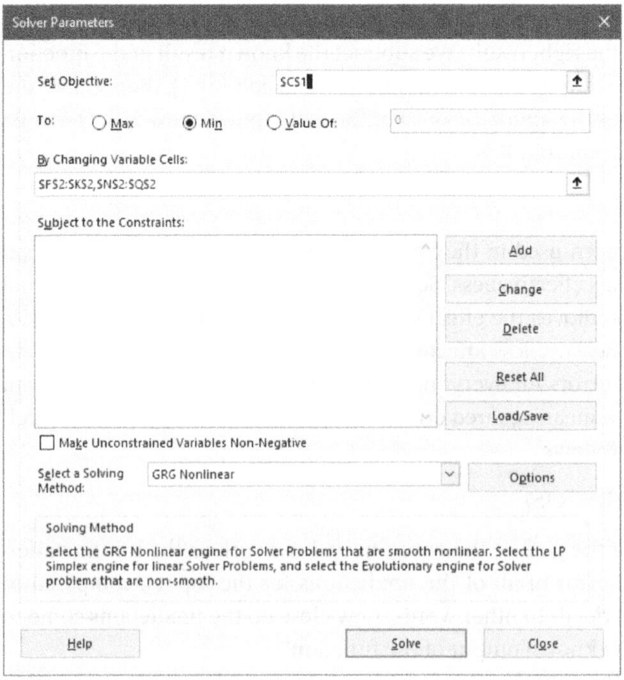

Figure 3.4 Excel solver settings.

method. We only care that it ends up choosing weights and biases that minimize the mean squared error across all input/output pairs and thus optimizes the predictions of the neural network. The solver takes only a few seconds to run on our little one-page deep neural network, and produces the weights and biases seen in Figure 3.1.

An important but surprising fact of both our one-page network and large-scale deep neural networks is that we must initialize the weights and biases to random values before beginning the training (solving) process. It may seem counterintuitive to do this, but if we look at the math being performed, we can see that if the weights and biases were set to the same number, or to zero, at the beginning then all nodes would calculate the same results initially, so it's as if the algorithm has nothing to work with and can't proceed toward a solution. You will be able to see this for yourself; if you don't preset the weights and biases to random numbers before running the solver, you may not get good results. There is a tab in the spreadsheet called Randoms that uses Excel's RAND() function to produce a set of random numbers for this purpose. Before you use the solver to calculate new weights and biases, copy the contents of cells A4–L4 in the Randoms tab, into cells F2–Q2 in the one-page deep neural network (DNN) tab to provide random initialization for the weights and biases.

As we've discussed in this chapter and others, neural networks are trained on one set of data, and then tested using another set of similar, but distinct test data. We expect the neural network to do better at predicting values from the set of data on which it has been trained, than on data it has never seen before, and we see that illustrated in the one-page network. Rows 5–35 are the training data, and as we've just described, the network is "trained" on these pairs of inputs and outputs, that is, a set of weights and biases are found that enable the network to predict the outputs from the inputs with the best accuracy possible, as shown by the minimization of the mean squared error for all of the input/output pairs. Rows 36–58 are test data. The input/output pairs are computed the same way as for the training data, but they are not used when the system is trained, so the network has never seen the test data inputs and outputs until after it is trained. We expect our little network to do better at predicting the data it used during training than the new test data, and this is what we see. In cell R2 we see that the mean squared error for the training data is 0.0022 and in S2 we see that the mean squared error for the test data is 0.0052.

The last feature of the one-page deep neural network is a simple plot (Figure 3.5) showing each input on the x-axis along with its known actual output value (gray line) and the value predicted by the neural network (black dashed line). Note that the plot shows both training data (inputs 1–31) and test data (inputs 31–54). For both sets of data, the lines match up nearly exactly, showing how well even a simple neural network can capture and predict a function from nothing more than a modest set of inputs and outputs.

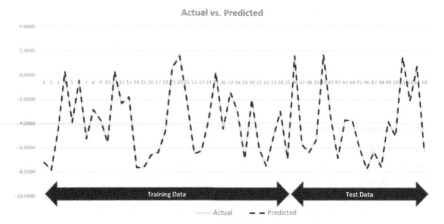

Figure 3.5 Actual vs. predicted values from the one-page deep neural network.

In the example above, we used a straightforward quadratic function, which while not linear, is quite regular. Let's give the little neural network more of a challenge. We change the function, in cell E5 (and copy it to E6–E58) to calculate the output, based on the input in cell D5 as follows:

$$= (SIN(D5) * (1/TAN(D5)))/SQRT(D5)$$

This is quite an odd function, and it might be hard to find a natural or even synthetic system whose behavior was described by it, but sure enough our little neural network does a reasonable job of "learning" it, as shown in Figure 3.6. Notice that the mean square error is larger – 0.0915 for the training data and 1.4005 for the test data, and the difference between the errors on the training and test data is much larger, meaning that the network is not doing as well on data it did not see during its training. The plot also shows more variance in the predicted values vs. the actuals, as seen by the divergence of the lines in a number of places (Figure 3.7).

To wrap up this chapter, we will describe the four significant differences between our one-page neural network and a "real" neural network that would be used in a production application. None of these four have to do with *how* the deep neural network works. The one-page neural network takes all the magic out of deep neural networks by revealing that they are nothing more than a great deal of simple arithmetic, arranged in clever new ways. Speaking of arrangement, there are several kinds of deep neural networks including the convolutional neural network (CNN), recurrent neural network (RNN) which includes the long short-term memory (LSTM) and gated recurrent unit (GRU) network, generative adversarial network (GAN), and the autoencoder network. All of these use the same principles of deep learning seen in the one-page neural network but connect neurons in differing arrangements.

The One Page DNN	Layer 1 Weights				Bias	Layer 2 Weights		Bias	MSE (Train)	MSE (Test)
0.66846042	1.11201595	0.696379701	1.132443158	0.9751615	1.164048886	-15.91607487 -6.474332459		-0.007503143	22.63189148	0.1048	1.3306

Input	Output	Layer 1 Nodes			Wa+b	Activation	Layer 2 Nodes	Wa+b	Sq Error	Actual	Predicted
1.7518	-0.1360	1.1710	1.9484	1.2245	1.9836	1.7083	3.2008	0.9939	-15.9185 -6.4743	-0.0015	2.2336	0.1411	-0.1360	2.2336
1.9500	-0.2851	1.3035	2.1588	1.3630	2.2082	1.9016	10.1099	1.0000	-15.9174 -6.4747	-0.0015	2.2383	0.2834	-0.2851	2.2383
1.1067	0.4255	0.7398	1.2309	0.7735	1.2533	1.0732	6.2415	0.9981	-15.8871 -6.4624	-0.0015	2.2609	0.0209	0.4255	2.2609
0.2511	1.9391	0.1678	0.2783	0.1755	0.2843	0.2449	2.3167	0.9102	-14.4834 -5.8938	-0.0014	2.2474	0.0988	1.9391	2.2474
0.3799	0.6650	0.2561	1.0898	0.6849	1.1097	0.9556	5.6559	0.9965	-15.8626 -6.4525	-0.0015	2.3151	0.0813	0.6628	2.3151
0.3750	1.5196	0.2507	0.4171	0.2621	0.4248	0.3657	2.6650	0.9471	-15.0760 -6.1324	-0.0014	1.4221	0.0095	1.5196	1.4221
1.2553	0.2770	0.8391	1.3961	0.6774	1.4215	1.2241	6.3231	0.9390	-15.9024 -6.4886	-0.0015	2.2594	0.0003	0.2770	2.2594
0.7902	0.7917	0.5282	0.8786	0.5523	0.8948	0.7705	4.7895	0.9318	-15.7866 -6.4215	-0.0015	2.4221	0.1366	0.7917	2.4221
0.3468	0.6005	0.6529	0.6618	0.6618	1.0722	0.9233	5.5060	0.9360	-15.8538 -6.4488	-0.0015	2.3278	0.0744	0.6005	2.3278
1.3269	0.2097	0.8870	1.4757	0.3274	1.5026	1.2939	7.2515	0.9393	-15.9068 -6.4703	-0.0015	2.2532	0.0015	0.2097	2.2532
0.2441	1.9640	0.1632	0.2715	0.1706	0.2764	0.2380	2.2646	0.9076	-14.4471 -5.8766	-0.0014	2.3088	0.1175	1.9640	2.3068
0.7010	0.9127	0.4686	0.7797	0.4900	0.7939	0.6836	4.3806	0.9676	-15.7213 -6.3949	-0.0015	2.5142	0.1588	0.9127	2.5142
0.6031	1.0509	0.4072	0.6774	0.4257	0.6898	0.5940	3.9589	0.9813	-15.6200 -6.3537	-0.0015	2.6567	0.1584	1.0509	2.6567
1.8873	-0.2265	1.2616	2.0990	1.3192	2.1372	1.8404	9.8222	0.9999	-15.9172 -6.4746	-0.0015	2.2386	0.2263	-0.2265	2.2386
1.8651	-0.2256	1.2606	2.0977	1.3183	2.1359	1.8392	9.8167	0.9999	-15.9172 -6.4746	-0.0015	2.2386	0.2156	-0.2256	2.2386
1.5743	-0.0028	1.0524	1.7510	1.1004	1.7828	1.5353	9.3868	0.9998	-15.9144 -6.4735	-0.0015	2.2425	0.0602	-0.0028	2.2425
1.5232	0.0396	1.0182	1.6941	1.0647	1.7249	1.4854	8.1520	0.9997	-15.9135 -6.4731	-0.0015	2.2430	0.0421	0.0396	2.2430
1.1322	0.3931	0.7565	1.2593	0.7914	1.2822	1.1041	6.3588	0.9383	-15.8906 -6.4637	-0.0015	2.2781	0.0151	0.3931	2.2781
0.1985	2.2005	0.1327	0.2208	0.1387	0.2249	0.1936	2.0754	0.8885	-14.1430 -5.7529	-0.0013	2.7348	0.2053	2.2005	2.7348
0.0340	5.4491	0.0227	0.0378	0.0238	0.0385	0.0332	1.3209	0.7831	-12.5646 -5.1108	-0.0012	4.9553	0.2751	5.4491	4.9553
0.9596	0.9775	0.6419	1.0703	0.6530	1.0936	0.6403	4.1769	0.9849	-15.6775 -6.3771	-0.0015	2.5758	0.1613	0.9775	2.5758
1.5503	0.0164	1.0363	1.7243	1.0837	1.7557	1.5119	8.2767	0.9997	-15.9140 -6.4733	-0.0015	2.2431	0.0514	0.0164	2.2431
1.4947	0.0621	0.9992	1.6625	1.0448	1.6927	1.4576	8.0216	0.9997	-15.9129 -6.4726	-0.0015	2.2447	0.0333	0.0621	2.2447
0.9332	0.6912	0.6238	1.0379	0.6523	1.0588	0.9100	5.4457	0.9957	-15.8497 -6.4471	-0.0015	2.3336	0.0798	0.6912	2.3336
0.2806	1.8926	0.1742	0.2899	0.1822	0.2952	0.2542	2.3605	0.9158	-14.5454 -5.9168	-0.0014	2.1886	0.0783	1.8926	2.1886
1.0832	0.4438	0.7231	1.2114	0.7613	1.2335	1.0622	8.1614	0.9379	-15.8646 -6.4613	-0.0015	2.2845	0.0254	0.4438	2.2845
0.5430	1.1619	0.3630	0.6039	0.3795	0.6149	0.5295	3.5757	0.9748	-15.5170 -6.3118	-0.0015	2.8016	0.1288	1.1619	2.8016
0.8120	0.7635	0.5428	0.9031	0.5676	0.9196	0.7919	4.8897	0.9925	-15.7992 -6.4266	-0.0015	2.4046	0.1269	0.7635	2.4046
1.6653	-0.0731	1.1132	1.8522	1.1640	1.8859	1.6240	8.8041	0.9996	-15.9157 -6.4740	-0.0015	2.2407	0.0985	-0.0731	2.2407
0.6423	0.3991	0.4293	0.7143	0.4489	0.7273	0.6263	4.1111	0.9839	-15.6814 -6.3705	-0.0015	2.5395	0.1605	0.3991	2.5395
1.4365	0.1117	0.9603	1.5977	1.0041	1.6268	1.4008	7.7545	0.9996	-15.9113 -6.4722	-0.0015	2.2470	0.0183	0.1117	2.2470
1.8343	-0.1923	1.2262	2.0401	1.2821	2.0772	1.7888	9.5793	0.9999	-15.9170 -6.4745	-0.0015	2.2389	0.1660	-0.1923	2.2389
1.2237	0.3017	0.8181	1.3677	0.6595	1.3926	1.1992	6.6053	0.9969	-15.9005 -6.4678	-0.0015	2.2622	0.0016	0.3017	2.2622
0.8132	0.7621	0.5436	0.9044	0.5684	0.9209	0.7930	4.8951	0.9926	-15.7998 -6.4268	-0.0015	2.4037	0.1294	0.7621	2.4037
1.6629	-0.0712	1.1115	1.8493	1.1823	1.8830	1.6215	8.7925	0.9996	-15.9157 -6.4739	-0.0015	2.2408	0.0374	-0.0712	2.2408
0.0335	5.4645	0.0224	0.0372	0.0234	0.0379	0.0326	1.3183	0.7889	-12.5578 -5.1081	-0.0012	4.9649	0.2436	5.4645	4.9649
1.3520	0.1776	0.9104	1.5148	0.9520	1.5423	1.3292	7.4125	0.9994	-15.9171 -6.4710	-0.0015	2.2509	0.0054	0.1776	2.2509
1.5253	0.0369	1.0196	1.6964	1.0661	1.7273	1.4874	8.1616	0.9997	-15.9136 -6.4731	-0.0015	2.2430	0.0428	0.0369	2.2430
1.2757	0.2576	0.8527	1.4188	0.6917	1.4448	1.2440	7.0167	0.9991	-15.9038 -6.4691	-0.0015	2.2574	0.0300	0.2576	2.2574
0.0087	10.7327	0.0058	0.0097	0.0061	0.0098	0.0085	1.2047	0.7694	-12.2466 -4.9815	-0.0012	5.4026	28.4105	10.7327	5.4026
0.8966	0.6593	0.5393	0.9372	0.6257	1.0153	0.8743	5.2776	0.9948	-15.8372 -6.4420	-0.0015	2.3511	0.0950	0.6593	2.3511
1.6448	-0.0576	1.0995	1.8293	1.1497	1.8626	1.6040	8.7099	0.9996	-15.9154 -6.4739	-0.0015	2.2411	0.0882	-0.0576	2.2411
0.3480	0.5590	0.6337	1.0544	0.6627	1.0736	0.9245	5.5158	0.9960	-15.8542 -6.4489	-0.0015	2.3273	0.0739	0.5590	2.3273
0.9564	0.5894	0.6394	1.0636	0.6685	1.0851	0.9332	5.5523	0.9961	-15.8566 -6.4499	-0.0015	2.3239	0.0705	0.5894	2.3239
1.4180	0.1278	0.9479	1.5771	0.9112	1.6058	1.3828	7.6686	0.9995	-15.9106 -6.4719	-0.0015	2.2478	0.0144	0.1278	2.2478
1.3268	-0.2509	1.2878	2.1427	1.3466	2.1817	1.8787	10.0025	1.0000	-15.9174 -6.4746	-0.0015	2.2364	0.2395	-0.2509	2.2364
1.5066	0.0523	1.0071	1.6756	1.0531	1.7061	1.4692	8.0760	0.9997	-15.9131 -6.4729	-0.0015	2.2443	0.0369	0.0523	2.2443
1.8704	-0.2158	1.2503	2.0802	1.3074	2.1181	1.8233	9.7448	0.9999	-15.9171 -6.4746	-0.0015	2.2387	0.2066	-0.2158	2.2387
1.3733	0.5703	0.6506	1.0825	0.6803	1.1022	0.9491	5.8235	0.9964	-15.8611 -6.4518	-0.0015	2.3175	0.0639	0.5703	2.3175
1.2132	0.3188	0.8103	1.3482	0.6473	1.3727	1.1821	6.7254	0.9968	-15.8930 -6.4672	-0.0015	2.2642	0.0030	0.3188	2.2642
0.9486	4.5294	0.0325	0.0541	0.0340	0.0551	0.0474	1.3979	0.8003	-12.7398 -5.1816	-0.0012	4.7105	0.0328	4.5294	4.7105
0.6636	0.3671	0.4436	0.7380	0.4638	0.7515	0.6471	4.2088	0.9854	-15.6848 -6.3801	-0.0015	2.5654	0.1614	0.3671	2.5654
0.1822	2.3041	0.1218	0.2026	0.1273	0.2063	0.1777	2.0005	0.8800	-14.0215 -5.7035	-0.0013	2.9056	0.3610	2.3041	2.9056
1.4987	0.0598	1.0018	1.6669	1.0476	1.6972	1.4616	8.0398	0.9997	-15.9129 -6.4728	-0.0015	2.2446	0.0345	0.0598	2.2446

Figure 3.6 One-page neural network with alternate function.

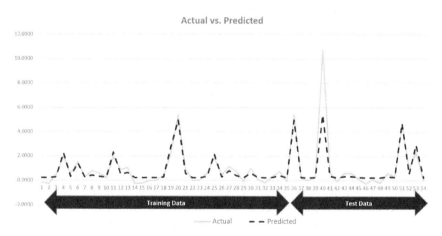

Figure 3.7 Actual vs. predicted values for alternate function.

The first difference between the one-page neural network and a production neural network is that the one-page neural network has only a single input. By having only a single input number, and a single output number, the vector of all inputs together is one dimensional, and when multiplied by the vector of weights, becomes only two-dimensional, and thus easy to show on a single flat page. In most applications, each input consists of several numbers, such as the features of a house, or even thousands of numbers, such as the pixels in an image. We still use the same matrix multiplication process, but the results are larger and harder to see at a glance.

Difference number two is that most neural networks have more neurons in each layer. The one-page neural network has five neurons in the first layer and three in the second layer. A production neural network will have more, but perhaps not as many as one might think. A dozen or up to a few dozen nodes in a layer are typical. Too many neurons in a layer and the matrix multiplication becomes time-consuming, which can make training the network take hours, weeks, or longer, even on a large computer. Some well-designed neural networks, when trained on large datasets, still take many hours to train, even on a multi-graphical processing unit (GPU) system optimized for deep neural network training. The one-page network is so small and uses just a small set of data, so it can be trained in seconds, even using Excel.

The third difference is that a production neural network will have more layers. Research continues into the optimal number of layers for various kinds of applications, but networks with dozens or even hundreds of layers are not uncommon, and these of course take more time to train. Note however that the one-page neural network can be extended to many layers by simply repeating the process described: add an activation function such as the sigmoid, to the output of the second layer, and use the result as the input to a third layer. There is no magic to additional layers – they all work the same way as the one-page network.

The fourth difference is that production neural networks have better and faster ways to train themselves. Like the number of neurons in each layer and the number of layers in the network, training algorithms are a subject of much experimentation and research. *Back propagation*, though developed in the 1970s became the core of the deep neural network revolution due to recent advances in computing power. No matter the approach, however, the objective is the same – find a set of weights and biases for each layer that minimize the overall error of the network. Better training algorithms simply aim to reduce the time it takes to successfully find those weights and biases.

As can be seem from these four differences, the one-page neural network does everything a large-scale neural network does, but at a tiny scale. Larger networks are simply more of the same. One can be sympathetic to the sentiment that the biggest obstacle to understanding neural networks is that they are called *neural networks* and thus conjure images of brain-like complexity and mystery and of humanlike intelligence. In this chapter, we've attempted to erase most, or even all of the mystery surrounding AI with deep neural networks, using terms and concepts available to everyone. The point has not been to minimize the amazing accomplishments of AI and neural networks and of those who build them. The accomplishments are real. AI and deep neural networks are succeeding at applications that are both amazing and more importantly productive. At the same time, AI and neural networks are not proceeding on a path toward humanlike intelligence. It's a paradox, but one that can be understand based on what we've presented in this chapter. On the one hand, AI and neural networks are becoming "smarter" in that they are able to perform more complex tasks successfully, even surpassing human performance levels. On the other hand, they are not becoming more intelligent in the human sense. Here's an analogy that may help.

Imagine we want to build an AI-based robot that can play baseball as well as a good human player. We start by developing a machine that can pitch a baseball harder and more accurately than any human pitcher. We also develop a machine that can run between the bases faster than any human player, and one that can reliably hit any legal pitch well out of any ballpark. Each of these capabilities surpasses human performance, but how close are we to building a superhuman, automated baseball player? The point is that there is much more to playing baseball than just the individual skills of throwing, catching, and hitting. Those building the robots know how throwing, catching, and hitting work, and can thus implement clever machines that surpass human performance at those tasks, even if in a different way: the pitching machine may use wheels and gears, rather than a humanlike arm. But these skills do not simply add up to being a baseball player. The problem here, and in AI, is that we can't describe exactly how a major league human baseball player does what he does – we can only copy the individual skills we do understand and implement them in computers and robots. It's why we can build a machine that can play chess better than any human player in the world, but still not be able to carry on a conversation as well as a two-year-old.

4

Intelligent Systems and the People they Love

There is a high expectation that intelligent systems technology will have a significant impact on economic growth in the United States and other developed countries. Accenture estimates that artificial intelligence will double the economic growth rate in many countries by 2035, but this won't happen automatically – it will happen through the design and implementation of new intelligent systems.

In this chapter, we'll explore one of the fundamental questions in the public consciousness about AI and intelligent systems: can computers think, or understand? People who are not specialists in AI technology may well assume that AI means building thinking machines, and by *thinking* they mean humanlike thought. Can computers be made to think in the way humans think about thinking? Can we simply assume that a computer that plays chess at a world championship level is thinking its way through the game in a way similar to how a human would play chess?

4.1 Can Machines Think?

It's important to note at the outset that we are not asking if *machines* can think. They certainly can – we are thinking machines. We are biological in nature, but there's no reason to think that only biological machines can think. What we are asking here though, is can computers, running software programs, be said to think, or to understand? It's one of the most considered and debated questions in the subfield of philosophy known as the philosophy of mind, and fascinating on its own, but the understanding of this question is also important to engineering.

Those who build AI systems may consider this entire discussion irrelevant – it is not necessary to have a philosophical answer to the question of whether machines can think in order to build AI systems using current deep neural network approaches. AI systems of this type are useful, regardless of our philosophical positions. At the same time, investigating the subject can illuminate our

Engineering Intelligent Systems: Systems Engineering and Design with Artificial Intelligence, Visual Modeling, and Systems Thinking, First Edition. Barclay R. Brown.
© 2023 John Wiley & Sons, Inc. Published 2023 by John Wiley & Sons, Inc.

thinking about computers, machines, and intelligence, and may also serve to make us less susceptible to AI fears born of illogic and fantasy.

To put this in a rather cheeky way, we might say there are several types of AI: deep neural networks, expert systems, reinforcement-learning systems, and a special category we might call M5-class systems. The name is taken from an original series Star Trek episode of the same name; M5 refers to a computer that was able to control the entire starship *Enterprise*, but which also had qualities like creativity, initiative, flexible decision-making, and exhibited emotions including guilt, shame, and even suicidal ideation. It is these M5 class systems that produce most of the fear of AI systems. The problem is that most people do not differentiate the types of AI in their thinking, in public discourse, media, and policy making. No one is afraid of a deep neural network that predicts the weather, but when lumped together with M5 systems, people worry. Is it necessary to point out that M5 class systems exist only in science fiction as of the early 2020s, and perhaps for the foreseeable future and beyond. The most productive and useful AI systems today are probably deep neural networks and reinforcement-learning systems and neither has the capacity for humanlike mental functions such as understanding, thought, and emotion. This is not an uncontroversial statement, but in this chapter, we'll make a case for it, and whether you end up agreeing or not, the journey will be worth it for the understanding of the issues and positions involves.

4.2 Human Intelligence vs. Computer Intelligence

In beginning to wrestle with the idea of an intelligent system, we should first spend a moment comparing human intelligence to machine or computer intelligence. It is clear from the progress of technology and the recent acceleration in the success of AI and machine-learning algorithms that computers can be made to act in ways previously thought to be possible only for human beings, which is to say, through the operation of human brains. A brain and its more abstract capability, which we call a mind are unique in all creation, at least at present. To a reasonable observer comparing a brain/mind to a computer running a program, the difference seems apparent. No matter how cleverly we make the computer perform, no matter how good it is at chess or go, and no matter how much of a conversation it can carry on, we sense that there is something different between what it is doing and what human, or even animal, minds are doing. In the remainder of this chapter, we examine this matter in some detail. Mostly the domain of philosophers who focus on the specialty of philosophy of mind, questions concerning brains, minds, and computers are fascinating. However fascinating the subject may be, those with a purely engineering or practical orientation may find this chapter somewhat academic, and you have the author's permission to skip it, and proceed to the next chapter. Before you go, at least for a moment, seriously consider the question, can a computer think? If you are satisfied with the answer you come up with, feel free to skip the rest of this chapter. If on the other hand, this seemingly simple question captures your imagination and suggests other questions including: What is thinking? What is a computer? Is a computer the only kind of intelligent machine? Is there any behavior that proves a system is thinking? Then press on through this chapter's tour of the questions of machine intelligence and the philosophy of mind.

4.3 The Chinese Room: Understanding, Intentionality, and Consciousness

The Chinese room argument, originally proposed by John Searle in 1980, is one of the most discussed and debated illustrations in the field of philosophy of mind. It attempts to show that a computer, running a program, is incapable of possessing understanding. It does so by analogy, proposing a thought experiment in which a person, who does not know any Chinese, is placed inside a closed room, equipped with books of instructions written in a language the person can understand, say English, for manipulating symbols, the meaning of which are unknown to the person. Other symbols are passed into the room, and the person follows the given English instructions to manipulate those symbols, producing a second set of symbols which the person passes back out of the room. The person is allowed the use of scratch paper, pencils, and unlimited time. Unknown to the person in the room, the input sets of symbols are questions in written in Chinese, and the English instructions written in the books are designed to produce appropriate answers to those questions, with the answers also written in Chinese. If the person in the room gets good enough at using the instruction books and doesn't make too many mistakes, the room will appear to be able to converse in Chinese, answering Chinese questions correctly in Chinese. Anyone outside the room would observe the room answering questions in Chinese just as a native speaker of Chinese would. The observer could plausibly infer that there is someone or something in the room who *understands* Chinese.

The Chinese room, if it had enough instructions, could potentially even pass the Turing test. The Turing test, named for its originator Alan Turing, claims that if a human being carrying on a conversation, cannot tell if his or her conversation partner is human or machine, then the test is passed and the partner, if a machine, can be considered intelligent or even conscious. The Turing test can be stated in propositional form as follows:

1. If X passes the Turing test, then X can understand.
2. A computer running a computer program can pass the Turing test.
3. Therefore, a computer running a computer program can think/understand.

While some may argue against premise (2), the Chinese room poses a situation that accepts premise (2) but rejects premise (1). Searle claims that the person in the Chinese room (he uses himself as exemplar) is doing exactly what a computer running a program would do, that is, manipulating symbols according to very specific prewritten rules. To him, the symbols are not even known to be symbols – they are "meaningless squiggles" (Searle 1980). Searle in the room has no way to relate the squiggles to any kind of meaning, and thus is limited to following the rules blindly, with no understanding of the meaning, and no way to gain any understanding. He could perform these manipulations indefinitely without coming to learn any Chinese.

Searle's claim is that even though the person in the room is functioning *as if* he or she understood Chinese and can convince observers that this is so (and thus pass the Turing test) he or she does not (and cannot come to) understand Chinese. For Searle, it follows that the claim that a computer running a program can understand Chinese is false, and the theory of *strong artificial intelligence,* which relies on this claim, is false also. Strong AI, according to Searle (1980), claims "that the computer understands… and that the program in some sense explains human understanding." Searle claims that the Chinese room argument refutes this claim about artificial intelligence.

In order to argue against premise (1), that if a computer passes the Turing test, it must have *understanding,* Searle must differentiate between something that appears to understand and something that in fact understands. If one assumes that there is no difference – that understanding is merely a matter of observable behavior, then according to Searle one is led to accept many sorts of objects as exhibiting understanding. Since a thermostat on the wall is able to respond to changes in temperature, it must "understand" temperature. Searle attempts to show that there is more to understanding than mere behavior – that understanding requires not just the ability to process meaning, but also the ability to manipulate symbols.

Searle's Chinese room argument suggests that *as-if understanding* (behavior that appears to understand) can be either the behavior of an understanding, thinking system (such as a human being) or the behavior of a system that merely *simulates* such understanding (such as a computer running a program). This difference is critical to Searle's argument for without it, it is compelling to assume that *understanding* behavior must be the result of an *understanding* system. The assumption that the appearance of understanding is equivalent to understanding is made by many of the critics of the Chinese room argument and without this assumption, objections to the CRA are substantially weakened. In what follows we will show that the distinction of apparent or simulated understanding vs. real understanding is quite clear to the developers of AI systems, even though it may not be acknowledged by Searle's philosophical opponents.

Searle offers the distinction that systems which appear to be understanding but are not, such as the Chinese room, are nonunderstanding because they lack the ability to handle semantics or meaning, and thus the squiggles remain meaningless despite the room's behavioral ability to answer questions in Chinese. Whether a system is capable of dealing with semantics or not may not be an easy question to settle by observation. I will argue later that there are criteria that can be applied – criteria which can be evaluated without making an a priori assumption about the system's ability to understand. These criteria are based on elements of systems engineering and computer science theory which explain how systems come to be able to appear to understand. A range of cases of understanding behavior will be considered, from systems that most would agree are not actually understanding, such as a wooden ventriloquist dummy, to the early AI programs such as ELIZA (which was able to carry on a rudimentary conversation in plain English), to those that are much more sophisticated, including modern AI and deep neural networks which are capable of exhibiting behavior that plausibly suggests they do have understanding. A clear example is IBM's Deep Blue computer which was able to beat the then-current world chess champion, Garry Kasparov. While it is plausible to assume that Deep Blue indeed understands chess, Searle (1998) questions whether it is even actually *playing* chess in the way a human being plays chess, that is, knowing the semantics of the game vs. simply following a program as the Chinese room does. As mentioned elsewhere in this book, Deep Blue relies on preprogrammed lists of moves for much of the game – the opening and the endgame, and it's hard to imagine that a human or machine player who plays by looking up moves from a list, is actually playing, let alone understanding chess.

First, we will consider the various objections raised to Searle's Chinese Room Argument (CRA), his responses, and the subsequent replies to those responses, focusing special attention on the systems reply, which is the most relevant to our purposes in this chapter. Then, we will consider in more depth the matter of simulated-thinking systems vs. actually thinking systems, and how these may be differentiated. Finally, we will look at how the Chinese room could be made into an actually thinking system, and how performing this transformation illustrates that in its original form, the Chinese room does not think, and thus, the main conclusion of Searle's CRA succeeds.

For the purposes of this chapter, we aim generally to support Searle's conclusion in the CRA, by answering, on his behalf, some objections raised to his responses to the system reply. Our position is that his replies, which one might claim are based on unsupported intuitions on his part regarding the nature of computers and programs, or on now-outdated understandings of these, are in fact correct, especially when we examine them with an understanding of how computers and programs actually work. Predictions made about the capabilities of computers running programs to learn, understand, and even *be* minds, have not in the 40 years since Searle posed the Chinese room, come true, and not for the lack of sufficient computing and memory capabilities, as was commonly thought, but by the fundamental nature of computer programs. Computers and even systems made up of multiple computers, despite their increasing power and speed, are progressing, not toward being minds, but toward being better and better simulators of minds. Our argument for this will proceed as follows:

1. We argue that the Turing test is not adequate to show that a system understands, since it is solely based on behavior and as-if understanding behavior can arise from sources other than actual understanding, specifically the simulation of understanding accomplished through computer programming.
2. We argue that the problem of other minds does not prevent us from distinguishing apparent or simulated understanding from real understanding. We use Searle's notion of underlying structure and modern systems engineering theory to show that while a complex system may appear to understand, and while in fact it may not be apparent to a casual observer how or why it does or does not understand, this is not a sufficient reason to assume that it in fact understands.
3. We will argue that the apparent understanding exhibited by the Chinese room does indeed rely on real understanding, but that this real understanding is not a part of the Chinese room system and is instead a function of a human mind – the mind of the programmer. To do this I will argue that a program, written by a programmer, is not and cannot be a mind, but acts as an agent of the programmer's mind. This reinforces Searle's main aim in the CRA which is to show that a computer, by virtue of it running a program, cannot understand.

In his original formulation of the Chinese Room Argument, Searle (1980) focuses his argument against strong AI on the question of whether the Chinese room *understands* Chinese. He takes understanding to be one of a number of possible cognitive states, perhaps the most fundamental, since it would seem that without it, other states are not possible. He focuses on the understanding of natural languages and how people, such as himself, can exhibit varying degrees of understanding of say, their native language, a second language they have learned later, and perhaps one with which they have no familiarity and cannot even recognize. He emphasizes that degrees of understanding are not appropriately attributed to devices that simply behave as if they understand. While we may speak of the adding machine understanding how to add numbers, this is metaphorical and not to be taken even as a low degree of the same kind of understanding human beings (and other brains) have. Thus, he begins to draw a sharp distinction between the behavioral appearance of understanding and actual understanding. His proof for this is the CRA itself – a system can behave as if it has perfect understanding of Chinese, and yet understand none of it.

This construal of understanding may not be satisfying to those who wish to argue, against Searle, that he has simply assumed that only systems with brains, such as human being and animals can possess understanding, and that without any better explanation of understanding, the existence of behavior indistinguishable from that of brain-based understanding machines is itself sufficient to guarantee understanding. This leads AI proponents to argue (or perhaps just assume) that the kind of understanding that appears to be going on in computer programs is the same kind that is going on in brains. The distinction between apparent and real understanding will be taken up at length later; for now, we note that Searle certainly means something more (and other) than behaving-as-if-understanding when he uses the term.

In the CRA, Searle treats understanding as an intentional state and argues that intentionality is a causal feature of the brain. He further explains in Searle (1983) that such causality is naturally attributed to brains with the capability of language but may also be logically attributed to animals even though they lack language. This brings into sharper focus what Searle has in mind about what it takes to attribute intentionality to a thing, be it human, animal, or computer. In analyzing how intentionality works, he emphasizes the directed-ness of intentional states – that they are *about* something, or *of* something. It is this directed-ness that is lacking in the computer running a program, since it is following a programmed set of instructions, and though it may appear to have intentional states, such as understanding, it does not, because it lacks the capacity to do so. As a more detailed analysis will later show, the computer is not even understanding its own programmed set of instructions, if we remain consistent about our use of the term understanding.

One might assume that Searle's analysis, so far, is simply out of date, that he is referring to traditional computer programming, and not modern techniques for AI such as machine learning with deep neural networks. As we show in Chapter 3, deep neural networks are doing nothing other than simple calculations, which could in fact be carried out in a spreadsheet (as we show) or by a traditional, procedural computer program. Neither the very vague resemblance of a neural network to the neurons in a brain, nor the term *neural network*, provide any support for the idea that a deep neural network is doing what a brain does, or has any more ability to understand than a computer program written in a more traditional programming language. In fact, programmers new to neural network programming are usually trained to program a deep neural network using ordinary computer programming including arrays, for loops and repetitive calculations, before moving on to using matrix mathematics and special-purpose neural network programming systems.

Searle's point is *not* that a machine cannot possess intentionality, but that a computer running a computer program cannot. In this chapter, I will argue that he is right on this point by showing that while a computer can be programmed to simulate intentionality or understanding, and may even fool human observers that it has intentionality, it does not, and further, that this fact is clear to those who create such programs. To a programmer, it is quite clear when they are "programming in" intentionality and when they are not. The temptation to attribute intentionality to a computer running a program stems, as we will show, from mistaken attributions of humanlike powers to sufficiently complex machines.

4.4 Objections to the Chinese Room Argument

A number of significant objections to the CRA have been raised since it was proposed in 1980. These fall generally into one of several themes. While the main focus of this chapter is the *systems reply*, other objections to the CRA are surveyed here to provide context and also because some of them will become relevant in the later analysis.

4.4.1 The Systems Reply to the CRA

More than any other significant objection to the CRA cited and answered in Searle's original 1980s article, the "systems reply" has resulted in the greatest amount of discussion since it was raised. It states that while it is true that the person in the Chinese room does not understand Chinese, the entire system, which consists of the person, the instructions, scratch paper, pencils, and the room itself, does understand Chinese. Searle's response to the systems reply is to modify the thought experiment and have the person memorize the instructions, do the work without the aid of scratch paper and pencils, and work outdoors so there is no room at all. Searle claims this modification changes nothing essential in the main argument and seems satisfied that this objection is quelled. Nevertheless, the systems reply and its derivative objections to the CRA have lived on for decades.

The primary proponents of the systems reply usually seem to proceed from the assumption that understanding *must* be going on in the room, because the room is able to answer questions put to it in Chinese. In order to show that this assumption is unwarranted, I will argue for the distinction between a system that has understanding and one that merely simulates understanding, thus defending Searle against this line of objection.

4.4.2 The Robot Reply

The "robot reply" from Yale, to which Searle also replied in the original article (Searle 1980), claims that by adding peripheral devices to the Chinese room and making it mobile and able to interact with its environment, it would gain the ability to understand. Searle's reply is that the CRA in its original form would apply equally to the robot – there is nothing to preclude the input symbols coming from cameras, sensors, and microphones which link the robot to its environment. Though a robot might provide more humanlike behavior in that it can move and speak, the robot still does not understand Chinese by virtue of its program.

4.4.3 The Brain Simulator Reply

The "brain simulator" reply, also stated and answered in Searle (1980), suggests that if a machine were constructed that simulated the structure of the brain in its computer program, it would then be capable of actual understanding, since it is presumably operating the way a brain does. Searle's response is that simulating the structure while running the program does not overcome the limitations of the program – that is, that it only manipulates symbols and does not carry the meaning of those symbols. Searle also suggests that the simulation of the brain's structure in some arbitrary medium is not enough to guarantee brain-like function, that is, understanding.

Searle could not have foreseen the dramatic success of deep neural networks that began 25 or so years after he wrote, but his response to the brain simulator reply is pertinent in today's world of increasingly complex deep neural networks. Networks with billions of parameters are not uncommon, but while the structure of a deep neural network in a limited way resembles the neurons and interconnections in the brain, there is still no way for these parameters to carry the semantic meaning necessary for understanding and consciousness.

4.4.4 The Combination Reply

The "combination reply" is next in Searle's original list of objections to the CRA and claims that the combination of the previous three replies would produce something to which we must ascribe understanding. If a robot with a computer in its head were programmed to simulate the synaptic operations of the brain and connected to suitable input/output devices so that it acted like a human being, it would be, as Searle admits, "rational and indeed irresistible" to believe this robot had intentionality, absent any further information. But given the information that the robot is operating by means of a program that carries no meaning about the input and output symbols, nor about the kinds of devices that provide those symbols, the assumption that the robot has a mind is unwarranted.

4.4.5 The Other Minds Reply

The "other minds" reply, minimized by Searle in the original paper, asserts that we only know that something has cognitive states by virtue of observing its behavior, else we run into the problem of other minds, and cannot infer that other people in fact have cognitive states. The claim is that if robots or computers exhibit the appropriate behavior, we must also attribute cognitive states to them. Searle's reply is that observed behavior is perhaps a necessary condition, but is not sufficient to guarantee cognition. It may look like a duck, and walk like a duck, but if we know that inside, it is not built like a duck, we must not be fooled into thinking that it has the cognitive states of a duck.

The problem of other minds is one well known in the philosophy of mind, and refers to the position that while one cannot see the mind of another person, one can know that other people have minds by analogy. Since the other person is like me in obvious ways, including appearance and behavior, he or she must have a mind like mine. Searle claims that more is necessary to infer the existence of a mind in another. I must know more about the structure of other people. If I know they have a brain like mine, then I may infer they have cognitive states like mine, but if they have a computer program instead of a brain, and it cannot be proven that the program does what a brain does, including producing cognitive states, then I cannot make the assumption of cognitive states, consciousness, or understanding.

As stated in his earlier arguments, the presence of a program of the kind described by Turing, posited in the CRA, and implemented in modern deep neural networks, is not sufficient to guarantee understanding and no amount of humanlike behavior will overcome this deficiency. An increasingly accurate appearance of human operation is not sufficient to indicate the presence of a mind. This theme will be taken up further in what follows regarding the nature of computer programs. Searle also suggests that we do not make the inference that other people have minds based solely on their appearance and behavior – we reason also by analogy with ourselves. I have a brain and my behavior proceeds from my mental states; others have similar brains and thus one can infer that their behavior also comes from their having mental states.

4.4.6 The Many Mansions Reply

Finally, Searle (1980) mentions what he calls the "many mansions" reply, named for its biblical reference to the multiplicity of possible kinds of systems. Here, responders from Berkeley claimed that the CRA applies only to analog and digital computers, and that someday devices may be built that indeed overcome the limitations of such computers and thus could be capable of understanding. Searle (1980) does not object to this reply, emphasizing that the CRA takes aim only at computers running computer programs.

The "many mansions" reply leads to an important point regarding Searle's CRA that seems misinterpreted in successive literature – the issue of whether one day a machine could be made that can indeed have understanding or consciousness. Searle states that the CRA is only applicable to the situation of a machine running a program that manipulates formal symbols. This is the kind of machine that digital computers are, and digital computers are the focus of most, if not all, AI development work. It is tempting to imagine computers to be capable, someday, of transcending this type of programming, or perhaps any type of programming, and somehow becoming intelligent, and capable of understanding. Searle's point turns here on what is meant by a *program*. Later, we will see how some opponents of the conclusions of the CRA use the term *program* to refer to what a native Chinese speaker is using in order to speak Chinese. This broadens the term to cover what is perhaps better described as *behavior* or *function*. When we instruct and educate our children, we do not refer to it as programming, and we don't think of what we are giving them as a program, in the computer program sense. In fact, often we want our children to learn to function in the world, without following some kind of program, or set of specific instructions. We want them to learn to think, to understand, not follow a program.

The point of the CRA is that understanding cannot come by way of a program, which Searle further clarifies as something that can be reduced to instructions that perform calculations and move information – programs that can be implemented in a Turing machine. Can other kinds of programs be developed someday? Perhaps, but if they are to run on digital computers, they ultimately must be implemented as sets of instructions and computations. In Section 4.5, the reasons for this will be described in a response to Haugeland (2002) regarding the nature of computer languages and programming.

Searle allows for the logical possibility and even the physical possibility of the creation of a kind of machine that can think, since as he notes, we are precisely that kind of machine. While we can be thought of as a digital computer, in that we can certainly run programs, follow instructions, and manipulate symbols like an electronic digital computer (albeit more slowly and with more errors), it does not follow that any digital computer can do what we can – understand, have intentionality, and possess mental states. The digital computer has only a computational (Turing) program; we have much more. Could a machine be created that has this "much more"? Yes, Searle replies, but it will not be like a digital computer running a computational program.

4.5 Agreement on the CRA

Before moving into the specifics of the systems reply and the arguments surrounding it, we consider the areas about which Searle and his opponents agree. First, as has been suggested already, Searle and his opponents agree that it is logically possible for a machine to have understanding and intentionality, for as Searle (1980) states, we are precisely such machines. Going further though, he allows that such machines need not necessarily be biological – that it is logically possible for a non-biological machine to have understanding and intentionality.

There are interesting research projects that are attempting to construct a synthetic brain, and while they have not achieved success yet, there is that possibility. One of the major obstacles to such a project is our currently limited understanding of how the brain works. It is difficult to build a machine that does what the brain does, if we don't know how the brain does what it does. As a metaphor, we understand enough of how the kidneys work to build an artificial kidney in the form of a dialysis machine, with prototypes now being created for a much smaller, implantable artificial kidney. But we don't understand the brain at this level, yet.

Next in the list of widely agreed topics, Copeland (2002), Hauser (2002), and others agree with Searle that simulation is possible – that the appearance of behavior that suggests understanding is not the same thing as a machine possessing understanding. As Wakefield (2003) points out, "we have a natural tendency to use mental language in regard to computers" but that this tendency does not mean we use those terms literally. The common tendency to attribute human terms to simple objects, such as "the calculator understands how to add" and "the thermostat wants to raise the temperature" should not be construed as asserting that these objects possess understanding or intentionality.

Further, there is general agreement that there *is* something special to understanding and intentionality – not everything possesses it. What the something special is, of course, a matter of great disagreement, but many do agree that things we encounter cannot necessarily be said to understand even if they exhibit complex behavior. Behavior is not enough.

There is almost a universal agreement on the idea that a human being can in fact do what a computer program can do, though perhaps more slowly and with more tendency to err. This assumption will come up again later in the patient human test used in conceptualizing artificial intelligence applications. There is little or no objection[1] to Searle proposing that the person in the Chinese room can in fact carry out what a computer with a program would do to answer the questions in Chinese. Of course, there is disagreement on the reverse, and the question of whether a computer running a program can perform what a human does in answering Chinese questions is at the heart of the strong AI hypothesis that Searle is arguing against.

On the nature of programs, there is general agreement that computer programs, as we know them today, are procedural and computational in nature, and are either completely or nearly completely reducible to universal Turing machines. New types of programs, as long as they run on digital computers and including those that use parallel processing, deep neural networks, reinforcement learning, adaptive learning, and the like, do nothing to overcome this limitation.

It is also acknowledged by most of those engaging in argument over the CRA, including Searle, that it is possible that other kinds of machines could be constructed, ones that are not reducible to Turing machines, and that these could possibly achieve understanding and intentionality. Searle's repeatedly

---

1 Copeland (2002) has raised the issue of Turing's O-machines which by definition can do things that Turing machines cannot, and thus may be capable of understanding, even if universal Turing machines are not.

emphasizes that the CRA is aimed not at these speculative, future machines but only at machines that run programs which are essentially procedural and computational, which is arguably all the machines and programs being used in AI in 2021 and in the foreseeable future.

In a way it is a paradox that we seem to have found, so far, that AI is most successful when we don't try to make it operate like we imagine a human operates, but find other techniques, such as deep neural networks and reinforcement learning, that produce humanlike results, but operate in a completely different way. No one thinks the human brain is multiplying large matrices in order to choose which product to purchase on Amazon.

4.5.1 Analyzing the Systems Reply: Can the Room Understand when Searle Does Not?

Cole (2004) and Bishop and Preston (2002) sketch the path of the systems reply starting from Searle's original statement and response. The first major move in responding to Searle is to claim that Searle's analogy of the person as the computer is incorrect. The claim is that the person is only the central processing unit (CPU) of the computer. The systems reply notes that the entire computer system consists of much more than a CPU and includes temporary and permanent storage, input/output devices, and so forth. The essence of this position is that while the person (the CPU) does not by itself understand Chinese, the entire system, which clearly speaks Chinese, must.

The systems reply may be outlined as follows:

S1. A system may have an ability that none of its parts possess.

S2. The Chinese room system consists of more than just the person in the Chinese room.

S3. The person in the Chinese room does not understand Chinese.

S4. So the Chinese room system might understand Chinese even if the person in it does not.

S5. A system like the Chinese room system can be capable of understanding Chinese, even if the person in it does not.

S6. From its ability to communicate in Chinese, the Chinese room system must understand Chinese.

S7. Therefore, the CRA is wrong and the Chinese room system understands Chinese.

Ned Block (2002) calls the system reply one of the "best criticisms" of the CRA. He identifies two components of the reply. The negative aspect claims that an entire system may exhibit a behavior that none of the component parts exhibit that behavior (S1 above). None of the parts of my body individually can be said to be able to ride a bicycle, but my body as a whole certainly does. Searle's claim, with which Block agrees, that the person in the Chinese room does not understand Chinese, does not entail that there is no understanding of Chinese going on at all, or that the room as a whole does not understand.

Block (2002), Kurzweil and Gilder (2002), and Rey (1986), proponents of the system reply, emphasize S6 and start from the assumption that there must be Chinese understanding going on in the room, since there is Chinese language conversation. With these two points in place, the obvious conclusion is that the entire system must understand Chinese. The other possibility – that there is no understanding of Chinese going on at all – will be taken up later.

Starting from the tacit assumption, supported by an appeal to intuition, that there *must* be an understanding going on (S6), and admitting that the person in the Chinese room does not understand (S3), Block proceeds to explain how it can be that the system understands when the person does not. This is the positive aspect to the systems reply, that the entire system, consisting of the person, program, scratch paper, etc. does indeed understand Chinese.

Searle responds to the systems reply in the original article (1980) by arguing against S2, S4, and S5. His original response to the systems reply takes aim at S2 and states that the CRA works as well if all of these components were internalized by the person in the room, making the person the entirety of the system. In this case, Searle claims, there is still no understanding of Chinese going on. Block's (2002) reply is that the role of the person as merely the "CPU of the computer" does not change by adding additional responsibilities to the person, such as input/output, maintaining intermediate notes, and storing the set of symbol manipulation rules. These components still exist, he says – they are simply concealed inside the person's own head. Block concludes that this proves that the system understands Chinese and thus the CRA is wrong and a properly accessorized computer can indeed understand Chinese.

Copeland (2002) responds to Searle's "outdoor CRA" by considering the system inside Searle. He asserts that Searle has not proven that if a system does not have a certain ability, then no part of it has that ability. Searle, in his original response to the systems reply (1980), claims that if he were to memorize the instruction book and work in a field, then he would still not understand Chinese, and thus neither would the system. Copeland's claim is that it may indeed be the case that while Searle is not aware of understanding Chinese, the entire system inside Searle may indeed understand Chinese. He provides examples including a muscle that may compute tensor equations to enable a player to catch a ball, while the player himself is not able to solve such equations. Copeland seems to be imagining that what is happening in the outdoor version of the CRA is that Searle is aware only of following the instructions he has memorized for how to manipulate the symbols, directing him to write down (or speak) the new symbols. He is not aware of any understanding of Chinese. Simultaneously, somewhere else in Searle's body (since his body comprises the entire system) Chinese is being understood; therefore, the system understands Chinese.

As with the Block's general approach to Searle's response to the systems reply, there seems to be an assumption that Chinese understanding must be going on, and the need is only to explain how it could be possible, given that Searle in the room (or outdoors) is not aware of any understanding. Copeland's argument, however, does not *depend* on this assumption. His argument is more generally about systems and their components. In any complex system, modules may have specialized functions and abilities – abilities which are not abilities of other modules, or even of a supposed "master" component. In this way muscles are capable of doing things the brain itself cannot do. Copeland thus shows that it is possible that something in the Chinese room could understand Chinese, while Searle, that is, his conscious thoughts, do not.

Copeland does not claim, of course, that a module in a system can perform *any* function. Thus, he has not shown that the system in fact understands Chinese, only that it is possible that it does. Searle counters that, considered in these terms, it is *possible* that Searle's pancreas is understanding Chinese, but Searle is unaware of this. It would seem that more evidence is necessary that something in the system is indeed understanding Chinese to draw the conclusion that understanding is happening, and Copeland does not offer any such evidence, other than the assumption Block also makes – that Chinese language responses to Chinese language questions indicate that understanding must be going on. Copeland has only shown that understanding might be happening in the Chinese room.

In sum, it seems that Block and Copeland, as well as others, take arguments for the possibility of understanding to be support for the functionalist position that understanding is in fact happening. Support for this functionalist theory is limited to an appeal to intuition, as if to say, if something is able to converse in Chinese, it is much more likely than not (perhaps even certain), that it understands Chinese. This would be more plausible if there were no possible explanation for how something can converse in Chinese and yet not understand. It is as if I assume my car understands the chemical equations of combustion because it is able to burn fuel effectively.

The CRA, however, offers exactly such an example, as do several other systems that we will describe later as "simulators" including the uncontroversial Aunt Bubbles machine described by Block (1990, 1995, 1980) which all seem to agree that it is possible for a system to perform humanlike conversation without understanding. The argument regarding whether it is more intuitive to assume that a Chinese conversing system is understanding Chinese or whether it is not, is at the heart of the systems reply as argued by Block and Copeland. However, they offer no support for their position on this. They simply present how it could be possible that a system exhibits a quality that none of its parts possesses, a well-known phenomenon in systems known as emergent behavior and described in another chapter.

Haugeland (2002) raises further concerns about the whole–part relationships implied in Searle's response to the systems reply. His analysis of Searle's argument questions Searle's premise, which in Haugeland's words states that "if the system understands Chinese, then Searle himself does too." Searle's claim, in his original response to the systems reply, is that if he were to internalize all the components of the room, thus becoming the entire system, he would still not understand Chinese. Haugeland's claim is that in this case Searle is not becoming the whole system as Searle claims, rather he is implementing "another mind." Put this way, of course, we would not expect Searle to understand everything this other mind understands. The picture Haugeland proposes consists of a system (Searle) implementing a mind (the entity that answers Chinese questions and understands Chinese). That Searle doesn't understand Chinese does not entail that this implemented system also does not. For our purposes, here there are two important aspects to Haugeland's argument:

a. Is the assumption of understanding warranted with respect to the entire system (the Chinese room)?
b. Does the introduction of a logical decomposition within the system make it more likely that understanding is going on?

Like Block and Copeland, Haugeland proceeds from the assumption that a Chinese-understanding mind is present as part of the Chinese room system. In fact, his conception seems to be that the only mind present is the Chinese-understanding one, and Searle is acting only as the implementing hardware and software. For Haugeland, we know that Searle does not understand Chinese, and we would not expect him to, just as we would not expect the implementing hardware and software of any system to understand what the system is doing. Since Searle does not understand, and the entire system *must* understand (per the initial assumption), it *must* be the program, or set of instructions in the Chinese room, that understands. Later, I will challenge the premise that understanding *must* be going on in the room. Without this assumption, there is much less (perhaps no) reason to assume that understanding is indeed going on in the room.

For Haugeland, it is the "system implemented" that understands. As in the arguments of Block and Copeland, there is no proof that understanding is going on, only the proof that *if* it is going on, it must be going on in other than the CPU (Searle in the Chinese room). Since there is no other component in the Chinese room that can seem remotely capable of understanding (book of instructions, paper, pencils, and walls), it must be that somehow the entire system understands

Chinese. Since it is implausible that these inanimate objects (printed instructions, paper, pencils, walls, etc.) are doing any understanding, Haugeland takes what would seem to be the only avenue open to him, which would be to propose that there is a mind in the Chinese room other than Searle's.

Cole (1991) argues, however, that the fact that Searle has a mind is irrelevant to the CRA. In the CRA, Searle's mind is only in use to follow the printed instructions and read and write the meaningless (to him) symbols. He is doing only what the computer running its program would do; thus, the fact that Searle happens to have intentional states (for example, his feelings about being assigned this job of working in the Chinese room all day) is unnecessary and incidental to the point of the CRA. We can therefore simplify Haugeland's point about there being another mind in the Chinese room as being a quest to find *any* mind in the Chinese room.

To maintain the analogy of Searle to a computer running a program, we can't take advantage of any of Searle's capabilities outside what a computer would have. Haugeland characterizes Searle's thought experiment as Searle asking "himself what it would be like if he were part of a mind that worked according to the principles that strong AI says all minds work on…" This assumes that Searle is operating as a mind in the Chinese room, but as Cole emphasizes, he is not, so it is nonsensical to ask Searle what it is like to be working in the Chinese room. It is equally nonsensical to ask a digital computer what it is like to be running the program it is running. Each is only mindlessly following the program it has been given.

Searle's emphasis is to ask if understanding is going on, and he is using the case of himself acting as the computer in the Chinese room as a rhetorical device. If he had meant that the Chinese room required a mind to be present to follow the program (i.e. to *understand* the program), and then proposed that we examine the question of whether that mind had understanding, we would have quite a different case. In this case in fact, since there is an assumption that there is a mind and thus the capability of understanding in the Chinese room, plus the apparent ability to understand Chinese, we would be justifiable in concluding that the room does indeed understand Chinese, for three reasons:

- It is unlikely that a mind known to be capable of understanding and producing understanding behavior is not using its mentality to produce this understanding behavior, without clear evidence to the contrary.
- We would assume by analogy that an understanding system, such as a human being, speaking Chinese, is understanding Chinese, since when we ourselves speak, we do it while understanding the language we are speaking, and we are aware of our own understanding.
- While it is possible that an understanding system could be speaking Chinese by following a program rather than understanding, this seems unlikely. It would mean that an understanding system is ignoring its understanding capabilities

and resorting only to following a program. This is actually the case with Searle in the Chinese room – he is not using his potential ability to learn Chinese; he is simply following the instructions given to him and manipulating the meaningless symbols. A similar case would be a school child who has learned that when he or she sees "2 + 2 = ?" the correct response is "4" without any understanding of the actual arithmetic.

4.6 Implementation of the Chinese Room System

The question of whether the system in the Chinese room is capable of understanding is of central importance to the systems reply, and perhaps even tips the balance between the intuition that the Chinese room is indeed understanding or is simply behaving *as if* it is understanding, without possessing understanding. Opponents to the CRA who support the systems reply leave this question unexamined and proceed from the assumption that the room system can understand, even if Searle does not, without explaining clearly how this understanding can be going on. For there to be an understanding, there must be a mind, and if Searle's mind is the only one in the room, where is the other mind? Haugeland proposes to answer this by his notion of the "implemented mind" – that is, Searle's mind (or perhaps the entire Chinese room system) implementing another mind: the one that understands Chinese.

Haugeland elaborates how this understanding is accomplished by the implemented system. Haugeland casts Searle's CRA as a modus tollens:

a. The hypothetical system understands Chinese.
b. If the system understands Chinese, then Searle himself does too.
c. Therefore, Searle understands Chinese.

Since the conclusion is false, claims Haugeland, Searle tries to show that the first premise is false by assuming the second is true. Haugeland sets out to show that this second premise is false in two ways. First, he claims that Searle's original response to the systems reply is inadequate in that in it, Searle, even though he has memorized the rule book (analogous to the computer program), has not become the whole system, but has only become the "implementing hardware and software." Haugeland goes on to describe how the Chinese room should be considered to be a system, implemented in "hardware and software" with the "system" being something separate from these that does the work of understanding. The system, he claims, understands, while the implementing hardware and software do not. In systems terms, the claim is that Chinese understanding is an emergent behavior of the system consisting of the person and the other materials in the Chinese room. But this may be the misstep. Is it clear that Chinese understanding is going on, or only Chinese question-answering?

4.7   Is There a Chinese-Understanding Mind in the Room?

Haugeland continues to expand on this notion of the system being something distinct from the hardware and software of the system, and next makes an analogy to multiple personalities. Imagining Searle's own mind (which does not understand Chinese) to be one of the minds in the system, he imagines another mind which he calls *Hao,* who is also in the system and who understands Chinese. Thus, for Haugeland, Searle's reply to the systems response fails, because while the Searle-mind in the system does not understand Chinese, the Hao-mind can and does.

Haugeland backs up from this self-admittedly bizarre construction to focus on the central question, "why, exactly, should we conclude that Hao doesn't understand the Chinese that he appears to be reading and writing?" Searle might feel that the burden of proof is the other way round, and ask, "why should we conclude that Hao does understand Chinese?" and in fact, more will be said about this later. Haugeland proceeds from the assumption, common, as it has been shown, to the positions on Searle's reply to the systems response, that understanding must be present *somewhere* in the room because the room appears to be reading and writing Chinese. By eliminating Searle, the rule book, the paper, and pencils as possible "understanders," the only remaining possibility is that there is another emergent mind in the room, implemented by these other components. This other mind is for Haugeland, the *system* in the systems reply.

There are several problems with this conceptualization. One is the difficulty in establishing a clear existential nature for this Hao-mind – it must not be physical, since we have eliminated all of the physical components in the room. It is best thought of as emergent behavior, not as an independent entity. Haugeland argues that the system (the Hao-mind) can indeed have the property of understanding, while none of the implementing components may have this property. Searle has claimed that this cannot be, because there is nothing in the system that is not in him and since he does not understand, neither does the system. Haugeland argues here that the word "in" is used in two different senses in this argument. To say that a system has understanding "in it" is to say that the system has the property or behavior of understanding, while to say that the system is "in" Searle is to say that the system is implemented by Searle. For Haugeland, Searle's claim that since the system is in him and he doesn't understand thus the system does not understand, fails.

Haugeland has attempted to show that even though Searle doesn't understand the Chinese that the room is reading and writing, that there may indeed be understanding going on in the room. Though this claim will be disputed later, even if it is true, how does it affect the point of Searle's argument? Searle's point is that he is doing exactly what an analogous computer program would do to speak and write Chinese. His mind (not some Hao-mind) is implementing the program, following the instructions, reading and writing the symbols, and so forth, and this same mind does not understand Chinese. That there could be another mind in the room that happens to understand Chinese would be incidental and unnecessary to the room's ability to read and write Chinese. A native Chinese speaker could be sitting beside Searle, in the room, watching everything that is happening, and understanding all of the Chinese questions and answers, but the mind of this native speaker is not necessary for the operation of the Chinese room as Searle has described it, and in fact, Searle can gain no understanding of Chinese by its presence.

We might take Hoagland's two-minded Chinese room and ask if we can somehow subtract the Hao-mind while leaving the room with the ability to answer questions in Chinese. It would seem that we could. If the Hao-mind is being implemented by the Searle-mind, we could replace Searle with a machine, perhaps a very simple computer, whose capabilities are limited to matching symbols and writing new symbols via a series of simple "if-then" statements. Constructed this way it would seem that the assumption that there is somehow a mind in those if-then statements, is unwarranted, and this is precisely Searle's point in the CRA.

4.7.1 Searle and Block on Whether the Chinese Room Can Understand

Searle, his supporters, and his various opponents all initially appeal to intuition on the matter of whether understanding is going on in the Chinese room. Searle's original argument is based on the intuition that understanding is *not* going on in the room, because despite the room's ability to produce answers to Chinese questions, Searle himself does not understand Chinese. His opponents begin with the intuition that understanding *must* be going on in the room, since the room is able to answer questions in Chinese. They proceed to explain how it might be that the room understands, even though Searle does not. To break this down, we might imagine there are several classes of systems among those that exhibit as-if understanding behavior:

I. Systems that everyone agrees do in fact understand (that is, are capable of mental states, including understanding).
II. Systems that everyone agrees do not in fact understand, despite their behavior.
III. Systems about which there is disagreement about whether they in fact understand.

Class I systems include human beings and animals, and nothing else that we know of. Searle states that there is no logical impossibility of this class including systems other than humans and animals, and even nonbiological systems, but so far there is no other example that has gained general agreement.

Class II systems include things from the toaster that seems to understand how to stop when the bread is done, to the thermostat (mentioned by Searle), to Microsoft Word which seems to understand spelling and grammar, to water which "understands" how to flow downhill. Hauser (2002) agrees that this attribution of mentality and understanding across all such phenomena and devices is unwarranted and agrees that there are systems that despite their behavior should not be considered to understand or have mental states.

Class III systems are the difficulty – systems about which there is widespread agreement as to their "intelligent" or as-if understanding behavior, but about which there is disagreement about whether they actually understand. In this class, we find complex systems, mostly based on digital computers running software programs. In working through the arguments as to whether such systems are capable of understanding, there is a tendency toward a false dichotomy in which these systems must either be considered *understanding* in the same way humans and animals are or must be relegated to the class of toasters, which have no understanding at all.

What's missing in most treatments of the systems reply is a recognition that systems can be programmed to *simulate* understanding, and therefore they appear to understand, but at the same time lack actual understanding. They may even be capable of doing what an understanding system can do, and in some cases, doing it better or faster, yet still without understanding. A clear example of this is the often-mentioned Deep Blue chess-playing computer, who can play chess better than virtually all human players and yet, as evidenced by how it operates, does not understand chess in the way a human player does. How a nonunderstanding system can beat an understanding one "at its own game" will be clear in the later section on the mind of the programmer.

Block (2002) underscores the perplexing situation of these Class III systems, stating that Searle's argument is that the person in the Chinese room is the CPU (central processing unit) of a system that is a "Chinese simulator, not a real Chinese understander." He thus acknowledges in agreement with Searle, that it is possible for a system to pass the Turing Test and yet not understand Chinese. However, Block goes on to indicate that he believes understanding is going on in the Chinese room. His admission and assumption are critical to the success of the systems reply. Revising premise S6 in the systems reply as mapped out earlier, in order to make this explicit, we have

S1. A system may have an ability that none of its parts possess.

S2. The Chinese room system consists of more than just the person in the Chinese room.

S3. The person in the Chinese room does not understand Chinese.

S4. So the Chinese room system might understand Chinese even if the person in it does not.

S5. A system like the Chinese room system can be capable of understanding Chinese, even if the person in it does not.

S6a. Systems that are able to communicate in Chinese (thus passing the Turing Test) are either Chinese simulators or real Chinese understanders.

S6b. The Chinese room system is not a Chinese simulator.

S7. Therefore, the CRA is wrong, and the Chinese room system understands Chinese.

Since unlike some other supporters of the systems reply, Block holds S6a to be true, he must show that S6b is true or the systems reply fails. Since the Chinese room is specifically designed to simulate Chinese understanding, that is, to fool observers into believing it is a real Chinese understander, and is in fact built exactly the way a simulator would be built, the burden of proof for S6b likely falls to those who wish to show that it is not a simulator at all, but actually understands Chinese. It would be ironic if those who design and build understanding simulators were to be shown, to their surprise, that they are building actual understanders. It would be as if a sculptor's clay figure of a horse came to life and walked away.

4.8 Chinese Room: Simulator or an Artificial Mind?

For the systems reply to success, Block must provide support for S6b and show that the Chinese room is not a *simulation* of a Chinese understanding system but is actually a Chinese-understanding system. Though Block must believe this for his argument to succeed, he seems to be of a divided mind about it. On the one hand, he uses phrases including "you implement a Chinese understanding system," "you are also implementing a real intelligent Chinese speaking system," and "The Chinese system also thinks…" but also speaks of "the Chinese speaker being simulated" and "a real [Chinese] speaker." Despite this ambiguity, or perhaps because of it, Block offers no evidence to show why the Chinese room system is anything other than a simulation. In fact, he later indicates, in agreement with Searle, that he is doubtful that the Chinese room system is phenomenally conscious. If the Chinese room is not a mind, capable of consciousness, then its apparent understanding is likely as-if understanding and not indicative of the presence of mental states.

On balance it seems Block treats the Chinese room system as a simulation of an intelligent, understanding Chinese speaking system such as a human Chinese-speaker. His extensive explanations about how one system can implement another, and how systems may have properties that none of their components possess, thus miss the point, and add nothing to the support of the systems reply.

Searle (2002) in his more recent remarks on the systems reply suggests that these attempts are evasive and that the original CRA works in the case of the entire system as well. The person in the room does not understand because he does not know what any of the symbols mean, and neither does the entire system know what any of the symbols mean. Adding additional components (books of rules, scratch paper, pencils, etc.) to the CPU (the person in the room) does not give it any increased ability to attach meaning to the symbols it is processing. Attempts to locate the semantic content within the system by resorting to constructions such as "the system" or an "implemented second mind in the room" fail because the entire principle of AI is that the program is constitutive of a mind, regardless of the hardware on which it is implemented. For Searle, it is this that results in the systems inability to understand and in accordance with the proponents of strong AI, the implementing hardware does not matter. Searle's point is that the program, or the system, does not become capable of understanding by virtue of the program.

The CRA makes clear an additional distinction important to the source of the Chinese room's ability to speak and understand Chinese. If we compare a native Chinese speaker speaking Chinese with another person following a program of instructions (as in the Chinese room), we can see that while they may exhibit identical behavior, there are several key differences. The first is that while the program-follower is clearly following a program (we know this because someone wrote the program in a language that the person can follow and gave it to him or her to carry out), the native speaker has not been given a program to follow. As noted earlier, we could broaden the term "program" to include what the native speaker is doing, but only at the expense of describing all behavior as programming. This broader meaning is clearly not what proponents of strong AI mean when they talk about programs running on digital computers. On the agreed meaning of a program, the native speaker is not "running a program."

To further support the conclusion that the native speaker is not running a program, we might imagine what it would be like if he or she *were* running a program. The person would have a set of instructions, in a form that can be understood and followed, and the person would be following those instructions. Assuming the person is paying attention to what he or she is doing, there would be awareness that he or she is following a program. If a condition occurred that was not foreseen by the creators of the program, the person would either get stuck and call for help, or could employ his or her conscious mind to figure out a way to proceed.

The second difference is that while a native-speaker acting as a program-follower may be a conscious, intelligent person who understands many things, these qualities are completely unnecessary for him or her to function in the manner required to carry out the program. A program-follower simply lacks the equipment to understand what he or she is doing with respect to the performance of the program. The semantics aren't there. In addition, the program-follower is exhibiting no causal powers in his or her execution of the program. The program is in charge; the program follower is simply along for the ride. The question of who started the program is a relevant one and will be taken up as we discuss the role of the programmer in Section 4.9.

The native-speaker, on the other hand, *must* be a conscious, intelligent machine, capable of understanding Chinese. We cannot imagine a native-speaker to be unconscious, or incapable of understanding, and yet be a native Chinese speaker. We cannot imagine a native Chinese speaker, speaking Chinese and yet not understanding, of if we do, we are actually imagining a situation where a native-speaker is made to answer some Chinese questions by following a program, before returning to the original conscious native-speaker state. In this case, again, the program-followers causal powers and understanding are suspended while he or she runs the program given, and then return when the program ends.

4.8.1 Searle on Strong AI Motivations

Searle (1980, 2002) makes the point that he suspects arguments such as the systems reply are motivated largely by researchers whose grants depend on the strong AI thesis. Even if Searle is right, the creation and programming of machines that increasingly successfully simulate as-if understanding behavior is a noble and worthwhile goal and will continue to produce important and useful systems. Such systems can easily surpass the performance of human beings in specific task areas, like weather prediction, missile guidance, or playing chess. The term *artificial intelligence* is apt if construed to mean, not the creation of artificial machines that embody real intelligence, but instead the creation of an artificial kind of intelligence: the kind that looks like and performs like real intelligence but is not.

Differentiating artificial machines with real intelligence from artificial machines with artificial intelligence, opens up a dual path in AI and cognitive science research, with one branch focusing on the creation of artificially intelligent machines, intended to do useful work and provide value to the world, while the other branch seeks to understand and perhaps one day reproduce brains and thus one day create artificial machines that are intelligent in the way humans are. To Searle's point about careers and grants and such, it may actually hinder this second path to believe that the best or even the only path open to the creation of minds is the programming of digital computers. Perhaps we should leave the digital computers to become artificially intelligent systems, while pursuing other approaches to create artificial brains with real intelligence.

4.8.2   Understanding and Simulation

Functionalism, the view that something can be determined to be conscious or not by examining only its behavior, at least in its purest form, has been rejected by both Searle and those objecting to the CRA. Searle claims that the CRA clearly describes a system that exhibits behavior identical to a conscious, native speaker of Chinese, but is not itself conscious or capable of metal states, especially understanding.

The rejection of functionalism is easily argued by posing examples of systems that seem to exhibit understanding – that is they behave as if they had understanding – but clearly do not. Block (1990, 1995, 1980) has posited the "Aunt Bubbles" machine which can carry on a conversation as well as any conscious human. Aunt Bubbles works, however, by simply looking up each response in a database, based on the conversation's progress so far. Any programmer with a month of experience could write the program to perform this look-up.

Everyone seems to agree that Aunt Bubble's simple (though large) set of if-then statements are not sufficient to produce mental states. Searle (1980) further raises the examples of simple electronic devices such as thermostats which seem to make decisions about the temperature and then activate heating or cooling systems, but which may consists of no more than two pieces of dissimilar metals that bend this way or that depending on the temperature, and which again all would agree understand nothing.

To make sense of this, Searle introduces the notion of simulation, and of the possibility of a system simulating another. The simulation of a rainstorm, he says, will not get you wet, nor will the simulation of a digestive system digest pizza. The question here is, is this notion of a simulated system logically well-defined, that is, can we uniquely determine whether a system is a simulation of a phenomenon, or the phenomenon itself? If we can approach such a definition, we can attempt to assess whether the Chinese room is a simulation of a Chinese-speaker, or an implementation of a Chinese-understanding system, complete with understanding, mental states, and even consciousness.

To begin, we suggest that a simulation has several important characteristics that distinguish it from the actual phenomenon:

1. A simulator is constructed and designed intentionally – simulators do not occur naturally.
2. A simulator is designed to simulate a specific, observed phenomenon, usually, but not necessarily a naturally occurring phenomenon.
3. A simulator is constructed out of materials that differ from the materials comprising the phenomenon being simulated.

4. A simulator does not, in most cases, produce the actual result of the system being simulated.
5. Whether the simulator achieves its goal of simulating the phenomenon depends on the understanding of the phenomenon by the designer and his or her skill in constructing the simulator.

The above list is meant to capture the ordinary use of the term "simulator." Simply stating these characteristics, however, does not show that the Chinese room is or is not a simulator, or whether various kinds of AI systems are in fact simulators and thus cannot be said to be minds, or whether they can be considered to be minds. Wherever someone is on that question, it is likely we can begin by agreeing to the following regarding simulators:

1. Agree that the concept of a simulator is sensible and without logical impossibility,
2. Admit that simulators exist, and
3. Agree that simulators can indeed appear to have understanding.

With these agreements in place, the question becomes, is the Chinese room a simulator? We begin by describing the characteristics commonly attributed to simulators and differentiate the simulation of something from an artificial version of the same something.

First, simulators are purpose-built, that is, they are built to study a phenomenon, understand it better, and perhaps help people adapt to, or even alter the phenomenon. A computer program that simulates tectonic-plate movement in the earth, or a simpler physical model using rocks and glue, could both be used to study the phenomenon of tectonic activity and perhaps better understand earthquakes. In neither case would the simulator be confused with actual tectonic-plate movement. A simulator of a rainstorm, the example mentioned by Searle (1980), constructed say using a computer program and color graphical displays of clouds, accompanied by digital readouts of humidity levels, barometric pressures, and rainfall volumes, does not produce rain, but is clearly intended to study rain.

Simulators are constructed precisely because the phenomenon to be studied is not fully understood. More precisely, the effects of the phenomenon may be well understood (the destructiveness of an earthquake) but the cause and the functioning of the system that produces them may not be. Simulators are built based on the full or partial understanding of the causes or effects of the phenomenon. This understanding is possessed by the creators of the simulation; otherwise, they could not construct a simulator.

What if, however, a system were constructed that actually produced rain, say in a large, tall building, by creating the actual conditions under which clouds form and produce rain (such as happens in NASA's Vehicle Assembly Building in Florida)?

This should more properly be called an *artificial* rainstorm, not a simulated rainstorm. An artificial system is one built or constructed, instead of naturally occurring. In order to construct an artificial version of something, we must understand the something well enough to build one. We can't, for example, create an artificial raccoon since we don't have the knowledge of how the raccoon produces the behavior we see. We can on the other hand construct a simulated raccoon that exhibits some or even all of the observed characteristics of the raccoon.

An even clearer example is lightning. Lightning was observed long before anything was understood about its actual functioning and causation. Now that lightning is fully understood, we are able to create artificial lightning easily by building up a static electric charge and providing a path for it to discharge to ground. Such a system produces artificial lightning – lightning indistinguishable from the naturally occurring kind. The ability to do this depends on a complete understanding of how lightning works. Artificial lightning turns out to be a lot easier to create than artificial raccoons.

From observing the progress of science over time, one can observe that when phenomena are first discovered, they tend to be studied through observation and simulation, but once fully understood, can be created artificially. Creating a simulation supports the understanding of the phenomena, while creating the phenomena artificially serves a different purpose. We might simulate a flock of geese flying in a V-formation in order to see how they achieve more efficient flight, but this is quite a different thing from creating a flock of artificial geese.

Simulation involves creating a system that exhibits or mimics one or more externally observable aspects of the system. Simulations are often referred to as "black boxes" meaning we care about how the system interacts with the world outside itself, and but we don't care how it works inside. For a simulation, it's doesn't matter how it works inside – its purpose is to show external interactions. Relevant to our discussion here, so little is currently known about how the brain produces mental states such as understanding, it is quite unlikely that we are accidentally producing artificial understanders.

If it can be shown that the clear differentiation of a simulated system and an artificial system exists in the arguments of those who suggest the Chinese room is capable of understanding, then the question becomes simply, is the Chinese room more likely to be a simulator of understanding or an artificial understander? In what follows, we try to show that it is the latter – that it simulates understanding without actually understanding.

Consider the first two systems: an AI program run on a digital computer, and a molecule-for-molecule artificial brain. We argue here that the AI program is a simulator of understanding, while the artificial brain can exhibit actual understanding. Mooney (1997) takes this up briefly questioning whether a computer simulation of the brain at the neuronal level could produce intentionality, that is, can a clear sense be given to the question of whether something is a simulation of a mind or an actual mind? With Dyer (1994), Mooney emphasizes that for a simulation to "be real" would require our scientific knowledge to be complete enough for us to understand the brain's interactions at a molecular level, and then to create a system with that has those interactions. Whether these interactions could be created in a computer program is hard or impossible to determine without knowing what they are. It is at least possible, and Searle would say likely, that when we understand such interactions sufficiently, the machine necessary to produce them artificially will need to be something other than a digital computer running a program.

The distinction between a simulated system and an artificial one is similar to Harnad (1989)'s distinction of a *c-implementation* created from a computer running software, and a *p-implementation*, which is an artificially constructed device based on knowledge of the phenomenon. Wakefield (2003) underscores this same distinction:

> For example, artificial flowers, artificial limbs, and the artificial Moon environments used by NASA for training astronauts are *not* genuine flowers, limbs, or Moon environments, respectively, but rather simulations of certain features of the genuine instances, whereas artificial diamonds, artificial light, and artificial sweeteners are *genuine* diamonds, light, and sweeteners, respectively, which are created as artifacts in contrast to the naturally occurring instances. *Artificial intelligence* thus potentially refers both to simulations of crucial features of intelligence and to creation of genuine instances of intelligence through artifactual [sic]means.

Copeland (1993) also uses this distinction, but conflates the two meanings into the single-term simulation, stating that some simulations are duplications of the real phenomena (what we call here artificial) and some are not (what we call simulations). Damper (2004), citing Copeland, proposes that a computerized automobile engine management system, which of course, actually controls the engine, would not be such a duplicative simulation. Not all systems are simulations, however. It is hard to see why this automotive engine management system would be called a simulation at all, except to bolster the argument that simulations might duplicate another system.

A clearer example might be to imagine a simulated driver vs. an artificial driver of the automobile. A simulated driver might consist of a computer program, or even a mannequin sitting in the driver seat, and it might be used to study the positioning of controls, the seat geometry, crash safety, etc. What it would not do is drive the car. An artificial driver would not be a simulation of anything, but would in fact drive the car. To create the simulated driver, we need to only know what the driver does, that is, move his or her arms and legs about the controls. To create the artificial driver, we must understand how to drive a car, how the actual controls work, and in fact, something about how the car itself works.

Searle (1983) approaches this from a biological perspective and draws the analogy with other biological processes such as digestion. "Mental states are as real as any other biological phenomena," he explains and are caused by biological processes. The reason that digestion poses no philosophical problem while mentality does, is that digestion is well understood as to its causes, effects, and processes. There is no seeming magic to it. Of course, there was a day when processes like digestion, circulation, respiration, and the like were not understood, and this led people of various eras to invent magical explanations for these phenomena. Respiration was thought to have to do with a person's spirit (as seen in the etymology) so when the breath stopped, the spirit was lost. Now that we understand respiration, we are able to build respiration machines that perform the respiratory function without requiring the involvement of spirit. One might observe that the understanding is the hard part – the engineering to implement the system is straightforward, once the understanding is there.

On Searle's account then, mental processes are a function of brains, much as respiratory processes are a function of lungs. Artificially created lungs, that is, systems that perform actual, artificial respiration are possible and theoretically, so are artificial brains – systems that perform actual mental processes. Naturally, such systems must be able to do what brains do, though there is no logical requirement that they must be constructed of the same kinds of materials. The difficulty is that we simply don't know very much about how brains do what they do, making it much more difficult or even impossible at present to produce an artificial brain, or thinking machine. It is certainly not enough for the machine to reproduce the apparent effects of the brain without reproducing their actual function. Simulated lungs consisting of expanding and contracting bags of air would not be doing respiration unless they were able to re-oxygenate the blood. A machine that can mimic the input and output of a brain, say by answering questions put to it in Chinese, cannot by virtue of this be said to be producing mental processes.

The clear distinction between a simulated system and an artificial one allows us to make several important observations about the Chinese room, and especially about the systems reply:

1. A simulated system attempts to mimic or demonstrate certain behavioral functions of the system, such as holding a conversation, so it should be no surprise that such a system can pass a Turing test, and Searle affirms this.
2. The actual functioning of the phenomena, how the brain understands, does not need to be known to simulate the *effects* of such understanding. All that is needed is to understand the effects. A simulator of planetary motion can be constructed without understanding what it is that moves the planets about.

3. It may be difficult or even impossible to determine whether a system is a simulation or an artificial instance of the phenomenon simply by observing its behavior. We must know something about how it was created. In the case of the Chinese room, this will lead us to the "mind of the programmer" to be taken up in the Section 4.9.

Given (3) above, a question as to burden of proof arises. If one observes something that looks like a smartphone on a park bench and out of its speaker is coming spoken poetry, should the observer more sensibly assume that the smartphone has become conscious and is composing and reciting poetry or that it is simply playing back a recorded voice, simulating human poetic recitation? The answer does not depend on the quality of the poetry, nor on the volume or tone of the voice. A better simulation does not move the system closer to being an artificial poet.

One might ask if a Turing machine can simulate a native Chinese speaker. In a sense the question is ambiguous. If we truly mean simulate, as characterized here, then, yes, since a native Chinese speaker takes in language and puts language back out, and Turing machines are certainly capable of language input and output. This is a hollow and trivial answer, since the same could be said of voice recorders and voicemail menus. If we mean, can a Turing machine be an artificial Chinese speaker, then the answer, for Searle, is no. Opponents to Searle on this point are unable to explain how it is that the Chinese room, a Turing machine, has understanding and is in fact a simulated Chinese speaker, when it is far simpler and more intuitive to characterize it as a simulation.

In the case of the Chinese room, we have a system that is created as a simulation, created to mimic the behavior and function of a native Chinese speaker. Like "blockhead" and other such systems, it was clear to its creators that it is a simulation of intelligence, not an intelligent system itself. It is curious why intelligence, understanding, or mental states would being claimed for such a system. The systems reply appeals to intuition based on behavior but forgets that there is another, and more likely explanation for as-if understanding, and that is simulation. Occam's razor comes to mind here, suggesting that we look for the simplest explanation for the phenomena we observe. Is it simpler to believe that a computer (or a person) following a set of rules for manipulating symbols of which they have no understanding, is somehow actually understanding anyway, based only on the observation that it behaves similarly to other systems (human) who do understand the symbols? Simulated understanding would seem to be the most likely explanation, rather than actual artificial understanding.

Even with all that said, however, it would seem that for any system to be able to answer questions in Chinese, there must be an understanding of Chinese going on *somewhere*. If the understanding is not in the simulation, then where is it?

4.9   The Mind of the Programmer

Only a computer programmer knows how dumb computers really are. Modern digital computers, running computer programs boil down, not only figuratively, but quite literally into machines that perform nothing more than the original operations proposed by Turing. Turing described a machine that can access a memory store (a tape in his early descriptions) and can change the value in a binary storage location from a "1" to a "0" or vice versa. Digital computers, as their name implies, do no more than this. The relevant question for purposes of the Chinese room is whether any program can be a mind and can have mental states such as understanding. Here we will explain that while computers running programs have become better and better at simulating understanding and may yet achieve persuasive as-if understanding and pass the Turing test, they are limited by the nature of computer programming to being simulators and not minds.

Since 1980 when Searle proposed the Chinese room argument, computers and computer programming have progressed in several ways. They have overcome some of the limitations in processing speed and storage capacity thought to be responsible for their lack of intelligence. Moravec (1997) maps out the progress of the computer, both in processing speed and memory, concluding that as of that writing, computers had reached the power level of a monkey's brain. What's notable, however, is that they were not able to exhibit the mental states, intentionality, or intelligence of a monkey. Of course, a computer (even well before 1997) can be programmed to simulate the behavior of a monkey and to, say, converse in sign language as well as a monkey can, but this is simulation, as differentiated earlier. In 1997, according to strong AI, it should have been possible to create a computer, running a computer program that operated as an artificial monkey brain. Then, a decade later, perhaps we might have been at the level of a toddler, owing to the increases in computer power. Moravec suggests why this did not happen. One possibility is that computer programming is just hard work and that we just haven't figured out how to write the right kind of computer program to implement a monkey-mind, but another more serious problem may be that computer programs simply do not have the ability to function as artificial (vs. simulated) minds.

Searle takes aim specifically and exclusively at digital computers running programs as the object of the CRA. Many of the objections to the CRA involve an appeal to intuition that such computers must surely be able to overcome their current limitations and one day think, understand, have at least rudimentary intelligence – perhaps even actual artificial consciousness. We have already seen one reason for this intuition – that computers have over the years been able to deliver more and more intelligent behavior, in many cases superior to what people can do. IBM Deep Blue beat the greatest human chess player at the time, and since then many computer programs have been developed that play chess better than any human, and they can run on a small computer. Any desktop computer can perform complex calculations and search information much faster and more accurately than any human. We have made the point that such behaviors do not necessarily indicate understanding of the subject matter involved, but are simulations of understanders. Deep Blue does not really understand chess; it is a clever simulation of a human chess player's understanding and playing of chess.

Admittedly, the possibility of simulation, however, is no evidence that the system in a particular case is a simulation. As we have shown, we cannot by behavior alone determine whether a system is a simulation or an actual understanding system (whether biological or artificial). Searle (1980) suggests that it is important *how* the system produces the behavior prompting a further investigation of how computer programs work. As Searle admits, there could one day be a new kind of computer, one that runs a new kind of "program" (if it could be called that) that might not have the limitations he claims for current programs. While "anything is possible," I will show here that with a more complete understanding of how computers and programs function, the limitations claimed by Searle are inherent in all current methods of programming running on digital computers, including all known artificial intelligence programming techniques. Even if this were not so, strong AI proponents claim that programs may be represented by a Turing machine, and these limitations certainly apply to any such machine.

Here, we'll consider the following relevant questions:

- Do computers operate by understanding their programs?
- Is following a program something different from understanding?
- Do human beings operate by following programs?

To address these questions, we must first examine what a computer program is, and how it works. To watch 1960s-era early science fiction TV and movies, one would conclude that computers operate much the way people do. A typical scenario is someone feeding loads of collected data into a computer and then asking the computer for conclusions drawn from the data. After some suitable whirring and clicking, the computer prints the answer on a small strip of paper. Another common mental image in the public consciousness is that computers are programmed by creating large collections of principles and rules, much the way a human teacher would teach a student. The computer is thought to be able to understand these rules and apply them to new situations. As the set of rules grows, the computer's knowledge and understanding grows as well, and eventually the computer is an expert. There is a class of AI systems programmed this way known as *expert systems*, but their limited success has caused them to be less popular in recent decades.

In addition to these popular perceptions of how computers work, we should note that as computers become more and more advanced, and able to be programmed in languages that increasingly resemble natural languages it becomes increasingly tempting to assume that the computers are understanding that nearly natural language.

None of this is how computers actually deal with programs. Digital computers were first developed using mechanical relays – which are simply electrical switches that can be activated by electrical currents sent to them by other relays. Early computers contained hundreds or thousands of such relays and were quite noisy as they clicked their way through calculations. These computers were programmed by connecting various relays together, using jumper wires. Later the relays were replaced by vacuum tubes and even later transistors and integrated circuits, but they are still enormous collections of switches. In modern computers, the switch is actually a small bit of silicon that exhibits an electrical charge, or not, representing a single binary digit (abbreviated as "bit"). Why is this important for our purposes with the CRA?

While it may appear that a computer can be programmed by giving it statements in nearly natural languages, such as "Add incoming.shipment to current.inventory" and "Store new.inventory in permanent.database" such statements are simply strings of symbols to the computer. In fact, these symbols are interpreted by another computer program (called a compiler or an interpreter) which, following its own set of instructions, translates them into more cryptic symbols such as "LOAD R1," "ADD R3," and "STORE R4." These symbols are not understood by the computer either – they form the input to yet another computer program called an assembler, which translates these symbols into machine language, which looks something like "0100100010010110." Does the computer understand these binary symbols? No. These are translated into those little electrical charges in the bits of silicon. To take it to the absurd, can we say that the computer understands these electrical charges? No – to affirm this we would be left attributing understanding to the thermostat and toaster also.

All this is to say that while it may appear that computers traffic in humanlike languages, this is an illusion – such programs are in effect, meaningless symbols, which are used to produce other sets of meaningless symbols. The CRA could thus be applied to each stage illustrated above – the modern computer is a sort of series of Chinese rooms, each feeding sets of symbols to the next, with no understanding at any stage.

Computers do not *understand* their programs – they follow them. A computer following a program is much like the child who writes "4" when he or she sees "2 + 2 = ?" but who does not yet understand arithmetic. The CRA suggests this by placing Searle in the position of simulating a computer, by dealing with symbols he does not understand and using only instructions he can follow. If the Chinese equivalent of "2 + 2 =?" is sent to the room, Searle finds the statement in the instruction book, that says, in effect, when you see "2 + 2 = ?" write "4." He does all of this without understanding what "2," "+," and "4" mean. If we change the Chinese room scenario to an English room and write all of the input questions and output answers in English, then of course Searle in the room would understand them all, but what's important is that he would not *need* to understand them all to be able to carry out the instructions. His understanding is incidental and unnecessary to his function in the Chinese room. More importantly, his understanding is not coming from the program he is following at all. Nothing in the program of instructions helps him understand. His understanding in the "English room" comes as a result of his abilities as a thinking, intentional human being, not as a result of his following of a program. Of course, this is precisely the situation with a digital computer, except that it has only the ability to follow the program, which is insufficient for understanding.

So, to the second question above, we must answer, yes, following a program is something quite different from understanding. To the third, we should answer yes and no. Human being can indeed operate by following programs or instructions, like the child answering "2 + 2 = ?" Following programs, however, does not encompass all of human behavior and ability. What remains? Intentionality, semantics, understanding, and other mental states are clearly a function of human beings, but not a function of the following of programs. Human beings can operate either by following programs or by using their intentional mental states. Computers operate solely by following their programs, without understanding.

In Section 4.4, we find that Searle in the Chinese room does not understand because he is merely following a program as a computer would. If we ask now what it would take to make the Chinese room understand, that is, to give it not only the ability to follow a program and produce Chinese answers, but also the ability to understand Chinese as it does so, how could this be done? Some of the critics of the CRA, especially those who support the systems reply, ask how it can be that the Chinese room produces Chinese answers to Chinese questions, thus exhibiting understanding, when Searle himself in the room clearly does not understand Chinese. We have answered that it is by following a program as a computer would, simulating a Chinese understander, and exhibiting as-if, simulated understanding, not actual understanding. If we dispense with the assumption made by these critics that understanding must be going on in the room, this suggests a more important question: How can questions be answered in Chinese with no understanding of Chinese anywhere? They answer is, of course, they can't.

The question becomes, *where* is the Chinese understanding that enables the Chinese room to answer the questions? If we look for it in the program, we do not find it, but examining the program leads us to the answer: the understanding of

Chinese is in the person who created the program. The person or team of people who created the instructions used by Searle in the Chinese room must of course understand Chinese in order to create those instructions, or a program. They are like the computer programmers who write programs in a language the computer can follow and cause the computer to exhibit as-if understanding.

Deep Blue, the IBM chess-playing computer, is an instructive example. Deep Blue was able to play good chess because a number of chess experts and computer experts combined their knowledge and understanding of chess and reduced this knowledge to a set of instructions that Deep Blue can follow. Deep Blue understands no chess. Deep Blue is, at its heart, a large set of alternating electrical charges in little silicon chips, the voltages following exactly its instructions. Through its instructions, Deep Blue "knows" what to do in any given chess situation. For example, Weber (1997) reports that it is programmed with winning combinations for all possible endgames where there are six or fewer pieces left on the board. By examining the data, Deep Blue can make the best possible move in any such situation, or at least the best possible move according to the experts that set up this information. Imagine if Deep Blue lost a chess game, which has certainly happened. Where is the shortcoming? It is not in Deep Blue, the computer, or in its ability to follow its program. It is not even in the program – the program operates perfectly the way it is written. We cannot say that the program made a mistake. Of course, the shortcoming is in the minds and actions of the programmers.

In the original Chinese room situation, let's imagine that a question is answered incorrectly. Who's to blame? Searle? No – the answers do not depend on his abilities, other than his ability to follow the instructions. Assuming he followed the instructions without making a clerical mistake, the wrong answer can be blamed only on the person or people who wrote the instructions. It is the programmer's understanding of Chinese that enables the Chinese room to function. If the room was asked, "How are you today?" it could only answer based on how the instructions enabled it to answer. The instructions might simply specify to answer in the Chinese equivalent of "fine" or they might specify that Searle should take his temperature and pulse, and then use these data to look up some further instructions and write some Chinese symbols accordingly. In any case, the way the question is answered depends on the programmer. Imagine if the Chinese room had a nasty tendency to be sarcastic. This could only be due to the programmer implementing sarcasm as part of the program.

There is therefore no mystery about how the Chinese room is able to exhibit as-if understanding, without itself understanding. The understanding is a function of the mind of the programmer. One might view the Chinese room as an extension of the programmer's mind, much as the hand or arm is an extension of a mind. The hand is able to carry out the instructions from the mind, and we are not confused about whether the hand understands how to play the guitar, or whether it is the mind. Transplant the hand of a guitar player to another person, and no guitar playing ability will be transferred.

To return to our thought experiment, let's again assume we wanted to make the Chinese room an understanding system, an artificial mind. How could this be done? We allow the programmer to enter the room with Searle. The programmer, watching Searle carry out the instructions, would of course understand what all the symbols mean and we could imagine him or her making notes about how to later improve the instructions to produce better answers. Weber (1997) described exactly this scenario with Deep Blue's programmers huddled around watching it play against Garry Kasparov. The Chinese room instruction-writer, or programmer, in the room with Searle, could also engage Searle in conversation, in English, about what is going on, pointing out for example that he is now working out the response to the question, "When did human beings land on the moon?" In this way Searle could come to understand Chinese, little by little as he begins to understand what the symbols mean, tutored by the programmer.

The important point here though is that this kind of learning requires capabilities beyond and separate from the program Searle is following. Over a long time of Searle and the programmer working together in the room, Searle may come to speak Chinese fluently. He could thus begin to answer questions from his own understanding, without needing to follow the programmed instructions. Of course, this is no longer the Chinese room scenario – the program is no longer being followed. We also note that nowhere in this process of Searle learning Chinese, does the program gain any understanding. If at some point we replaced the program-following Searle with another person, any understanding of Chinese Searle has gained by working alongside the programmer, goes with him. The behavior of the room is unchanged by Searle's understanding of Chinese, or its loss.

AI proponents might object at this point and describe how programs have been developed that can indeed "learn." The program in the Chinese room could learn also, if it were programmed to do so, but again, it would have no ability to understand what it has "learned." For example, it might be programmed to ask how the weather is outside, and being told it is raining, remember this answer in case it is asked for a weather report later. In doing this, it has not come to understand what "weather" or "rain" means. Computer programs, including deep neural networks, learn by example, in ways described in Chapter 3, and do not learn in the way humans learn. Our knowledge of how the brain learns is still very limited, so it is certain we are not able to program computers to do what humans do when we learn. Even in the cases of systems that learn from data, such a deep neural networks, or systems that learn by repeated attempt based on a reward function, such as reinforcement learning systems, the point holds. These systems learn to perform a function based only on the data, goals, and constraints provided by the programmers.

4.10   Conclusion

The point of this chapter has been to show that the systems reply to the Chinese room argument fails in four major ways:

a. It assumes that understanding must be going on in the Chinese room, when in fact, there is no basis for such an assumption.
b. It fails to acknowledge that as-if understanding behavior can be the result of a simulated system vs. actual artificial understanding.
c. It fails to acknowledge the limitations of programs, whether followed by humans or machines to have or gain understanding.
d. It fails to acknowledge that the source of the Chinese room's ability comes from the understanding abilities of the programmer, not anything in the Chinese room.

This chapter has reviewed some of the important ideas in the philosophy of mind, related to the idea of machine intelligence and computer intelligence, which as we've shown are two different things. If we imagine the likely future of intelligent machines, we can envision two parallel paths continuing for some time into the future. One is the path of practical simulated intelligence. In this path, we have technologies that include deep neural networks, natural language processing, reinforcement learning, and all other AI technologies in use today. These technologies will produce useful and valuable applications that do apparently intelligent work for us, such as summarize texts, answer questions, write reports, classify images, play games with us, drive our cars, and control complex systems like aircraft, satellites, ships, and military systems. They will also become useful to us as assistants, giving us the weather or turning on our lights with a voice request. Products like Amazon Echo and Google Home will continue to develop and evolve to be more powerful and more useful. On this path, there is no chance that any of these systems will become intelligent as we think of intelligence in human beings. They are simulators. They will beat us at chess or go, without knowing what they are doing, or understanding anything about the game. They will fly our planes without even knowing they are in the air. They will drive our cars for us, with no awareness of the lives they are endangering, or protecting.

On a parallel path is the quest to develop intelligent machines – machines that are intelligent in the way we are. They will have mental states, emotions, intention, and eventually consciousness. There is nothing impossible about this. One way it could be done is to take a living brain, say one from a small animal to start, and figure out how to connect it to input and output devices. It would be a "brain computer" and would not follow what we know as computer software programs. It might appear to be an electronic system, with keyboards, screens, voice input and audio output, but inside it would be powered by a brain. For a dramatic version of this story, see *Spock's Brain,* the premier episode of the third season of the original Star Trek series, where Mr. Spock's brain is stolen and used to control complex machinery on a foreign planet. In philosophy, the term *brain in a vat* is a thought experiment used to reason about human mental states and the environment uses exactly this configuration – a disembodied brain, connected to other devices and sensors.

Depending on the capabilities of the brain used, the resulting machine could have the intelligence of a mouse, a cat, a monkey, or even a human. If using a brain from an animal or a recently deceased human makes us uncomfortable ethically, perhaps we can learn to grow a brain in the lab through a cloning process. If we grew a human brain in the lab and connected it to electronic inputs and outputs, there is no reason to think that it would not behave in a human-like way and be able to learn the way humans learn. It might start out with the mentality of a newborn, but could be able to grow, learn, and develop. What its legal and societal status would be is an interesting question. Would it be considered alive? Predictably, it would have the status of a pet, a living thing that we can buy and sell, but which we are obligated to care for in a reasonable and compassionate way.

Another way that an intelligent machine could be developed is to build a synthetic brain, a device that does what a brain does. It does not *simulate* the behavior of a brain the way a computer chess program does – it actually carries out the same processes that occur inside a brain. As we've said, developing a synthetic brain is difficult, mainly because we do not know how a brain works. Though we may figure this out someday, it would seem that the shorter path is to simply build on what nature and evolution have provided and use actual or cloned brains. While this may sound creepy, humans have always used naturally occurring components for our own purposes, even ones from living things. We make knives of bone and tooth, we wear leather, we harvest wood from living trees to build our homes, and we raise animals for food. So why not raise brains to power our new kind of intelligent machine? But beyond the pure scientific fascination of seeing if we can do it, there is the question of whether this line of research would even be useful. If we today have computers that can handily beat us at chess, of what use is a brain-based machine that though conscious, must learn chess the way humans do, and would, by definition, be limited to the playing abilities of a human brain? At least, it would give better interviews before and after the match.

From what we've said in this chapter, admittedly at great length, the one thing it seems we do *not* need to worry about is digital computers, running software programs, however speedy or clever, becoming intelligent in the human sense, developing intentions, setting their own goals, planning for our demise or even becoming conscious or self-aware. They just can't.

References

Bishop, M. and Preston, J. (2002). *Views Into the Chinese Room: New Essays on Searle and Artificial Intelligence*. United Kingdom: Clarendon Press.

Block, N. (1980). What intuitions about homunculi don't show. *Behavioral and Brain Sciences* 3 (3): 425–426.

Block, N. (1990). Consciousness and accessibility. *Behavioral and Brain Sciences* 13 (4): 596–598.

Block, N. (1995). The Mind is the Software of the Brain. In: *An Invitation to Cognitive Science*, p. 377. United Kingdom: MIT Press.

Block, N. (2002). Searle's argument vs. cognitive science. In: (ed. Preston).

Cole, D. (1991). Artificial minds: cam on Searle. *Australasian Journal of Philosophy* 69 (3): 329–333.

Cole, D. (2004). The chinese room argument. In: *The Stanford Encyclopedia of Philosophy (Winter 2020 Edition)* (ed. E.N. Zalta). https://plato.stanford.edu/archives/win2020/entries/chinese-room/ Retrieved 14 May 2022.

Copeland, B.J. (1993). *Artificial Intelligence: A Philosophical Introduction*. Oxford: Blackwell.

Copeland, B.J. (2002). The Chinese room from a logical point of view. In: (ed. Preston).

Damper, R. (2004). The Chinese room argument – dead but not yet buried. *Journal of Consciousness Studies* 11 (5–6): 159–169.

Dyer, M. (1994). Intentionality and computationalism: minds, machines, Searle, and Harnad. In: (ed. Dietrich).

Harnad, S. (1989). Minds, machines and Searle. *Journal of Experimental and Theoretical Artificial Intelligence* 1 (1): 5–25.

Haugeland, J. (2002). Syntax, semantics, physics. In: (ed. Preston).

Hauser, L. (2002). Nixin' goes to China. In: (ed. Preston).

Kurzweil, R. and Gilder, G.F. (2002). *Are We Spiritual Machines?: Ray Kurzweil vs. the Critics of Strong AI*. Discovery Inst.

Mooney III, V. J. (1997). Searle's Chinese Room and its Aftermath. *Center for the Study of Language and Information Report No. CSLI*, 97, 202.

Moravec, H. (1997). *When Will Computer Hardware Match the Human Brain?* Carnegie Mellon Robotics Institute.

Rey, G. (1986). What's really going on in Searle's "Chinese Room". *Philosophical Studies: An International Journal for Philosophy in the Analytic Tradition* 50 (2): 169–185.

Searle, J.R. (1980). Minds brains and programs. *Behavioral and Brain Sciences* 3 (3): 417–457.

Searle, J.R. (1983). *Intentionality*. Cambridge University Press.

Searle, J.R. (1998). *Mind, Language and Society*. New York: Basic Books.

Searle, J.R. (2002). Twenty-one years in the Chinese room. In: , vol. 2002 (ed. Preston).

Wakefield, J.C. (2003). The Chinese room argument reconsidered: essentialism, indeterminacy, and strong AI. *Minds and Machines* 13: 285–319.

Weber, B. (1997). The contest is toe-to-toe and pawn-to-pawn. *New York Times* (May 10).

Part II

Systems Engineering for Intelligent Systems

5

Designing Systems by Drawing Pictures and Telling Stories

For in Calormen, storytelling (whether the stories are true or made up) is a thing you're taught, just as English boys and girls are taught essay-writing. The difference is that people want to hear the stories, whereas I never heard of anyone who wanted to read the essays.

(C.S. Lewis, *The Chronicles of Narnia*)

One of those most fundamental and essential aspects of engineering and building systems is achieving full agreement with all the stakeholders on *what* is to be built and *how* it is going to work for its users. Stakeholders include not only everyone who has a vested interest in the system being developed including the intended users but also those who will fund, install, maintain, manage, and rely on the system. Gaining stakeholder agreement sounds simple enough. Can't the system architects, designers, and engineers just talk with all the stakeholders and potential users for a new system, ask them about their needs, wants, and expectations regarding the system, and merge all this information together, resulting in a clear statement from which they can work? It turns out there is much more to it.

5.1 Requirements and Stories

In the 1960s, the practice of writing lists of specific requirements was developed to capture user needs and describe system functionality. When the development of a system is be contracted by an acquiring company or organization to another, the list of requirements is often the most important part of the communication between the acquiring and providing organization. It's not uncommon for a military or defense organization to provide thousands of requirements to the company that will build the new system. Requirements are often called "shall statements" because the traditional beginning of each requirements statement is "the system shall…"

Engineering Intelligent Systems: Systems Engineering and Design with Artificial Intelligence, Visual Modeling, and Systems Thinking, First Edition. Barclay R. Brown.
© 2023 John Wiley & Sons, Inc. Published 2023 by John Wiley & Sons, Inc.

Lists of thousands of requirements though can fail to produce a complete picture of the system that is understandable to everyone, leading to unnecessary and resource-wasting change and rework throughout the systems development lifecycle. There are a variety of reasons that this simple ask-users-and-write-down-requirements process yields less than optimal results and these reasons fall into several categories:

- Unclear input
- Ambiguous input
- Contradictory input (both actual and unintentional)
- Overwhelming amount of input
- Overly specific "requirements" may be found to be unfeasible or undesirable

Requirements can simply be unclear as stated. English is notoriously imprecise, and what the writer of a requirements statement had in mind might not come across in the terse format of a shall statement, without context. Requirements can be ambiguous in that more than one meaning is possible, so without additional information or understanding, assumptions are required to interpret the meaning. Large sets of requirements can include contradictions, either because the various groups of stakeholders who contributed to them disagree, or due to oversight and error. Large sets of requirements are overwhelming when considered as a list of independent items. It would seem humanly impossible to read a list of several thousand statements and come away with a cohesive picture of the system, its intended purpose, and desired operation. Requirements may stray into implementation details. When a requirement from the customer states that the system should be implemented in a certain way, the supplier must either follow the requirement even if the implementation method is not the best choice or challenge the requirement through formal processes. Most guidance on the best ways to write requirements suggest that requirements should never include implementation details, but sometimes the line between requirement and implementation is a fine one, and implementation details do creep into requirements sets.

The strength of the practice of writing requirements in shall statements is also its weakness. There is no widely agreed way of writing shall-statement requirements – no formula, no syntax, no prescribed language. Some have been proposed, but forcing requirements to be written in a specific syntax, or even turning them into something resembling high-level software source code, reduces their ability to be written and reviewed by all stakeholders. The likely result of making requirements syntax more structured and specific is that stakeholders would turn back to writing narrative descriptions and leave the translation of their work into requirements to a specialist, introducing the potential for error in the translation.

5.2   Stories and Pictures: A Better Way

In this chapter, we will focus on techniques useful in the early stages of any systems development project, whether it be designing a new home or building a spacecraft. The techniques are most valuable when the system has not yet been designed – when it's just an idea, or a need. Looking to stories, movies, and visual art for inspiration, we'll explain several techniques for understanding, modeling, and conveying the needs, functionality, and performance of a new system.

Our approach will acknowledge and include the familiar technique of developing use cases in the MBSE process, but we will show how the idea of use cases can be expanded to larger-scale systems and systems of systems. Expanding from the idea of individual use cases, we will introduce a way of combining more general business processes with use cases to produce a more flexible way of describing a system's use in an operational context. To that we will add the concept of timeboxes, which allow for the flexible modeling of loosely connected processes and systems.

As we proceed, we'll explore the techniques professional storytellers and filmmakers use to create compelling stories, images, and scenarios in very little time and with few words, including storyboards, movie trailers, and recaps. Some insights from the science of the relative passage of time and even from the fantasy of time travel will be applied to help create compelling stories of the systems we seek to engineer.

5.3   How Systems Come to Be

This is a book about engineering and designing systems. In this chapter, we focus on presenting a philosophy and approach for how systems can be conceived and designed. As engineers, we tend to start from the system itself. We take a system that we can see and break it down into parts and proceed to think hard about the parts. Breaking things down into parts becomes our fundamental approach to designing new systems. Notice, however, that systems don't come into being simply because somebody started putting parts together. Every system that's ever been designed and built had a beginning in someone's mind. It began as a thought or an idea, and that thought was embedded in some kind of story.

Perhaps there was a need or problem that arose in a situation, and it was thought that a system could meet that need. The system was a response to the need in the situation. Not all situations or even problems require a system, but all systems are conceived within some kind of situation and story. Every system is surrounded by a context, a situation, a human condition, or a problem that brought the system into being. The same dynamic is true even in the very small situation of a person we'll call Lisa, working alone in her home workshop and creating something new to solve a problem at home, or perhaps some new device just for fun. If we watched Lisa build the new device or system through a video camera, what we would see would deceive us about her creative process. It would look like Lisa simply walked into her home workshop and start putting pieces together, trying out mechanical approaches, or writing some code on a computer. What's deceiving is that much more is going on in the unseen world of Lisa's thoughts. Many questions occur to us that are not answered by observing Lisa's actions: what is she trying to accomplish? What is her purpose? Why does she have that purpose? What need is she trying to satisfy or what opportunity is she trying to exploit? Who will use what she's building, and why – what are their requirements, needs, and expectations? And why do they have those needs?

Before we know anything about what Lisa is going to build or how it will be engineered and constructed, there is a rich story of how her project came to be. The context and the need are always there. Ask any creative person, maker, or home improvement DIYer about their projects at home, and you'll get the need and the story first.

If it's just Lisa working on the project, it's easy for her to hold the context and the need in mind and then begin work. As she works, various approaches are tried, and there may also be a progressive refinement of the context, and of the need. All of this can happen in Lisa's mind, with no worries about communicating the needs to anyone. She has complete freedom to change her mind about any aspect of the project without impacting anyone else's work.

Let me offer a personal example. I've been thinking lately about how I can create an easy way to load a special large kind of tandem bicycle (actually a tandem tricycle) into a van. As I'm thinking about this need, I'm thinking at the same time not only about mechanical devices that could help but also about what are my real needs. I'm thinking, for instance, that when I load the bike into the van, I also want to leave room in the van for other things. How can I maximize the available space left over after putting the bike in there? Perhaps I can store the bike as high as possible in the van to be able to utilize the space under it. I'm also thinking about the weight of the bike and how I may or may not be able to lift and use my own muscle power as part of the system to load it. My own strength is both a resource for the design, and I can use my own strength as part of the system, but it's also a constraint on the system – I can only lift so much safely.

I'm also thinking about the complexity of the construction of the loading apparatus because I don't want it to be too complex to build. If it's too complex, I'll probably never build it. All these considerations are there together in my mind as I'm working on this design. Now to someone looking from the outside, all they see is me physically trying things: putting pieces of wood or metal together, ordering a pneumatic cylinder to see if that will work, and a fair amount of blankly staring into space, lost in thought. Though it may seem chaotic from the outside, the think-try-think-do process is an intuitive, obvious, and very efficient process for anyone starting with a personal need and trying to design and build something to meet that need.

Perhaps unfortunately, one-person-designed-and-built systems are rare in the commercial world. They happen at home and in small shops with small systems, but in large companies, it takes more than one person to conceive, design, and build something. As soon as we add people, we add the need for communication and the need to work together. The more people we add, the more complexity and difficulty there is to the engineering process, and it's easy to see why. The moment I communicate something to someone, they begin to think and work, based on what I have told them. What if, in the next moment or the next hour or the next day, I start thinking differently? The other person started with what I said before and is going from there in his or her thoughts. We now have my changed thinking and the other person's progressed thinking, which was based on my previous thinking and communication. It would not be surprising if we are already far apart in our thinking, and this is just two people and a single communication. It's impractical and inefficient to share everything I'm thinking all the time with another person and for them to share everything with me, so getting out of sync is bound to happen. Small teams of under say seven or so people have an easier time of this if they communicate very frequently, work together in the same physical space and grow to know each other.

As is often the case, science fiction provides us with some dramatic examples of alternate worlds and processes. The Borg, a race from the Star Trek universe, is a fully integrated hive-mind, with every newly assimilated drone (formerly an individual person) gaining complete, continuous, real-time access to the thoughts of the entire collective. The Borg's technological advancement and ability to rapidly adapt to changing situations is thought to be a function of this integrated mind, admittedly at the loss of almost all individual identity and independent thought. Nevertheless, it's intriguing to consider how efficient a group of engineers could be if they were literally able to think together with the colleagues all the time.

Back in the real world, frequent communication has been found to help groups of engineers work together more efficiently. In agile methods for system development, frequent, short meetings called stand-ups are held simply so people are forced to communicate with each other often, hopefully cutting short the impact of any divergence of thought among team members.

Many important systems are large and complex and require the efforts of dozens, hundreds, or even thousands of engineers and designers. The communication overhead can become overwhelming to the point that either sufficient communication does not happen, or communication consumes much of the time and effort that might otherwise go into the engineering work. The usual approach, developed in the decades since the 1960s space program, is to boil down the thinking of a person or a group into a compact work product, such as a document or a drawing, and pass that along to other people or groups who need the information. No document or drawing can capture all of the thinking, however, so there are gaps, omissions, misunderstandings, and questions, and of course, a document or a drawing is at best a fixed snapshot of thinking at the time.

The question we'll focus on here is how can we improve the quality of communication? How can communication be made more complete, with less potential for omission and the resulting need for guesswork by the recipients. The approach we suggest here is that of telling more complete stories about the need and purpose of the system – communicating not with more words, but with more information, not with more drawings, but with more meaning.

An analogy from the military environment will get us started. When giving commands to those below in the chain of command, a commander has two approaches: give only explicit detailed instructions for exactly what to do or include communication of the *commander's intent*. Commander's intent has become a term that refers to the purpose, the general objective, and the main idea behind the mission or set of instructions. By communicating commander's intent, the commander has a better chance of the mission being accomplished because the commander calls on the troops to think, as well as follow instructions. When the given instructions no longer fit the intent, they can be modified by troops who are guided by the commander's intent. The commander's intent is a more complete story about the mission including not just the next steps but also the broader context, the reasons, and the purpose of the mission and the most important intended outcomes. In business, the principle of *management by objective* is similar. In what follows, we focus on how we can better communicate the broader picture, including context, intent, and even a vision for the system we are engineering.

When a system requires more than just a few people to design and build, the engineering process is different from that of an individual or a small team. It is here that we want to focus because large complex and intelligent systems are most often designed by groups and teams. How does a large group of people think together about the context for the new system, about the needs, wants, and expectations that the system will meet for its customers and users? With the individual working alone, we have seen that this thought process is a very fluid, interrelated, even nonlinear kind of process. As soon as a need is brought to mind, it suggests a certain approach and then as that approach is considered, it leads to thinking about a certain design. Imagining that design could easily lead to the recognition of the impracticality of that design. Thinking might then back up and reconsider the same approach could be implemented with a different type of design, or/and that if that doesn't work, we might back up to reconsider the need. Maybe our understanding of the need is wrong. Needs are not always obvious or clear, so this is a very iterative process, occurring in the single engineer's mind.

The question we want to address in this chapter is, what kind of tools can we use to enable this fluid process across a larger group? That the process is fluid is very important. Trying to force the process into a linear step-by-step sequence has risks of its own. We may succeed in describing and even demanding such a linear process, but the systems produced will likely be nonoptimal. In the development of large-scale engineering systems, it often occurs that if we try to write down all of the specific needs and requirements for a system at the beginning, before we are allowed to think about designs approaches, or what's feasible in materials and in techniques, we may end up with a set of requirements that sounds great, but are either impractical or simply not what we would have stated had we known then what we know now. We may wish we could go back and change some of the requirements, knowing that it could result in a much better system.

I remember talking with a group in a large defense company many years ago, who told me about how, when late in the design process for a new system, a new approach was conceived and discussed. Everyone agreed that it would save money and be a better way to design the system – the systems engineers, the customer, and the program managers were all for it – but it simply could not be done because so much work had been done already based on the original approach. This is a situation we can only avoid by trying to learn more, sooner. If we are somehow able to time-travel to the future of the design process and learn now what we will learn then, we can take advantage of it now. We want to find ways to discover superior approaches and ideas much earlier in the engineering process, when there is still the opportunity to change the design without too much rework or wasted effort. We'll call this a very desirable phenomenon an *early discovery*. Eventually, of course, engineers discover every necessary approach, design concept, and flaw in the system as it's built, but as we've illustrated, discovering such things late in the system development process is expensive. In the history of systems engineering, there are countless stories of project teams investing lots of time and money pursuing a design, creating detailed drawings, specifications, and prototypes and even building the system, only to discover that something doesn't fit, doesn't work, or isn't what the users really needed. Discovering these kinds of things early will save time and money.

5.4 The Paradox of Cost Avoidance

As an aside, there's a paradox here that causes difficulty in implementing new techniques that save time and money. Saving time and money in a systems development process is known as cost avoidance; it's time and money not spent. If that money was not spent though, how do we account for it, and show the CFO that money was actually saved? If the project were something that the company does over and over again, then we can easily compare the time and money spent on doing it one way with the cost of doing it another way. But systems development projects are often unique and hard to compare with other projects. This is a hard problem. The best we can do is to be honest with ourselves about the likely rate of late discovery, the magnitude and nature of the discoveries, and the typical cost to rework or make corrections to the system. If as a company, we can admit that we typically spend about 20% of the program's total cost correcting and reworking things that came up during the system development lifecycle, and that about half of those things could have been discovered earlier in the process and corrected at low cost or even prevented altogether, then we can reasonably say that our early discovery methods stand to save the project about 10% of its overall cost through cost avoidance. It's important to notice that we don't claim savings when compared to the projected cost of the system, since we know that projected cost is unrealistic by 20%. Claiming savings based on the projected cost is too high a bar and may cause organizations to forego useful improvements in engineering processes when they can't show savings when compared to the overly optimistic projected cost. Organizations must have the nerve to be realistic about the cost and how long a system is likely to actually take to develop in order to evaluate savings. An innovation that causes systems to be developed in only 105% of the originally projected time and budget, may not sound exciting, but in real terms, it may be an actual savings of 15% of the projects actual cost.

A number of approaches have been developed to try to accomplish this early discovery of better design approaches. The two main ones that have proved to be effective are agile methods and modeling. They are not mutually exclusive and can be used together. In an agile approach, instead of trying to lock down all of the requirements in detail at the beginning of the project, a more general set of requirements and needs is developed quickly, and then work is begun to implement some small aspect of the needed functionality quickly, in a short time interval called a sprint. Typically two to four weeks in duration, a sprint aims to create something that works, and is demonstrable, but which performs only a small part of the overall functionality of the system. Ideally, sprints are arranged to address the most uncertain or difficult part of the functionality first, thereby progressively reducing the technical risk in the project. Agile methods have proved to be a successful approach, but they are not the focus of this chapter and this book. Let's look at the other primary way to accomplish early discovery – modeling.

A model of a system is an abstraction that represents aspects of the system in a way that can be analyzed, but which is much faster, easier, and cheaper to create than actually creating the entire system. When architects create a mock-up model of a new building using foam core board and hot glue, they are doing it not only to show off their creation but also to work out details that aren't apparent from a flat drawing. When systems engineers spend time thinking about how a system will be used and documenting use cases, they are likely to discover new things about the needed system. When designers create a first prototype using slow 3D printing, it might take hours to produce a single part, but what can be learned is well worth the wait compared to the risk of manufacturing a batch of 500 only to find an issue that forces the scrapping of the entire batch. When a descriptive model is created using the SysML modeling language to precisely describe the interfaces between a system and other systems with which it will interact, the model enables everyone to agree on the exact way these interfaces will work, before time is wasted building systems that end up not connecting correctly. It is scarcely possible to overestimate the time and work that can be saved by employing this modern version of measure twice, cut once. Modeling enables the team to do it right the first time because, at the risk of piling on the aphorisms, there may not be time to do it right, but there is always time to do it over.

5.5 Communication and Creativity in Engineering

Both agile methods and modeling, in addition to their other benefits, help facilitate communication among team members. Agile does it by forcing frequent communication and synchronization through stand-up meetings and short sprints. Modeling does it by forcing the team to make design decisions explicit and clear and expressing them in a form that is understandable by everyone, without ambiguity. Complete communication of this type is easy with a small team, working closely together, but as teams grow and people become more specialized in their roles, it becomes more difficult. It may be tempting to ignore the need for complete communication, organizing a large team by simply dividing up the work, sending teams off to do their own part, and asking for regular reports on each term's progress. A requirements team may be assigned to collect and document all of the requirements and then deliver them to the design team, who can then begin the design.

A linear process, however, is only efficient when nothing is ever discovered along the way that might cause us to want to go back to an earlier stage. If a discovery during design causes us to want to go back and challenge some of our assumptions made in the requirements phase, that's great for the eventual final system design, but we may have wasted time and money proceeding into design based on requirements assumptions that were invalidated based on later learning.

Since learning throughout the process is inevitable and even desired, it's better to use a more iterative process which accommodates evolving designs better.

Again, we see this very intuitively if we watch one engineer work on something in isolation, conceiving, designing, and building alone. For examples of this dynamic process in action, one need only look at the *maker movement*, and the many great guides for making creative things. YouTube provides a way for video-makers to earn money through included ad placement, so there are many of them showing the creations and their processes on YouTube, hoping to develop a large audience. I will highlight one of my favorites from whom I've learned a great deal. Mark Rober's very popular YouTube channel showcases not only his fascinating builds but also his engineering processes. A mechanical engineer who has worked for both NASA and Apple, Rober brings his knowledge of engineering to his entertaining projects. In his inspiring class, *Creative Engineering,* he teaches his process, and it's very much what is described above – an intuitive, iterative process that considers the needs first, then experiments with different design approaches, often causing the re-evaluation of some of the needs and initial assumptions. Rober encourages his students to first consider the scene within which a new device or system will exist. Considering the scene, or context, within which a system will be used is vital, not only for small builds that someone will complete in a garage workshop but also for large complex systems, defense systems, and spacecraft. It all begins with the scenario within which the system would be used.

5.6   Seeing the Real Needs

Peering into the scene of nineteenth-century New York City streets, we would see a city where horses, carriages, and horse-drawn trams were the primary modes of transportation. In a dense city like New York, we would see many, many horses, so naturally horse manure was an enormous problem. The good thing about horse manure is that it's biodegradable, and quite useful for fertilizer, but if there's enough of it, it's still very difficult to deal with. Ask people about their needs in that situation, and what you'll hear about is that people would like the city to be manure-and-odor free. Even then people wanted their personal transportation to meet the zero-emission standard.

Henry Ford famously said that if he had asked for what people wanted in a situation like that, they would have asked for a faster horse, with less emissions. Henry Ford and many others could see the situation, but that's different from envisioning a city full of automobiles. At the time, any observer would see people, horses, and carriages moving through the city, and a great deal of horse manure everywhere. That's the situation within which some system could be built to improve the world. There are many different approaches to such a system, and we know from history that various approaches to manure removal were tried. It's hard to imagine what it was like to consider solutions to this problem, at the time of the problem. A natural inclination is to keep the current situation mostly intact, and simply bring new technologies to it. One approach would be to remove the manure more efficiently. Another would be to eliminate the need for removal by eliminating the source of the manure. It occurs to us to replace the horses with something mechanical, and in fact early automobiles were referred to as horseless carriages.

This direct substitution is one way of seeing a solution in this situation, but if we didn't know already about automobiles, what other transportation options might have been considered? In fact, other options did and do exist and have been in competition ever since. Electric trams, cable cars, trains, and subways of various sizes have been used in cities across the globe. It may be tempting to think that all solutions were considered, but it may take some real creativity to imagine new solutions. In his 1940 short story, *The Roads Must Roll*, Robert Heinlein offers a unique approach to transportation. Science fiction stories are wonderful for their ability offer concepts and solutions outside our current ways of thinking. In Heinlein's story, we see a completely different solution to urban and intercity transportation: the roads and the sidewalks are moving conveyor belts, traveling at various speeds. The slowest one is moving at perhaps 5 mph, making it easy for someone to simply step onto it from a stationary platform. The next road over is moving at 10 mph, and since the difference between 5 and 10 mph is small, a person can easily step from the 5-mph road to the 10-mph road. The next road is moving at 15 mph. A person can simply walk from slower roads to faster roads in order to travel at up to say 100 mph, which might be the fastest road that's moving. There are stationary platforms and seats and even buildings, stores, and restaurants on these roads that move with the road. It's an entirely different concept for transportation. Is it more or less practical than other approaches? That would take a different kind of analysis – here we are aiming to imagine possible solutions.

Another creative solution to personal transportation that is also a sci-fi standard is personal jet backpacks. If we could simply strap a backpack on everyone's back, everyone could fly to their destination. Air traffic control might present a challenge, but perhaps that could be managed. Tony Stark, aka Iron Man took this approach. Interestingly, in the Iron Man story, the innovation that made his flying suit possible was the arc reactor, a small but extremely powerful source of energy. He had to invent a power source because that's the main problem with jetpacks is that it's very difficult to carry enough power to propel yourself off the ground and fly for any significant distance. The challenges are power, control, and safety, but power seems to be the hardest problem.

In the early stages of engineering, a complex or intelligent system, we face the question of how to describe the scenario into which the system will be introduced. How we communicate about the scenario among the stakeholders of the system is important in order to gain widespread understanding and agreement, and also to avoid biasing the solution in some way. When the scenario is more complex than just something someone can say in a few words, we need a way to communicate about it succinctly.

5.7 Telling Stories

Though it may sound like a contradiction in roles, we're suggesting that engineers learn to tell good stories. Storytelling has been around in the human race for a long, long time and likely predates both written and spoken language. Cave paintings have been discovered than seem to be tens of thousands of years old, and they tell little stories – stories of hunting animals, stories of danger and conquest, and stories of love and family. Storytelling and drawing pictures are one of the earliest forms of communication. We use stories to explain things to each other, to teach, to illustrate, and to instruct. Spiritual teachers use stories – Jesus used parables, Zen masters told stories, and Shamans tell stories to help heal.

There's something very fundamental and essential about stories to human beings. As C.S. Lewis describes in his book, *The Chronicles of Narnia,* children in the mythical land of Caldron learn to tell stories instead at a young age. They grow up through their school years telling stories instead of writing essays, as other children do. The big difference, he points out, is that people like to hear the stories while few would like to read the essays. We like to hear stories. When someone opens a conversation with, "Let me tell you a story," they have our rapt attention.

Stories are still with us, albeit in more developed and technological forms than in the cave painting days. For us, stories become novels, novels become movies, and movies become franchises. Star Wars is a story that has been with us for over 40 years, developed in numerous films, and they continue to be made. Epic sagas and stories follow patterns. Joseph Campbell, a historian, philosopher, and student of mythology, describes the hero's journey as a fundamental archetypal story that we see again and again in many, if not all of our great stories, including Star Wars and its hero Luke Skywalker. Campbell's hero's journey is like a template for a story where the hero has a humble beginning, has the sense that something more is possible and is then affected by some dramatic life event that leads to his or her discovery of a mentor, guide, or guru. The mentor leads the budding hero through a training process, an ordeal, or a challenge, and once the hero has come through that challenge, and is on the other side, is eventually transformed into something greater. The journey of personal transformation from Luke Skywalker to Jedi Master is a story that has been told countless times in countless settings.

System stories also have patterns. A story is not just any series of events. It has a flow to it, a beginning, a middle, and an ending. Stories are able to compress time. It's intriguing to examine how much time can be compressed and still communicate effectively. One of my favorite movies is *The Hunt for Red October*, a cold war movie about a new, fictional Soviet submarine, with a nearly silent and undetectable drive system. The movie lasts only 212 minutes but tells a complete and compelling story of this submarine. We learn about its capabilities, the kind of missions it was designed for, and its limitations. We see into how the Soviets view their "old adversary," the United States Navy, and how the Navy learns to detect and locate the "invisible" submarine, and to counter its effectiveness. We learn many useful things about this submarine, by seeing it play its part in the story. The movie is not a documentary about the submarine but tells a story that puts the submarine in a starring role. One might assume that more would be learned about a system like the *Red October* submarine from reading requirements, specifications, and designs, but weeks of pouring over papers might not be a match for what is learned by seeing the submarine in action, in a complex and compelling story, compressed into just a few hours.

Storytellers who create movies are masters at compressing a story into a short time. First, the story, which takes place over many days, months, or even years, is compressed into a novel that takes perhaps 20 hours to read. In fact, *Red October* began as a novel by Tom Clancy which in audio form requires about 19 hours of listening. The novel is then transformed and compressed into a screenplay, which will be far shorter, perhaps 15,000 or 20,000 words, where the book will be well over 100,000. Reading through a screenplay takes around two hours.

Shooting the movie involves time compression as well. A typical shooting ration for a film is around 10 : 1 meaning that for every minute in the final film, 10 minutes were shot. In shooting the director captures everything needed to tell the story, usually in several ways, from several angles, with variations that might end up being shown for only a moment, or left out of the final product altogether. In shooting the director assembles the building blocks needed to tell the visual story. Next comes the process of editing, a seemingly technical task, but actually a highly creative and artistic process, where the story is told as fully as possible, given the limitations of the eventual movie's length. Sometimes movies include deleted scenes where some of what was not used can be viewed. We can see how many decisions and compromises are made during editing, when a movie is released later in a "director's cut" edition, which uses more of the shot footage, including material that is interesting and useful to the story, but was not deemed essential for the theatrical version of the movie. Sometimes watching the deleted scenes or the director's cut actually makes the story richer and more complete, by revealing background details, character motivations, and less-important events that were dropped in favor of something more important in the regular edit. Those who want the rest of the story will enjoy the longer director's cut version. Through the editing process, the movie is cut down to a theater length which is somewhere around two hours, plus or minus an hour. The movie makers are able to compress that entire novel into a movie. The movie does not contain every detail in the novel, which leads to the common refrain of readers who later see the movie: "the book was better." Perhaps, but it takes 10 times as long to get through. The extra words are needed because the novel doesn't have visuals to help tell the story. As the audio book company, Audible used to say, "the pictures are in your head." Spending the time to create the visuals in the film is paid back in the compressed time it takes to understand the story.

It would be interesting if we could make a movie of every system that we wanted to build. If the United States Navy wanted a new kind of submarine with a new kind of drive system that was silent, and very hard to detect by sonar, perhaps they could make a movie like *The Hunt for Red October*, illustrating how this submarine would work, how it would function in different kinds of battles, defensive situations, or reconnaissance missions. If somebody wanted to build a real *Red October*, it would be quite effective to have the entire engineering team sit down and watch the movie. Everyone would get a very direct, visceral emotional feel for what the new submarine should be and how it will be used.

Immersed in the movies, we suspend our disbelief, allowing ourselves temporarily to step into a world where this submarine does indeed exist. We're able to see and explore the world where the new submarine exists, long before the submarine is actually built. In most movies, many of the fundamental concepts that make the story possible are impossible, like traveling at multiples of the speed of light, necessary for every interstellar space travel adventure. But at the boundary of the currently possible and the science-fiction impossible, is a space of innovation, of invention, of what could be possible, if we dare imagine it. Are the big doors on the side of the *Red October*, which hide the silent caterpillar drive, an impossible fantasy, or an inspiration for research into new kinds of submarine drive? Innovation requires both imagination of the kind the novelist, screenwriter, and director must have, as well as the creative inventiveness of the engineer. Is a caterpillar drive possible? Maybe or maybe not, but the question never gets asked unless the idea is imagined, explored, and dramatized in a story. It's no accident that science fiction has inspired the invention of new systems. Those who tell science fiction stories inspire engineers to create new systems.

We probably can't make a theatrical movie about every new system we might want to create, though it might be worth considering. Capturing the vision for the new system in a compelling and dramatic way could have a significant unifying effect on those who will be involved in creating it. Movies require a lot of money to create but with various simpler technologies, we may be able to achieve a similar effect and show how a system could work in a dramatic fashion. Returning for a moment to the process of move creation, let's go one more step. After the movie is made another film is created – the trailer. In three minutes or less, movie makers try to condense the entire movie into a form that tells just enough of the story and gives a feel for what the movie is like, but leaves out some of the more important details that will be revealed in the movie, in order to build suspense and a desire to see the complete film. Watching the trailer for *Red October* after seeing the full movie, one is struck by how much can be learned about the submarine, and the story in those two minutes. There is a remarkable richness and density of information in movie trailers.

The making of a short video to describe a new kind of system is seen in other science fiction stories. In the second Star Trek movie, *The Wrath of Khan*, Captain Kirk, Dr. McCoy, and Mr. Spock watch a very short video, summarizing a proposal for an engineering system. In just one minute and forty seconds, Carroll Marcus, the lead scientist, presents the entire vision of a new system called Genesis in a dramatic and compelling way (search "Star Trek the Wrath of Khan Genesis Project Summary" on YouTube to view this film within a film). Before we talk about some specific tools and techniques that can be used to capture stories of systems, let's look more closely at the process used to develop a movie. There's much to be learned here about the creative process, and we can apply it to how we develop and convey the stories of our systems.

5.8 Bringing a Movie to Life

Movies are developed in a series of seven stages:

1. Development
2. Financing
3. Preproduction
4. Photography
5. Postproduction
6. Marketing
7. Distribution

Our purpose here is not to describe the moviemaking process in detail but to see how these stages apply to the stories that we want to tell about our engineering systems.

The first stage, development, is where the ideas begin to come together. It's possible that the idea for the movie is brand new, or perhaps it comes from a book, or from real events or a person's life. Movies will sometimes identify themselves as "based on a true story" where the true story of the person's life has become the basis for a movie. In the stories we tell about the complex or intelligent systems we want to build, our "movie" is always based on a true story. The true story is the context, the situation into which we intend to introduce the new system. A new kind of maritime rescue system might be based on the true stories of rescue situations that have happened in the past. The story might show how those situations could have happened if the new system had been available.

The early story concepts can be in a number of forms, and they may be much longer than what we need at this stage. A novel has much more content than we'll need for a movie. To base a movie on a novel, there must be a sifting and sorting and getting to the essence, the core, the heart of the idea. The team struggles to find what's important in the novel – and what's important to show in the movie. It's not just a task of summarization. The best-selling novel Dune by Frank Herbert has been made into movie three times since it was written in 1965. It's fascinating to see what each moviemaker thought was important in the story and how much time and detail is given to each event. The latest movie adaptation, in mid-2021 spent a great deal of time on Paul and his mother Jessica's acceptance into the Fremen community, and in particular on Paul's duel with Jamis. Other adaptations spent more time on the position and mystique of the Spacing Guild Navigators.

The same kinds of choices must be made when telling the story of an engineering system that is to be developed. If we have a system that is going to involve many people in its use and deployment, such as a large defense system, spacecraft, satellite, or vehicle, there may already be a lot of words, a lot of information written about the system, including wishes and desires that people have expressed, market analysis, and even thoughts and projections about trends in vehicle design or available technologies. There may be the equivalent in size of a lengthy novel to consider as input in this development process. The systems engineering team must do the same kind of sifting, sorting, condensing, and combining to bring the story down to something that's compact – a small, early version of the story.

In the second stage of movie production, financing, the story must be pitched to someone who can fund the project. We have to take what we've developed as the short form of the story and shorten it much more so we can present it to someone quickly. Movie pitches are often very, very, short, perhaps only a single sentence. If the pitch goes well, it leads to longer discussions, but it's that very short pitch that must get the deal going. We go from a very large body of information, the novel, to a very small pitch presentation. When the pitch succeeds, then we move on to the next stage which is called preproduction. In preproduction, we have to make sure that the screenplay has been successfully developed from the original novel or from the original concept. As mentioned above, the screenplay is not only much longer than the pitch but also much shorter than the original body of information, the novel, or the collected information on which the film is based. In our engineering projects, our "screenplay," may be our early description, models, and diagrams of the system context – not of the system itself at this point in the process but of the problem area or need for the system. We will see in Chapters 6 and 7 how this work could include the context and the usage views of the system, including things like a context diagram and a use case model.

In preproduction, moviemakers also think through and plan out what is going to happen later in the production phase. They work out the casting and the locations of all the shoots, as well as costumes and props. All of those things have to be planned out so that they all come together in an organized and efficient way.

Everyone involved knows that flexibility will still be required and things may change as they go. But that's not a reason not to plan. In the production stage, the moviemakers make it all real, shooting the photography, as the actors do their craft. It's very much like an engineering project – all of that actual engineering is done efficiently and in the right order, minimizing wasted time and the rework of items built (or shot) incorrectly. The postproduction phase for a movie is extremely important to the final product. It's hard to imagine making a movie without postproduction, shooting film or video, perhaps with just one camera start to finish, and putting a title on it and releasing it. It would be a movie, but would certainly not be the best way to tell a story. In postproduction, we may add special effects that were not added with the actors when the photography was done they're added later. We add explosions, we had a scenery in the background, the actors may have done a shooting against a green screen or may have done motion capture and CGI-generated characters are put into the movie. Even more important to postproduction than effects, is editing. Editors, together with the director, must both arrange and reduce the shot footage to just what is going to be the theatrical version of the movie which might only be 100–120 minutes. They work very carefully to figure out where to cut from one scene to another scene so that the meaning and flow work, but so that the movie doesn't bore or confuse viewers. In recent years, moviemakers use more frequent jump cuts from one scene to another scene, asking viewers to following multiple story threads happening in different places and times. The editing must be done so that the audience is not confused but is intrigued and sees events happening at multiple times in multiple places, as part of an integrated story.

The next stage in moviemaking is marketing. Once the final product is created, it must be promoted so that people find out about and want to watch it. Part of marketing is where the movie trailer is created, a miniature version of the movie intended to intrigue people and make them want to see it. The trailer is a sort of a preview but not a summary – it must leave out some of the best parts to entice people to want to invest their time and their money in watching the movie. Systems engineers face the same challenge in developing systems and need to be able to communicate about the system in an accurate and compelling way to people who have not yet experienced it. The story of the system must be told, both before it is designed and built and afterward. In marketing a movie, the emphasis is not on the specifications, but on the story. Moviemakers don't lead with the details of the actors, the timeline, scenes, locations, and specifications but instead tell a story. The pitch is a very short presentation of the story before the movie and the trailer is a very short presentation of the story after the movie is made; they are different but are the same story. The pitch is a story of what the movie is going to be like, looking forward while the trailer is a very condensed view of the movie, telling the full story because it's already been developed, but leaving out some details to not give too much away. Both versions of the story are short but communicate the entirety of the vision at their respective places in the moviemaking process.

The final stage in moviemaking is distribution, where moviemakers figure out how to get this movie out to lots and lots of people. Distribution used to be a very straightforward process, since movies went only to theaters through established channels. For a system, distribution of the story of the system to all relevant stakeholders is a key to setting expectations and making the system attractive and successful.

What we can learn from the moviemaking process that we can apply in engineering is that the story is paramount. The story is front and center, at every phase of the moviemaking process. It's the story that is the heart of the movie. The movie is not about the characters, the actors, and the scene settings – those exists only to serve the telling of the story. If you talk to a moviemaker about what they are focused on, no matter what part of the moviemaking process they are engaged in, it is the telling of the story. The story is what's important. A movie may be unsuccessful, even though it has great actors and acting, beautiful scenes, and splashy special effects, if it lacks a good story. Conversely, a movie with a great story can overcome deficiencies in acting, writing, and photography, at least to some degree.

Stories can change. Sometimes in the process of making a movie, the story does change, hopefully making it a better movie in the end. Occasionally, a moviemaker will release a movie offering several different endings, or may distribute different versions of the movie with very different endings. In the movie *Little Shop of Horrors,* the original movie made in 1960 has one kind of ending, and the musical remake of the same movie in 1986 had a different ending. Very different. One might even say an opposite ending. The reader will have to watch both to find out the specifics.

While the story is central and paramount, it may not be fixed and immutable. The story can change; the story can have flexibility to it, but it is still the story that is paramount. When a product or a system is purchased or delivered, the customers are buying a story. DIY homeowners are buying not a drill, but a story within which the drill will accomplish things they want to accomplish. It's like they've already told the story, and there's a hole in the story for a drill, so to speak.

We learn from the filmmaking process that the story can be told effectively, in different ways, in different forms, for different audiences, or for different purposes.

The story of a system may be told originally in its long form, like the original novel, perhaps produced from lengthy analysis reports, background research, intelligence gathering, and market research. Then, the story must be told in a very short form as a pitch for it to be approved and funded. Then the story must be told as a "screenplay" with storyboards, in order to guide the work of the photographers and director. Later, the story of the system must be told after the system is developed and built, for demonstrations and promotion, along with a new very short form of the story in the "movie trailer." All of these are expressions of the same story, but told in many different forms.

5.9 Telling System Stories

While there are similarities between the storytelling of moviemaking and the telling of system stories, systems engineers and moviemakers will likely use different tools and methods. Nevertheless, it is intriguing to think of creating an actual pitch, movie, and trailer for a system just as Dr. Caroll Marcus did for Genesis. Whether we actually create a film or not, there are ways to incorporate storytelling in the systems engineering of complex and intelligent systems. We never escape the story. We ignore it at our peril. When the teams lose sight of the story, the project can lose focus, stray into feature-creep and needless enhancement, or even fail to give enough attention to critical functions and qualities of the system.

Stories have both a visual and verbal aspect which can be used together or separately when telling system stories. Though many people are fond of saying that they are "visual" rather than "verbal" we continue to rely on the verbal for the earliest and most fundamental stories, in both movies and systems. The early forms of the story in a movie are all verbal. The novel, pitch, and even screenplay do not typically have complete pictures and drawings. The single drawing on the cover of the novel may be suggestive of the tone of the story but is far from a full portrayal. Eventually, in the completed movie and in the trailer, the story tends to be told more through visuals.

In the novel, all the visuals are imagined by the reader, which is why the novel requires so many words. The pitch also may or may not have visuals. In its earliest and shortest form, it may not have visuals at all. In its more expanded form, it may have some early sketches or drawings to give a general idea of what the movie will look like. But good illustrations can be difficult to do and can require the skills of an artist. As we get into preproduction, the storyboards become much more detailed illustrations since the reason storyboards are done is to help communicate the story among the people who are doing the production. The visuals make sure that the writer's intent has been communicated successfully between the writers, the directors, and the actors, so that all understand the story, and are all playing their parts in the overall creation.

The actual video or film photography becomes the set of building blocks used to communicate the story. Different kinds of shots are used to tell the story efficiently. There are *establishing shots* where a picture of a building or setting is shown, for a moment, and then in the next shot, we are in an office conference room and see a few people meeting. The viewer makes the connection that the office is in that building, even though the viewer was not taken in the front door of the building, up the elevator and down the hall. By establishing the building with a single shot, and then jumping to the conference room, the viewers understand exactly what's going on – that we are seeing a meeting in that building. With just a few visuals, the movie communicates a lot. In the novel, the same event might have required several paragraphs describing the building, the setting in which the building exists, the conference room and the people present but, in the movie, we can do it in just a few seconds, with just a couple of visuals.

Part of postproduction and editing is limiting the visuals to only what is needed. It's always intriguing to watch the deleted scenes from a movie, when they are included on the DVD release version, since they clearly show what could have been shown, but was left out. Some scenes have an extended version where the extended version of the scene which might take three minutes, but the version of the scene that made it into the movie might only be one minute. It's fascinating to see what the editors eliminated, and to second-guess them about their decisions. Did they succeed in telling the story completely, but still eliminating two-thirds of the content? The editors must consider how to tell the story most economically. In some cases, the moviemakers are able to release a *director's cut* or *extended version* where they have the freedom to make the entire film longer, perhaps as long as they want and include the longer versions of scenes, previously deleted scenes and other material that was edited-out in the shorter theatrical version. Editors often need to make the movie as compact as they can, but the director may get this opportunity to make it as full and lush as he or she can, so there is a slightly different purpose guiding the editors of each version.

In the marketing phase, visuals are used too. The trailer is cut from the actual movie, so therefore everything in the trailer. In almost all cases, is actually in the movie. The trailer presents a condensed version in order to be attractive and gain interest. Trailers, as can be seen through a careful watching for a familiar movie, sometimes present things out of order, that is, not in the order that you would see them in the movie, or not in the order that they happen in the story. Sometimes the dialog is rearranged as well. In a trailer, bits of dialog might be used out of sequence, or to create a moment of "conversation" that doesn't appear in the film. A line may be heard over a visual and that arrangement may not be what happens in the movie. The creators of the trailer have their own skills of being able to evoke and create the feel and enough of the content in the movie in a very, very, short time to achieve their purpose.

Marketing also creates visuals for the movie in the form of movie posters. Movie posters are their own kind of art because they generally are not a just a still photo taken from the movie, and are not even a particular scene from the movie. Movie posters are different. They tend to be artistic renderings of the main character or characters, along with an image of a scene or object from the movie, put together to evoke the right kind of feel. The original movie poster for Star Wars (see https://www.starwars.com/news/7-things-you-didnt-know-about-the-original-star-wars-poster) gives quite an adventurous, almost romantic kind of feel for the movie, almost like a fantasy story, aimed presumably to attract people to the movie, even though the style and feel in the movie is somewhat different. Again, we see that the purpose for a visual is not to describe something accurately, like an engineering drawing, but to evoke and communicate the spirit of the movie. One could be forgiven for interpreting the poster as indicating a romantic relationship between Luke and Leia, but late in the movie, we learn that they are siblings. There is romance in the movie, but not between them. The poster takes some liberties to communicate in a single glance. The movie poster is not trying to be literal, but it's trying to be evocative of the feel of the movie, and is fundamentally trying to sell the movie.

The essence of all marketing is testing. A typical approach would be to create, or at least sketch multiple potential possible movie posters and test different ones to see what works the best in terms of generating response from the target audience groups. In the distribution phase, we also have ads for the movie, including print ads that will appear in newspapers, digital ads for websites, or poster-sized ads or on the sides of buses. Different images will be tested as ways of communicating enough of the story to be compelling.

Imagery is very important in many phases of movie development, and we'll see how images work in telling the stories of engineering systems. We may not be using photos, actors, and scenes in telling the stories of engineering systems, but we can learn from moviemakers how to tell stories, and how to illustrate stories. Even simple drawings can be visually compelling, sometimes even borrowing an image can be compelling. If we are designing a system, and there's a close analog in a science fiction story, perhaps we can use the reference, not to say that we are doing what is depicted in the science fiction movie, but to evoke the idea.

5.10 The Combination Pitch

One of the most effective ways of pitching a movie is by comparing it to the combination of two existing movies. This is called a *combination* description. In the 1960s when Star Trek was being pitched by Gene Roddenberry, he described it to network executives as "Wagon Train to the Stars." At that time, Wagon Train, a western-themed show set in post-civil-war America, was the number one show on the network, attracting big-name guest stars and cast members, and followed a group of travelers only a long journey from Missouri to California, but focused on the personalities, lives, and relationships of both travelers and the people they met along the way. By referencing the show with the notion of interstellar travel,

Roddenberry could pitch his new show in a single phrase. Anyone would quickly grasp the idea of a long-term voyage, with interesting people, and situations along the way, translated from the old west into futuristic space travel. It worked. Twilight has been described as Jane Austen meets Dracula. One might describe *Under Siege* as "*Die Hard* on a ship." Combination descriptions can convey a lot of information in a very short phrase by borrowing and combining the theme, storyline, and general tone of one work with another. They are also tools for creativity – imagining an established movie in a new setting or new time period can create something unique. Anyone care to make *Star Wars* under water, or *Titanic* in the desert?

President Reagan's strategic defense initiative from the 1980s was colloquially known as the Star Wars initiative, evoking images of high-tech defensive weapons in space. Using comparisons and combinations from existing stories, even when fictional, can be a very effective way to communicate the story of a system. Mark Rober, mentioned earlier, calls one of his most popular YouTube videos the *squirrel ninja obstacle course*, referencing the popular ninja competition television shows, which themselves are a reference to actual ninjas who were legendary for the physical agility, speed, and extensive training. In the ninja warrior television shows, competitors are challenged with an elaborate and creative obstacle course. A single word, like ninja can refer to a pattern, a theme, even a whole world. One must be sure that the listeners are familiar with the world that is being referenced. It does no good to refer to a ninja obstacle course if somebody has never seen any of the ninja warrior competition television shows. If a reference to ninjas conjures only warriors fighting to the death with swords in all black costumes, there is no connection to the squirrels in Mark Rober's backyard.

5.11 Stories in Time

Another major thing we can learn from the moviemaking process is how the stories and pictures we use in describing systems are connected in time. In our lives, we experience time in a linear fashion. Life happens to us one thing after another. Sometimes time appears to move fast to us, and sometimes it appears to move slowly, but it always moves forward linearly. We don't experience multiple things in multiple places at the same time; we are in one place experiencing one thing and then we go somewhere else and experience something. In movies, however, the storytelling is not limited to a linear process. Moviemakers are free to go forward and backward in time and to show things that are *happening* at the same time without *showing* them at the same time. They can show a scene here and a scene there, and the movie viewer will understand that those are happening at the same time. It's often important to do this to tell the story completely because in the world, things do happen simultaneously. One nation is doing something over here and another nation is doing something else over there and eventually they end up in battle. To understand the story, we need to see those preparations in both nations even though they're happening at the same time. The movie maker may take us back and forth between the preparations and in good moviemaking, we will understand that events are happening simultaneously.

Movies also make extensive use of flashbacks. Superhero movies, like the Marvel series, often do this to show origin stories. We are shown the superhero in the present times doing super-heroic things. Then we might flash back to a much earlier time, before the superhero was even a superhero: before Peter Parker was bitten by the radioactive spider that turned him into Spiderman, or before Bruce Wayne put on the cape for the first time. Movies can easily do this, and they usually signal it in one of several ways. Sometimes they will use a different actor, a much younger actor, to show we are now in a much earlier time, or sometimes through visual effects, they will just make the actor appear much younger and tell the story that way.

When we tell stories of systems in the engineering world, we have to find ways to do this sort of time-bending to tell better stories. Many of our engineering modeling techniques are a bit too literal, and do not allow easy representation of flexible timeframes. We often think of showing a flowchart or SysML activity diagram, and these show time in a linear forward fashion. They may show simultaneous events, but only as branches within a single, larger flow. If we have two independent flows but simultaneous flows of time are harder to depict. It's ever harder to show how two independent timelines connect and relate to each other in time. We'll look at some techniques later for how we can overcome this limitation, and think more like a moviemaker, focusing on presenting the whole story in a compelling way, using a new kind of diagram.

5.12 Roles and Personas

The third thing we can learn from moviemakers is about roles in movies. It's important for the characters to be established. We know from the first time an actor appears in a movie, that he or she represents that particular character role. It would be quite confusing in a movie to have the same actor portray two completely different character roles. Actors may even get a bit stuck with role they played for years. Leonard Nimoy struggled to clarify his identification with the character Mr. Spock, in his first autobiography, *I am not Spock,* and then his later book, *I am Spock.* In the same way in describing engineering systems and the stories within which they exist, we should consider the roles of people with respect to the system. In most complex systems, there are multiple users and multiple *kinds* of users. Each of these users may have different set of needs, different attitude, different set of skills, different purpose, different desires, and different expectations, with respect to the system.

In marketing, we refer to these as personas, and use them to characterize and describe different kinds of customers for which we are building a product or a system. A persona is a specific prototypical person of that type. Instead of simply saying that our product is meant to apply to men between the ages of 35 and 42, marketers will build a persona and conjure a specific functional person to help them better relate to the customers they are trying to reach. In this case, they might create a persona named Jose. Jose is 38 years old and grew up in Chicago, they invent. He went to university and earned a degree in business, and so on, sketching a complete background. By describing Jose, they can better imagine what he will like and be attracted to, what marketing approaches are most likely to work. Jose is a typical potential customer, with fictional specifics to make him more real to us.

Systems engineers can do the same thing when describing a system and its users. We identify roles of people with respect to the system – later we will call them actors when we discuss use cases. We'll think about those roles and the real people they represent, considering their needs, their wants, their expectations and other attributes about them as we design the system. Users are not all the same, and by considering each person's use of the system, and designing the system to fit his or her needs, wants, and expectations, we can make the larger system consisting of the users and the system we are developing, more intelligent.

For example, one category of users of our system is what we might call casual users, meaning that they use our system or our product only occasionally. For these users, we will try to make the system very intuitive to use, so that they don't have to remember too much about the system between their infrequent uses, or pull out and read the manual every time. Even very intelligent users will be annoyed if they have to read the manual every time they use the system.

In accommodating different kinds of users, it's often best to reveal advanced functionality of a system only progressively, meaning that when a user first approaches the system, the user sees basic system functions and is able to use the basic functions of the product right away, very intuitively and easily, and offer the user additional capabilities only later more advanced functionality is needed. As the user grows in familiarity with the system, more ability is gained to access more advanced functionality and understand it without being confused or uncertain. Software designers are often wise to present only basic options on the initial screens. Placing advanced functionality behind a button that literally says, *advanced,* signals users that they must make a deliberate choice to go into the advanced area, and only and then, are made aware of the available advanced features. More advanced and experienced users want efficiency with the system, so designers make available shortcuts, compound functions, and more comprehensive user interfaces to best meet their needs.

In this chapter, we've spent some time looking at how our oldest forms of communication, stories, are still one of the most complete and compelling ways to describe and understand complex and intelligent systems. We looked at how movies are developed, and how their visual aspects are matured and refined. We paid particular attention to the length of the story and how long it takes to tell, showing that a story can be told in a single image like a movie poster, a movie trailer of only a couple of minutes, or in its longest form, the original novel, with words and no images at all.

In Chapter 6, we'll look at how these ideas can be employed to communicate and tell the stories of systems we are developing.

6

Use Cases: The Superpower of Systems Engineering

Systems Engineering is a transdisciplinary and integrative approach to enable the successful realization, use, and retirement of engineered systems, using systems principles and concepts, and scientific, technological, and management methods.

(International Council on Systems Engineering Website, https://www.incose.org/about-systems-engineering/system-and-se-definition/systems-engineering-definition)

6.1 The Main Purpose of Systems Engineering

Before we describe how *intelligent* systems can be designed and engineered, we should look first at the way large, complex systems are conceived and built. The complexity of technological systems made possible by miniaturized electronics, computers, and software started increasing dramatically around the time of the US–Soviet space race in the 1950s and 1960s. Spurred by national pride and competition with the Soviet Union, the United States became very creative in pushing technology to its limits and managing the engineering of complex systems. The Soviets and engineers in other countries did the same. The field that emerged came to be known as *Systems Engineering.*

The development of large, complex systems such as spacecraft, defense systems, satellites, and self-driving cars faces two major challenges, both of which systems engineering attempts to addresses. First, the system must be organized, architected in a way that facilitates effective design and construction, and second, the work of the dozens or hundreds of engineers engaged in the project must be organized in a way that facilitates productive work. Systems engineering tells us how to think about and organize the system itself, by breaking it down into subsystems,

Engineering Intelligent Systems: Systems Engineering and Design with Artificial Intelligence, Visual Modeling, and Systems Thinking, First Edition. Barclay R. Brown.
© 2023 John Wiley & Sons, Inc. Published 2023 by John Wiley & Sons, Inc.

then sub-subsystems and components, designing and engineering each, and then integrating these designs back into the larger system.

Systems Engineering also solves the problem of how to organize large teams of people to design and engineer large, complex systems, and the solution is the same. Organize teams to design, engineer, and ultimately build the sub-subsystems, then organize teams to integrate those into subsystems, and a team to integrate those into the overall system. Sometimes the hard part of engineering large, complex systems is not solving the engineering problems and design challenges, but organizing all the people and ensuring efficient, accurate, and timely communication among them. Communication difficulties are one of the reasons that large development projects of all kinds often end up behind schedule and over budget.

What we want to focus on here is how the systems engineering process begins, specifically how the system's *requirements* are developed, used, and communicated. To set the stage, consider that if you ask almost any systems engineer, "What's the most important part of systems engineering?" the answer will invariably be "Getting the requirements right." If you read the explanation of why failed or delayed projects met their fate, usually the culprit will be identified as missing or incorrect requirements. Requirements thus become the scapegoat for almost any kind of system development project failure or delay.

Is it sufficient to simply blame the requirements and move on, or is there more insight to be gained? There is often the assumption that what is mainly needed in the development of large, complex, and ultimately intelligent systems, is to get the requirements right at the start. Before we can examine that assumption, let's look at what requirements actually are.

6.2   Getting the Requirements Right: A Parable

The term *requirements*, when used by systems engineers, really has two meanings, though they are usually treated as one. First, requirements refer to the facts about a system that must be true for the system to be accepted and for it to function in the way the users (or acquires/purchasers) specify. The keyword here is *specify*. For a requirement to be a requirement, it must be specified. Something that the users need, but which is not specified as a requirement, is just that – a need – and it will usually become an unfulfilled need if it is not translated into a requirement.

In the situation where one organization, such as a government agency, is paying another, such as a defense contractor, to design and build a system, it is literally true that needs not expressed as requirements are likely not to be met. Requirements are the currency used to "sell off" the system back to the acquirer. If there are 5000 requirements, then as each is shown to be met, using a process called *verification*, the acquiring organization pays the contractor an appropriate share of the cost of the system. Ensuring that the system meets the user's actual needs is a process called *validation* and is performed separately – gaps in validation must then be made into new requirements in order to be implemented.

The second meaning for the term *requirements* refers to the form of a requirement. The classic and most widely used form of a requirement is what's called a *shall statement*. It's as simple as it sounds. A statement is made about the system using the verb *shall*. For example, "The system shall have the capability to transport twelve soldiers at a speed of 35 mph over rough desert terrain." In a typical systems engineering process, the idea is to determine all the requirements up front, before any design activities begin, in order to fully express the needs of the acquires, sponsors, and users of the new system. The assumption is that a fully detailed and complete set of requirements can be determined as the first step in the systems engineering process, before any design work is done. We will challenge that assumption in a moment.

A great deal has been written about how to write good shall-statement requirements (a great example is the Guide for Writing Requirements, published by INCOSE, incose.org), but here we want to address something more fundamental. Is writing the kind of precise, unambiguous, complete requirements statements called for by systems engineering guidelines an easy or hard thing to do? It can be either, depending on the system's level of technological risk and uncertainty.

If the system one is describing has little technical uncertainty, and little experimentation and innovation are needed for its engineering and development, then it's easy to write many high-quality requirement statements. For example, imagine a company that designs security systems for office buildings. If their systems don't change much from customer to customer and use the same components, technology, and design patterns, and if most technical aspects of the system are well known, then there is little *technological risk* and *uncertainty*. In practice, this means that little research, experimentation, and analysis will be needed to reach a final design. The design of the next customer's security system may require a few more sensors than the last one, or a different arrangement of motion detectors and door locks, but there won't be cases where we are forced into the lab to experiment with a new technology or evaluate new possible technical approaches to achieving an important system function.

The term *risk*, used in this way, doesn't refer to a probability – the chances of failure in the system or the project developing it. Think of risk as a special kind of uncertainty. We don't know how we are going to design a certain part of the system, and it isn't just a matter of deciding. We'll need to do some hard thinking, some experimentation, perhaps construct some prototypes using different designs, run computer analysis and simulations – in short, do some real engineering, to figure out how to design some aspect of the system. High technological risk and uncertainty also bring risk to the project's schedule and budget. It's hard to estimate in advance how long it will take to figure out these uncertain aspects of the system. Project managers and engineers can and must make guesses, based on their experience with past engineering challenges of a similar perceived difficulty, but they can't be sure.

For a system with low technological risk and uncertainty, writing requirements is pretty easy. In fact, we can mostly copy the requirements from the last time we did a similar system, make a few changes and we are good to go. In fact, even if we leave out an important requirement, it is likely that this won't cause a major problem because everyone associated with designing the system is already familiar with the system and knows how it should be engineered. Leave out the requirement that the security system must have a battery backup? It is highly likely that someone will include one anyway, simply because those systems *always* include a battery backup. For systems that are already familiar to the engineers, and which need only variations and enhancements that carry low technological risk and uncertainty, requirements can easily be written since they are describing a system that is mostly well known.

When a system has areas of technological risk and uncertainty, however, requirements become much more of a challenge. It is traditional to think of requirements as independent of the design or implementation of the systems. Requirements are intended to be statements of need, not of design. It becomes difficult to write requirements, beyond the obvious and trivial, without assuming some aspects of the design of the system. Requirements provided to a government contractor for the construction of a defense system often run into the many thousands of shall-statements. It's clear that the government has something very specific in mind, but the provided requirements statements masquerade as pure statements of need, not of design and thus they fail to communicate all of the thinking of the government agency needing the system. This is an important insight, so we will illustrate it with a parable.

6.2.1 A Parable of Systems Engineering

The army of a large, powerful country sought to acquire a fleet of new light armored vehicles. Since many people had input into what the vehicle would be, and what it should do, many meetings were held and many discussions conducted. Often in these discussions, specific scenarios were discussed about how the vehicle could be used in particular missions or battle situations. There was both discussion of technical aspects and also imaginative, creative storytelling about how the vehicle might have provided an advantage, had it been available in past battle situations. The various meetings and discussions also debated some possible approaches to the vehicle, with some options emerging as preferred and others discarded for a variety of reasons based on justifications and rationale discussed at the time. Sometimes lengthy discussions resulted in single specific conclusions, such as the time the group decided that the vehicle should have redundant engine-cooling systems. It would be almost impossible to overstate the number of meetings, participants, expressed viewpoints, and debates resolved in these early conceptual discussions.

As the discussions were brought to a close, partly because they had converged on a set of decisions, and partly because there was just no time left for more discussions, the task was assigned to a group of systems engineers to finalize the requirements. They worked diligently to write a set of shall-statement requirements based on the conclusions that had been reached in the many discussions and meetings. As the list of requirements grew into the thousands, the team faced the challenge of making sure everyone's interests were represented. The weapons team, for instance, had discussed the weapons systems at great length, and the requirements team reduced these lengthy conversations into succinct shall-statement requirements.

When the set of requirements was completed, they were sent in a document to all participants in the discussions for review and approval. The weapons team, for instance, was free to review all 4000 requirements, but of course paid more attention to the weapons requirements and other teams acted similarly. Changes were sent back to the requirements team and incorporated and a final document circulated for final approval.

With the requirements work complete, the document was ready to be put to use. Through a request for proposal (RFP) and competitive bidding process, the contract to build the vehicles was awarded to a leading defense contractor. The complete requirements document was sent to the contractor as an attachment to the contract. The requirements would be used as input to the contractor's own systems engineering, design, build, and test process, and ultimately as the basis for the customer's acceptance of the completed system.

As the contractor's systems engineers began their work, they poured over the lengthy requirements document, and having read this book, knew the importance of trying to understand the system fully by developing and describing the use cases. They worked to understand the requirements and to imagine the use cases and mission scenarios the customer had in mind. For example, many of the requirements had to do with being able to fight a close-quarters battle, so the team imagined that the customer planned to use the vehicle in an urban setting. Nothing in the requirements stated this specifically, and they wouldn't have – it's too general to be put into a shall statement. The contractor's systems engineers were left to sit back and try to imagine the usages, mission scenarios, rationale for various decisions and many other important aspects, based solely on the shall-statement requirements.

The irony, and the main point of this parable, is that the original customer team, through all the discussions about the intended usages for the vehicle, had already done much of the work to describe helpful and informative use cases, needs, scenarios, and rationale, but were not able to communicate all of this valuable information to the contractor using only shall-statement requirements, leaving the contractor to attempt to re-create, probably imperfectly, the thinking of the customer based only on the requirements.

A better approach, which is already being employed in some similar situations is for the customer to begin the systems engineering work using model-based principles including the development of use cases, and providing that information to the contractor as part of the contract. Shall-statement requirements are still typically used, but ideally are carefully traced to the flows of events specified in the use cases.

The point here is not that developing requirements is bad, or that requirements are not a useful form to help describe a system. The point is that even though requirements are thought of as pure statements of need, from which designs can be derived later, in practice, they aren't. Requirements evolve along with the design of the system. Requirements written in advance of all design activities, especially in systems with some technological risk and uncertainty, are destined for change or worse, mislead teams and waste time. Let's consider a familiar example – building a home.

6.3 Building a Home: A Journey of Requirements and Design

If you wish to build a new home, perhaps you will enlist the services of a builder. Let's say this builder tells you, "You know, I'm not just a builder – I'm also a systems engineer. As a systems engineer, I know that we need to get your requirements before we do any design work on the house. So, what are your requirements?"

You respond with a kind of blank look, so the builder continues to prod. "Let's start with a very basic requirement for the house – the size. How many square feet should the house be? Oh, and let me mention that this number must be exact, and any change to it later will be very costly. As a systems engineer, I was trained that fully detailed requirements must be gathered at the beginning and then later if they need to change, we use impact analysis to determine the cost of the change."

You protest, "But I don't know the exact number of square feet in the house – how could I know that precisely now? I can make a guess, but I would hate to be held to it. Can I perhaps tell you this 'requirement' later in the process?" You, the future homeowner, have rightly noticed that even a requirement as seemingly simple as the size of the home cannot be sensibly determined this early in the process. To do so would be a guess.

Fortunately, the builder's architect is listening to this conversation and chimes in. "Could we begin by determining some of the requirements that you are certain of now, and then later proceed to deriving things like the total square footage? How many bedrooms do you think you'll need? How many bathrooms?" To these, you can make sensible responses with little uncertainty and little risk of change later.

The builder is happy with the progress, so the architect continues, "OK, and how big should these bedrooms be?" You reply, "Well, I'd love for the master to be $12' \times 21'$ and the other bedrooms to be $12' \times 14'$ but what would that do to the price of the home?"

To determine that, the builder explains, they would need to sketch out a rough floor plan, with the dimensions of the other rooms, because since this is a two-story home, the sizes of the rooms upstairs affect the dimensions of the rooms below them, and the whole house, the roof and even things like the number of windows and the amount of concrete needed for the foundation. The builder and his or her architect run off to make some preliminary designs, calculate some rough costs and bring back some options for your review. This is a sensible process but isn't the get-all-requirements-up-front-then-do-all-the-design approach that the traditional systems engineering process would seem to dictate.

The point here is that it is natural for those needing the system (the customer) and those designing the system (the builder, contractor, or designer) to work together collaboratively throughout the development process. The customer brings knowledge of his or her needs, goals, ideas, and preferences (I've always wanted a huge garage/workshop), while the designer brings knowledge of technologies, materials, standards, and good design principles (always put the stove *between* the refrigerator and the sink). The knowledge of the designer should not be underestimated. A good designer will not allow a customer's expressed desires to result in a bad design. I once saw a very high-end home built with an elaborate curved driveway and drive through garage, but no parking anywhere on the property other than that driveway – a very inconvenient arrangement if more than one person is to live in that large home.

Together the customer and designer work through the design of the home, from the general (numbers of rooms) to the specific (locations of windows) and arrive at the optimal solution, given all factors including of course, cost. For a small system, like a single-family home, where the customer and designer can sit together and discuss back and forth, a good design can be reached that meets the customer's needs. For larger and more complex systems, we'll need a different method, one that allows many people to participate in the process.

It may be tempting at this point to assume that the only method needed is to simply ask the customers to write down all of their requirements. What the customer *thinks* are the requirements for the system is of course important, but there are several reasons why assuming the customer is giving the designer the *right* requirements and *all* of the requirements, is a flawed approach and will likely result in an inefficient design process. It's a tempting assumption to make, and one that is made in many current approaches to systems engineering, but it doesn't produce a successful system. Let's examine the reasons.

First, customers (including all of us for various products and systems) are generally bad at *knowing* what they need or want. Human beings, thanks to the gift of a large prefrontal cortex, are able to think conceptually and endlessly about what we want. For those of us who have an adequate supply of food, shelter, and other basic necessities, it may be that the vast majority of our thoughts concern trying to answer the question, "What do I want?" It's not an easy question to answer in life, and it's not easy when it comes to a particular system. To illustrate, let's return to the matter of building a home. What if the prospective homeowner is a single person or couple with no children and no plans for them. Yet they want a home with large living, dining, and kitchen spaces, along with special rooms like a home office or workshop. Should they design a home with these spaces, but only two bedrooms (a master and a guest bedroom) and perhaps only two bathrooms? This may be all they need but a 3000 ft^2 home with two bathrooms is not a good design. It will be inefficient in layout (especially if multifloor), and will be very difficult to sell later, since it is what we call in engineering a "weird" design. It may not even appraise for what it will cost to build, due to market valuation. Does the couple "want" more than two bedrooms? Yes and no. They may not need to use those bedrooms as bedrooms, but they need them for the design to be good and to build wisely. An expert designer will guide them in this way, and they may end up with more bedrooms and bathrooms than they "need" but with a design that is optimal and wise, when all things are considered. The wise designer or systems engineer takes in the customer's expressed needs and wants, but then works with the customer to determine the real needs, based on the designer's expertise in that type of system.

The second reason that we can't simply ask customers to describe their needs and requirements and then take what they give us as "the" requirements for the system is that customers (all of us at times, remember?) are generally bad at *describing* our requirements and needs. The most famous saying about this inability is when Henry Ford reportedly said, "If I asked people what they wanted, they would have said a faster horse," and I'll add "with less emissions." There is doubt whether Ford ever actually said this, but we can appreciate the sentiment. People (us) will always have a hard time imagining that of which we have no experience. It's very hard to describe your need for a car, a smart phone, a flat screen TV, social media app, or even the ideal romantic partner if you've never experienced one.

Let's consider an example even closer to home. My guess is that since you're reading this book, there is a good chance you have a technical background, and if so, you are probably a regular consultant to your friends and family about technology products. When your friend Emily asks you to help her choose a laptop computer, it's natural to ask how she wants to use it. "Well," says Emily, "I need to do email, web surfing, social media, look at cat videos, etc." This is not very helpful, since all laptop computers will do those things. As a good friend, you will start a dialog with Emily, asking her questions you know are important to ask but which she wouldn't know to answer without your prompting. She doesn't know, for instance, the trade-offs of screen size, weight, resolution, battery life, and price that drive the laptop choices – if she did, she wouldn't be relying on you. She doesn't know whether she needs a faster CPU, more memory, or a higher screen resolution, nor how much each of these will cost, nor how much each would increase the enjoyment and utility of the laptop, if at all. As you may have noticed, this is a hard conversation to have with Emily, or anyone, and many of the questions may be frustrating to her since she doesn't know how to answer them. She doesn't know how to describe her intended use of the laptop in such a way that you get the information you need to make the right recommendations. You don't know her needs, preferences, constraints, and budget. So, you have a dialog, which my colleague in Quality Management, Larry Kennedy refers to as a "requirements interrogation" suggesting that a fair amount of assertive, even uncomfortable, probing may be needed to get the information you need to make a quality recommendation. Once again, if only two people are involved, then a dialog will do. With more complex systems, the number of people involved on both the customer and designer sides increase, and we need to move beyond a simple dialog.

What we need is a way to capture all of the best thinking about *what* we want the system to do, while avoiding the temptation to describe how the system will do it or the actual implementation. We'll describe this much more in the next chapter, but here let's begin by looking at the difficulties inherent in a traditional requirements elicitation process.

6.4 Where Requirements Come From and a Koan

In a traditional requirements elicitation process, those who want a new system, a new defense system for instance, as the stakeholders (including the purchaser, users, strategists, tacticians, engineers, and many others) to write down a large set of statements stating facts about the desired system. Stakeholders can't actually describe the system they want. To do so would require that a system already exists, either in reality or in someone's imagination, to which the stakeholders can refer.

In some cases, this is true – if someone is describing a new car, it would be natural to refer to cars that already exist. It's much less true is the system is something new.

In writing down requirements, stakeholders are free to write in any format, but in recent years have adopted the shall-statement requirement as a preferred format. Requirements are meant to be statements to which the system must conform and are used not only to guide the development of the system but also to test and finally judge the system as to its acceptability. Requirements, it is said, should capture the needed functionality and performance characteristics of the system, but should not include the specification of a particular design or implementation of the system.

Here we must face two common problems in developing stakeholder requirements, and these difficulties apply whether the requirements are being developed by the customer or by development team, or a combination of both. In the defense systems world, these two groups are quite separate, with the customer being say a military agency and the developer being a government defense contracting company. In the auto industry, the customer may be represented by a marketing or product management organization, and the developer is an engineering and manufacturing organization within the same company.

The first difficulty in developing requirements is, how do we know when we have written down all the requirements, that is, what constitutes a complete set of requirements? This is no minor question as missing requirements are very often cited as the source of significant problems later in the development process or even of overall project failure. Numerous reports from the US government and other sources in a variety of industries list poor requirements management as one of main causes of project failure. It's easy to see why it is tempting to blame the requirements. Whenever a project or a system design failure occurs, one can usually imagine a requirement that if it has been recognized, written down, and understood earlier in the project, could have prevented the failure.

Say an automated car drives into an obstacle. See? There should have been a requirement that the car not collide with such obstacles. But wait, negative requirements are usually disallowed because they are hard to verify and test. OK, we can write the requirement that the car must detect and avoid all obstacles. Still no good – too general. It's impossible to test the system's performance with *all* obstacles. We only need the requirement that the system detect and avoid the particular kind of obstacle that caused the collision, but the trap is obvious. It is impossible to write requirements statements to cover all possible situations and usages that might arise, including the ones we would never have expected. The situation is worse than that, though.

If a requirement is a statement that must be true of the system, it seems impossible even to write all possible true statements about any system, no matter whether the requirements writer is the customer or the developer of the system.

While writing requirements for a car, for instance, should we write a requirement for the minimum turning radius (the minimum space required for a U-turn without backing up)? Certainly, we wrote a requirement that the car must be able to steer (or did we – perhaps we just assumed everyone would know that), but does the customer actually have a requirement for a turning radius? Offhand, I don't even know the turning radius of the vehicles I have owned, so I'd have a hard time specifying a reasonable turning radius, and I sure don't want to specify something that will dramatically increase the price of the car. At the same time, I don't want the car to have the turning radius of a Greyhound bus. I just want a "normal" turning radius. I can't write something as vague and impossible-to-verify as "normal" in a requirements statement, so perhaps it's best I not specify a turning radius at all. Then later if the car turns out to be unacceptable to customers because of the large turning radius or overly expensive because of the engineering required for a very small turning radius, at least my ill-conceived requirement won't be to blame. We arrived at a perverse incentive to avoid specifying certain kinds of requirements.

What happens in practice it seems is that people, whether the system customer or system developer, write the requirements that are important to their interests and leave other requirements unstated for one of three reasons:

1. They don't know how to express the requirement correctly
2. They aren't aware of the need for such a requirement
3. That aspect of the system is not of interest to them

In an ideal process, how many requirements should we write and how detailed and specific should they be? If the "requirement for the requirements" is *not* that requirements should contain all true statements about the system (of which there are an unlimited number), then what constitutes a complete requirements set? There is no obvious answer.

A subtle but perhaps more costly difficulty in requirements development is that it is difficult or impossible to write requirements about aspects of the system that are truly new and innovative, or about which feasibility is not yet known. If I'm writing requirements for a conversational home assistant, can I write a requirement that the system should be able to understand verbal commands from any room in the house from a single microphone location? I don't know at this point if that is even possible with current microphone technology. What I really want to write, but which requirements-thinking prevents me from writing, is something like, "experiment with various microphone and voice processing methods, and then choose a reasonable compromise between performance and cost."

The purpose of a requirements development process is not to make an arbitrary number of true statements about the system, but rather to give clear direction to the designers of the system, so that they can proceed effectively with the design and implementation of the system. What is the real difference between requirements and design, though? A statement like, "The system shall be able to travel at highway speeds on the US Interstate system," is phrased as a requirement, and it may indeed *be* a requirement, meaning that the system must conform to the statement. But is the statement not also a statement of design? One might argue that the statement constrains the design to a rolling vehicle with wheels, when perhaps the designers might envision something that flies or hops. Perhaps the requirement, they imply, should be more like, "The system shall be able to transport 1–5 people from one place to another." Well, if we are an automotive engineering company, we need not concern ourselves with flying, hopping, or matter-changing transporters. We build cars that travel on highways (at least for now). Requirements often must dictate some aspects of design – there is no hard line between requirements and design. What is important is to be *aware* of the lines being drawn. When a requirement constrains the design, the team should consider if this is appropriate, and if so, it's fine to include it as a requirement. Requirements that constrain the design of a system are often referred to as *design constraints*.

A common explanation of the difference between requirements and design is that requirements express *what* the system needs to do and design expresses *how* the system will do it. In a large system, however, these are not unambiguous. If I'm at the systems level for say a military ground vehicle, the *what* to me could be, the system must have the capability to attack a specific kind of target at a specified range. The *how* from my perspective is that the vehicle will carry an appropriate vehicle-mounted gun. To the lead systems engineer for the vehicle weapons subsystem, at the next level of detail, the *what* is that the vehicle needs a certain class of gun, and the *how* is the design and specification of that gun. Now to the mechanical subsystem engineer, at the next level down, the *what* is the need to carry, turn, aim, and stow that specific gun, and the *how* is the specific design, materials, and parts to accomplish that. One engineer's floor (their *how*) becomes the next engineer's ceiling (the *what*).

Returning to the matter of requirements and design constraints, let's consider a software system example. What if a requirement were to state, "The system shall store the model data in a relational database"? Is it appropriate to constrain the design to a relational database or should we allow for the possibilities of using a graph database, or even something other than a database to store the model data? It's a decision worth consideration. In some situations, the database to be used is already identified and decided, and so it can be included as a requirement. If not, then such a decision can be left to the engineer/designer at the next level down.

We have come to the point where we can see that requirements statements, by themselves, are not adequate to fully describe the system in way that engineers at all levels can work together to arrive at a design that meets all stakeholder needs.

The main reason is that, at each level of detail, each *level of abstraction*, decisions are made without full knowledge of their impact. This is not a criticism – it is inevitable. It is like the homeowner being required to state the square footage of the home before any design is done – it's risky and uncertain.

6.4.1 A Requirements Koan

Zen masters often use koans to aid their students in becoming enlightened. A koan is a sort of riddle, but one that does not have a specific answer. The student's task is to contemplate the koan rather than to simply provide an answer. The student's response to a koan provided by the master demonstrates the level of enlightenment attained at that point. Famous Zen koans include, "what is the sound of one hand clapping?" and "If a tree falls in the wilderness, and no one is there to hear it, does it make a sound?" Here, we offer a Zen koan for requirements: "When is a requirement a requirement?" The question suggests that what are stated as requirements in a document may not be the actual requirements. As we've described, some "requirements" may be statements about the system that are not essential, but simply descriptive. If a requirement states that a vehicle must have four wheels, the statement is probably not intending to require that the vehicle have four wheels but is describing familiar vehicles. The vehicle could meet the requirements by having three wheels, or even no wheels. It is true that the vehicle needs some means for traveling on existing roads but having four wheels is a "requirement that is not a requirement." After wrestling with the koan for a while, the Zen requirements student may arrive at an understanding that a requirement is a requirement when it is a requirement.

To move beyond the limitations of using shall-statement requirements on their own, we need a way to describe what the system needs to do for the stakeholders and how it is to do it, with sufficient detail, but without describing a particular design. We need to work out enough detail about the system so that designers have a complete concept of it and can perform detailed design with full confidence that what they are designing will meet stakeholder needs. Use cases give us a way to do just that.

6.5 The Magic of Use Cases

In 1999, I joined a worldwide consulting company called iXL, one of several large firms devoted to creating something new for large corporations – websites. No one knew at the time what the Web would become (perhaps we still don't know), but everyone knew a corporation simply had to have a website. So iXL and a few other similar companies grew quickly by acquiring small Web design firms and consolidating them into a consulting and Web development organization with at the time, over 3000 employees worldwide. I joined as a project manager, knowing little about developing websites, but something about managing software development projects.

My first weeks at the company were planned to be a nice smooth integration, with me joining just before Christmas, working a bit, taking the holidays off and then easing into some real work in the New Year (that is of course if Y2K didn't halt civilization as we knew it). Things didn't work out as expected. Y2K turned out to be a paper tiger, but just after I left for Christmas vacation, I was told we had a consulting project that was "on fire" in London, meaning the client is not pleased with the progress on the project and is about to throw us out, resulting in unfortunate financial consequences for the firm I had just joined. I was to fly over immediately after the holidays, to try to put out the fire and get the Web development project on track again.

For a few years prior to 1999, I had been studying the work of Ivar Jacobson, one of the "three amigos" (along with Grady Booch and James Rumbaugh) who developed what came to be known as the "Rational Unified Process." A key part of this process, contributed by Jacobson, is the concept of the "use case" (Jacobson 1993, Jacobson et al. 1994). A use case, wrote Jacobson, is a sequence of events performed with a system that yields a result of value. He goes on to state that taken together, the uses cases describe all of the functionality of the system. I remember my surprise and then enlightenment when I read that. I had always held the mental image that complex software-driven systems had lots of functions and capabilities and that certain ways of using the system could be represented in various kinds of flow charts, but I had never seen a compact way to represent *all* of the functionality of a system. Since use cases tell the stories of system usage, a complete set of use cases describes all the functionality of the system, and even better, organizes that functionality in the way that is most important to users – how they use the system to accomplish their goals. There is simply no functionality in the system that is not part of one or more use cases.

Working with the customers in London, I used the concept of use cases to capture the needed functionality and behavior of the new website over the next week or two, set the developers to work designing and building, and the project was saved. As part of my reward of course, I was sent next to a larger "on fire" project in the United States, and I used exactly the same use case methods to get that project moving again.

From there, an even larger problem project, this one worth over $5 million to my company, was put in my lap to fix. In this case, there had been months of discussions and working sessions between iXL and the customer and many documents had been produced and reviewed, but there was no clear alignment on how the new website should work nor on what it should do specifically for various types of users. There was no way for the website software developers to get to work on a design since they didn't have any clear concept of how the system should work. The diverse collection of requirements-like statements scattered throughout the documents did not coalesce into a clear picture of the needed system.

Once again, I employed my now favorite go-to method: use cases. Printing out all of the documents produced so far in the project (a stack over a foot high), I literally set them aside, proclaiming them as background for what were about to do. In a workshop process that I'll describe in some detail shortly, we were able to get clear direction from the customer team, including everyone from the administrative assistant to the VP, merge it with the background knowledge of our consulting team, and produce a joint work product containing a set of use cases on which everyone could agree. Agreement was easy since everyone had participated fully in the creation of the use cases.

For over 20 years now, use cases have been my superpower – the one tool that always seems to cut through the confusion in an engineering or development project, focus everyone clearly on what the system needs to do, and provide clear input to the designers and developers of the system. Use cases apply equally well to systems that consist only of software, like websites or business applications, and also to engineered systems like cars, satellites, spacecraft, defense systems, and electronic devices of all sizes. Use cases are simple and easy to read and understand so everyone involved in the project can be familiar with them. Use cases are in essence stories, and everyone loves a story, so they are much more fun to read than a collection of shall-statement requirements. As stories, they are more expressive, and evoke more of a total sense of how a system works for its users. The purpose of the system in the first place is to do things for its users, and use cases describe those things in clear, specific, story-like terms. As we've described in Chapter 5, stories have a strange, almost magical power for human beings, and use cases tap into that.

For thousands of years, people have told each other stories, both factual and fictional, to communicate complex ideas about products and systems. Long ago, stories were communicated by oral tradition, then later in writing through books, and now through movies, TV series, books, and books made into movies. Take for example, a robot. How do you know what you know about robots? For me, it was stories like "I, Robot" by Isaac Asimov, the robot from the TV series Lost in Space and then later R2D2 and C3PO from the Star Wars movies. We learn what robots do from these stories. We saw robots interact with humans, working with them or against them, supporting their human creators' goals, or rebelling, making their own decisions and even dominating, enslaving, or killing humans. It is these stories that give us the raw material for the current debates about artificial intelligence, and its possibilities, risks, and dangers. We have always experienced the world against the background of our own stories.

We use stories in engineering, when we describe to users how they should use our new product. When Apple introduced the iPhone in 2007, it bucked the trend of describing digital products in terms of processor speed, memory, storage capacity, and screen pixels. Instead, they told stories. Apple TV commercials told a single story, or as many as 10 mini-stories in a 30-second spot – "there's an app for that" is really, "there's a story for that!" Who can forget the early iPhone commercial showing a person standing on a street corner, thinking about pizza, and using an iPhone to find a restaurant, check the reviews, and get exact walking directions?

Before moving on to show how to create and make use of use cases in the systems development process, let's consider why we are focusing so much on use cases in a book about designing intelligent systems. Use cases, as we have said, are about how the system is used by its users, not about how it works inside, so they keep the team focused, at the outset, on what the users need and want, not on the technology to be used in creating the system. It's a real danger – technologists tend to focus on their preferred (or newly discovered) technology, and on finding ways to apply it. A newly minted practitioner of AI machine learning algorithms can often be found figuratively or literally wandering the hallways of his or her company, looking for collections of unexploited data to which to apply the technology – a solution in search of a problem.

In the quest to make systems more intelligent, it may seem natural to focus on "intelligent" technologies, such as artificial intelligence and machine learning, and try to find ways to include them in new system designs or in updates to existing systems. As we have described in Chapter 1 however, making a system more intelligent, means making the system act more intelligently for, and with, its users. The focus should be on the users and how they use the system, not the inclusion of a particular technology. Use cases give us precisely this focus and enable us to describe, in detail, how the system will work for its users, while explicitly prohibiting description of the underlying technology. Intelligence in systems is in the eye of the beholder, and what matters is how the system works (to repeat) *for its users.* Designing an intelligent system starts with doing the hard work to figure out how the system could be the most intelligent and capturing that in the use cases – describing how the system should work for the users, but not including the technology used to implement it. Of course, it helps to have some sense of what technology is capable of, so we don't simply describe systems with matter transporters and anti-matter propulsion systems.

How much about the available technology do we need to know to write good use cases that are feasible and implementable? Sometimes, quite a bit but either extreme of knowing too little or too much about the technology can cause difficulties in the development of use cases. If the team developing the use cases knows too little about the available technology, they risk describing ways of using the system that are impractical and can't be implemented with current tech, or could miss opportunities to exploit the technologies of which they are unaware. On the other extreme, if the team includes technical specialists who know the technology well, it's possible for useful discussion about capabilities, needed system behavior and use cases, to be overshadowed by discussions of technological capabilities.

What's needed is a balance. Some amount of technological expertise and guidance applied to the development of use cases in the early stages can be helpful to keep use cases practical, but often it's best to keep the technologists out of the use case development process and let them review first drafts after they've been created. This way, the technologists avoid frustration with the lengthy and often chaotic discussions by users about how they want the system to work but can still point out technical infeasibilities early enough in the process to allow for easy correction.

6.6　The Essence of a Use Case

A Use Case is, literally, a case of usage of a system. It is a very simple concept, and yet one that can trip people up, especially engineers who know a great deal about how the system works inside. Use cases are about *using* the system, not know what it's doing inside. Use cases answer the question, "What does this system do for its users?" which is probably the same question as, "Why did the users acquire this system in the first place?" Users and customers buy systems to use them, not to admire their design or intricate functionality. In fact, haven't we all seen systems that are so beautifully complex in design that they are difficult to use effectively? Use cases focus on what's most important to the most important person in the systems development process – the customer. With complex technological systems, the customer may not understand the design, the internal workings, or the esoteric materials, but every customer and user should understand the desired usages that led them to acquire the system in the first place.

Use cases are stories of the usage of a system. There are as many use cases as there are distinct ways to use a system. In Chapter 5, we explored the power of stories, and with use cases, we bring that power to conceptualizing a new system in a concrete way, by describing the system in a set of stories, each one telling how a system is used to accomplish a goal for its users. Let's begin our exploration of use cases with an example.

The owner of a car probably purchased it with the idea of using it, unless the intent is to add it to a collection or donate it to a museum. A typical usage of a car is to commute to work, and we can imagine a story being told of how the driver of the car begins the commute by approaching the car, then interacting with the car in variety of ways including unlocking it, starting it, setting it to drive, driving the car to work, parking, and then later reversing the process to return home. The story ends with the driver parking and securing the car at home. Describing this story may seem like a useless exercise, but that's because we are already familiar with cars, and how they interact with drivers. In systems engineering, we are normally working on a system that is new, perhaps even novel or unique, so telling the stories of system usage is a surprisingly useful way to begin to reason about how the system should work for its users, long before we proceed to designing the system.

While a use case is the story of a single usage of the system, the story may have some variations. Perhaps on some days, the commuter stops at a fueling or charging station on the way to work or on the way home or stops at the gym (or a gym with a charging station – very efficient). These variations can be shown in the use case. It's still the same use case and still achieves the same goal, but with some variation in how the story unfolds. Of course, things can also go wrong during a use case. What if the driver runs out of fuel along the way? That possibility can be described in the use case as well – perhaps the car should warn the driver when the fuel level gets low or assist the driver in calling for an emergency fuel (or charging) delivery.

Use cases are independent of each other; one use case cannot depend on another. Use cases stand alone as stories of the usage of a system. They are not like chapters in a book, where one chapter leads to another. They are more like entries in an encyclopedia, each story complete in itself. Any use case can be performed with a system at any time, providing that the system is in a state that makes that use case possible. We call this a *precondition*, and it represents the state a system must be in for the use case to start. For example, in order to perform the use case *close garage door*, the garage door must be in the *open* or *stopped* state (if the door is moving, it can't be closed, only stopped). On a shopping website, the use case *checkout* can only be performed if there are items in the shopping cart. Apart from preconditions like these, use cases can be performed at any time and in any order.

Later in this chapter, we'll look at some good practices for writing use cases, but most of the benefits of using use cases can be produced by simply writing the narrative story of each use case in plain, unstructured text. Use cases are intended to be read and understood by everyone involved in the project, including the project sponsor (who may know little about the details of the system), the engineers creating the system, and the users. The best language for writing a use case is plain English (or another natural language). Jargon and technical terms should be included only if they are sure to be understood by everyone who will read the use case, and as we said, that's *everyone*. Think of each use case as a story that starts when a user takes an action with the system to accomplish some goal. After that first action, the story proceeds with the system doing things and the user doing things in a back-and-forth kind of dialog, including interactions with other systems. The dialog is not one for one – the system may do a few things, then the user does something, then an outside system does a couple of things, then the user does a thing or two, and so on, until the use case is finished, and the user's goal achieved.

If the use case is *withdraw cash from an ATM*, the use case would start with the user inserting an ATM card. Then, the ATM asks for the type of transaction, the user indicates cash withdrawal, the ATM processes the transaction with the user's home bank, ejects the user's card, then dispenses the cash and receipt, and the use case ends. Even this brief description of the use case is useful to reason about the basic design of the ATM. Should the ATM ask for the card first, or ask for the transaction type first, then ask for the card? Should the ATM eject the user's card and make the user remove it before providing the cash? These and other fundamental design choices can be considered as the use case is written and reviewed, long before the machine is designed, or any software written.

Use cases are for everyone involved in the systems development process and can be used in a number of ways throughout the development lifecycle:

- Customers utilize use cases to describe what they want the system to do for them.
- Project and program leaders utilize use cases to come to agreement and understanding between the customer and development team about what the system is to do.
- Business analysts and systems engineers utilize use cases to describe how the system is to function.
- Business analysts and systems engineers utilize use cases to describe how the system is to function.
- Software designers utilize use cases to design the system.
- Subsystem engineers utilize system use cases to understand their subsystem's role in the overall system operation.
- Team leaders and scrum masters utilize use cases to divide up work and implement the system iteratively and incrementally.
- System test and verification engineers base their test cases on use cases.
- Documentation writers utilize use cases as an outline for their user guides.
- Training course designers utilize use cases as the content outline for their courses.

Use cases are *complete* usages of the system. This is an important concept to understand and one that causes many practitioners of use case modeling to end up in trouble. Use cases are not just functions, or operations or various bits of behavior of a system. Use cases are complete usages of the system. Think of it as a user approaches a system, starts using it for some purpose, then finishes that usage and stops and goes away. The user has used the system for a specific purpose, has presumably achieved some important goal and then can either stop using the system, or begin using the system again to achieve another goal. Use cases often directly reflect the goals of the user.

If the concept of use cases as complete system usages is not understood, it is easy to end up with a lot of "small" use cases, which are not really use cases at all, and to lose the picture we are trying to create of what the system does for the user. For example, when trying to list the use cases of an automatic teller machine (ATM), it might be tempting to think of "Enter PIN" as a use case, since it is certainly something the user does when using the system. Entering the PIN is not however, a *complete* usage of the system. The user's goal is not to enter the PIN, the user did not come to the ATM in order to enter a PIN, and entering the PIN, on its own, does not accomplish anything meaningful for the user.

6.7   Use Case vs. Functions: A Parable

As a parable to illustrate why it's so valuable to identify use cases in designing a system, let's imagine the design of a GPS navigation system. If you remember, in the old days before smartphones included good navigation apps, many people had standalone GPS navigation devices in their cars, and some professional drives and others still find them useful. GPS navigation systems are small, disconnected devices, about the size of a small can of soup. People called them GPSs, but the Global Positioning System (GPS) is actually the set of satellites above the earth that provide the signals received by navigation devices. All these little "GPS" devices did was give navigation guidance – they were truly one-trick ponies. Imagine in our example that we are going to design a new GPS navigation product, a standalone product, meant to attach to a car's dashboard with a suction cup. In order to design a device like that, there are two fundamental ways we could go about conceptualizing the design.

One way we could proceed is to try to imagine all of the different *functions* that that device would need to carry out. We would list functions such as identify destination, find nearest business, avoid highways, plot route to destination, and reset route after deviation. We imagine all these little disconnected functions, and then take that list of functions and give it to the software and hardware people and have them build a system that performs all the functions. This approach would be successful in that it would produce a usable product. The product would do all the things that we want it to do, but there is a big potential disadvantage to this approach. We don't know how that product is really going to work in the user's hands, or the sequence of actions a user would perform. We may have a general idea, but we haven't thought through the actual flow of a user using the product. It might be that the product ends up with a button on the control panel for every one of these little functions, so to navigate somewhere, the user has to know how to select each function in a specific order, which could be very confusing for a user. We've probably all seen products that were designed exactly this way.

They were designed with individual functions in mind and the user had to puzzle out how to put those functions together into some useful sequence of actions that produces the result they want. While the function-based design approach does work, it has some disadvantages if our goal is to produce a product that works smoothly and intuitively for users – one that meets their needs, wants, and expectations.

The other approach is to start out by identifying and describing the intended usages of the GPS system – the use cases. Instead of asking what functions the system should be able to perform, we ask what usages it should make possible for users. For our GPS navigation system, we would answer the question in much the same way we would for the military vehicle we'll consider in Chapter 7. We think about how the system would be used, and what goals it would accomplish. The first use case that comes to mind for a GPS navigation system, of course, is *navigate to a destination*, and it's pretty easy to imagine. A driver is sitting in a particular location, and the system knows that because it's a GPS system after all, and the person requests the system navigate to a destination. The system responds by asking for the destination, and the person enters the address, and so on. As described in Section 6.10, we describe the sequence of events for that use case in a step-by-step story style. During this use case, many things could happen that change the story a bit. If the driver gets off course, or a road is blocked, the GPS must put the vehicle back on course. All these variations are part of that one complete usage.

The designers and builders of the GPS system would work from that use case and would design the product to work smoothly for that particular use case. Of course, there would be other use cases for the GPS, such as find the nearest McDonald's? Some current GPS systems will do that – it's a useful use case finding the nearest of a certain kind of business. Another use case might include finding a route to my destination that avoids highways. A more unusual case might be to find a Hilton hotel that I can reach in about five hours. To set the scene, imagine I'm on a long road trip and I want to stop for the night in about five hours, and sleep at a Hilton Hotel. It's a logical need, though not an everyday one, and it's certainly something one can expect a GPS navigation product to be able to do. To my knowledge, there are no GPS navigation systems or apps in the world today that will do that. To me, that's very interesting – why do no products include this ability? Some of them will find a Hilton Hotel along my chosen route, but they only look out perhaps 10 mi or so – but I want to stop in five hours. Many GPS navigation systems can show me the Hilton hotels in a particular city or area, but showing hotels near Dallas, which I'll reach in about five hours, doesn't mean those hotels are anywhere near my route. What I really want to ask the GPS system to do is to find me a Hilton on my route that I can reach in about five hours, and no products do that today. Why not?

Nobody has ever thought of that use case and used it as part of the input to the design of that product. Thinking about use cases is a very powerful way to design functional and easy-to-use products. What company tends to design its products this way, so that if you use the products the way they intend, they work very smoothly? You can't deviate too much from the way they intend. But if you use them the way they intend, they work very smoothly. Of course Apple is the one that comes to mind there. In fact, the early ads and television ads for the iPhone from Apple were not about specifications, such as how much memory or how fast the processor was, or even the resolution of the screen. Instead, the ads showed a person standing on a street corner, looking at the screen and thinking, that he wants pizza, and he puts *pizza* into this little screen in his hand. The screen shows the nearby pizza places, along with ratings, menus, prices, locations, and directions. The entire ad was a use case – the *find pizza nearby* use case. So it's very powerful and compelling to think from use cases, and use them to communicate clearly what a new product or system will do. In our example of the GPS navigation system, identifying all the possible use cases of the product is a superior approach than listing all the individual functions.

We can also see that there is a relationship between the functions we listed in the first approach, and the use cases that we listed in the second. It's very likely that the system will need all those functions in order to carry out the use cases. In a model based systems engineering approach, we identify the use cases first, and then figure out the functions that the system needs to be able to perform to accomplish those use cases. That's the heart of the process of MBSE. This will be explained more when we get to the R in CUSTARD: Realization in Chapter 7.

6.8 Identifying Actors

Before going into detail on how to identify and describe use cases for a system, we will look at how to identify and describe a system's users using the concept of Actors. We'll look at actors in even more depth in Chapter 7 when we describe the entire model based engineering process.

Describing a system from the user's or acquirer's perspective is the most important perspective, since the users and acquirers are the system's reason for being. We must first identify the system we are trying to describe. Of course, it's simple enough to name the system, but identifying it fully means describing the system's boundary, that is, where the system ends and the rest of the world begins. The most straightforward and simple way to do this is to identify our system's *actors*. An actor is someone or something outside the system that interacts with the system. An actor performs some interaction with the system – think of them as "inter-actors."

For any system of any kind, there are two main categories of actors:

- Human users (people who use the system)
- Other systems that interact with our system

The most obvious actors are the people who will use the system, once it's built and deployed, to achieve some objective. An automatic bank teller machine is used by a bank customer. A shopping website is used by a shopper. A Wi-Fi-enabled light bulb is used by the resident of a home. A missile system is used by a military officer. But there are always other less obvious actors. The teller machine must communicate with other banking systems. The shopping website will be used by order processing staff and will connect to a credit card processing system. The light bulb must connect to a Wi-Fi access point in the home. The missile system is used by command personnel, and must connect to GPS satellites and battlefield sensors.

To correctly identify actors, we temporarily consider the system-of-interest to be at the center of the universe, looking out at everything else. Outside the system (surrounding it, to continue the visual metaphor) are things that touch, interact with, use, or communicate with the system. Those things can be people or other systems. It's a deceptively simple notion: every system divides the world into three categories – the system itself, the things that interact with it, and everything else. The actors are the things outside the system that interact with the system. To further develop the idea of actors, we'll explore these notions of *outside the system* and *interact with the system* in more depth.

6.8.1 Actors Are Outside the System

The key here is that actors must be outside the system. Some examples will make this clear. Returning to the example of a car, who are the actors for a car? First, we consider the human users of the car which are the driver and the passengers. The driver and passengers should be represented as different actors, since they interact with the car differently; however, we need only one actor for a passenger, since in most cases, we can consider that all passengers other than the driver interact with the car in the same way. Now if the front seat passenger and the back seat passengers will interact with the car in different ways – say, for example if the car has a back seat entertainment system – then perhaps we should have an actor for front seat passenger and another for rear seat passenger.

6.8.2 Actors Interact with the System

Interaction is usually easy to identify. A human actor may press a button on the system or a nonhuman actor may send a message over a wireless data connection to the system. But is everything that touches a system, an actor? A car touches the road on which it drives, but is the road an actor with respect to the car – is there an interaction? The decision of whether to consider the road an actor depends on whether we are building functionality into the system to interact with the road. Certainly at a molecular level, the tires are interacting with the road surface, but from the perspective of system functionality the car doesn't interact with the road the way it does with say, the GPS satellites or emergency call center. To shift the example though, consider rain-sensing windshield wipers. There is certainly an interaction with the environment, specifically the rain, that starts the wipers, so it would make sense to show the rain as an actor.

6.8.3 Actors Represent Roles

Actors aren't actually representing human users themselves; actors represent the *roles* played by those humans. We don't show an actor called Emily, the owner of the car, but instead show an actor that represents the driver (who could be Emily) and another that represents a passenger (who could also be Emily if someone else were driving that day). The idea of an actor being a role that a user plays with respect to the system helps reduce the number of actors required to fully represent how a system works for its users. My friend Gene is a driver, passenger, and mechanic for his cars at various times, but we don't need a special actor for Gene or his special combination of roles – we just need actors to represent the roles individually – driver, passenger, and mechanic. It doesn't matter that some people will play one, two, or all three of the roles at different times. Consider a university where there are professors and students, and some people are both. We need only actors to represent the roles of professor and student, even though various people will play the roles individually or in combination throughout the life of the system. When a person is acting in the role of student, it doesn't matter if that same person also sometimes acts in the role of professor – roles are independent.

Identifying actors for a system is no mere formality nor is it an exercise in documentation of the obvious. It is in identifying actors that we start to conceptualize for whom the system is intended, who will use it, and with whom it will interact. Later, we'll specify *how* it will interact with these actors, but for now, the question is *who*. Since actors are either human users or nonhuman systems, we'll consider each in some depth.

6.8.4 Finding the Real Actors

It has been amazing to me over my years of practicing use case modeling, how little thinking goes into correctly identifying actors and how obvious this lack of thinking can be when the final system is put into actual use. An example that comes to mind is a home theater surrounded with sound system. Such systems can be professionally designed and installed, but many are do-it-yourself variety, which consists of a TV, audio/video receiver, speakers, and source components such as streaming devices and DVD/Blu-ray disc players. The enthusiast relishes the task of choosing each component, connecting them all together and making the system produce a wonderful home cinematic experience. To others in the household, the system looks like a lot of boxes, buttons, and confusing remote controls. Frustrated cries of, "I just want to watch TV!" can be heard when the enthusiast is away from home. If the designers considered the nonenthusiast user as a valid actor – let's call it the "casual user" actor – then different design ideas might emerge. Perhaps the system would have a second, simpler remote control, that allowed switching between the main options like "TV," "Netflix," and "Play Disc" defaulting the more detailed settings to those set up previously by the enthusiast. The casual user and the enthusiast are different actors because they have different needs and use the system differently.

At first glance, a list of the actors for a system used by a family would include mom, dad, son, daughter, guest, etc., but this list doesn't work to describe the different roles played by users of the system, and so it isn't helpful in designing the system. Mom, if she's so inclined may be more of an enthusiast user and Dad might be the casual user, or they both might be enthusiasts. A son could be an infant, and not a user at all, or a son could be an enthusiast or casual user – same for a daughter. Just because son and daughter are different people, they aren't necessarily represented by different actors. What we are after is to identify all the different *kinds* of users of the system, where *kind* means a different role that someone plays with respect to the system. Each role represents different *ways* of using the system.

Appropriately identifying actors is vitally important to getting the system right. Miss a type of actor or unthinkingly conflate two different actors into one, and the system could be great for some users and terrible for others. Enthusiasts will be frustrated by the simplistic controls and limited options of a system designed only with casual users in mind, and casual users are easily confused and frustrated with complex options that delight enthusiasts.

The fields of *design thinking* and targeted marketing use a very similar notion – profiles. When designing a new business process, marketing initiative, or product launch, designers identify some fictional but realistic customers or users that represent types or classes of users, giving each a name, background description and even a fictional photo. As design proceeds, designers ask themselves, "will this work for Sam (the 35-year-old technology enthusiast) and also for Marty (the 22-year-old recent business college graduate)?" Designers and marketers call these profiles, and they are a great way to personalize the team's thinking about the people for which the system or product is intended. Profiles are described the same way one would describe a real person – they have a name, photo, family background, education, interests, hobbies, and a profession. The idea is to make each profile as realistic and concrete as possible. The team ends up talking about the Marty profile, just as if Marty were a real person: what would Marty expect from this product? how would Marty like it to work? what would make Marty want to buy this product? what would make it better for Marty? The team checks their new product idea or design against all the fictitious profile-people to make sure they don't design something that only works for some types of users. Identifying human actors for a system takes a similar approach by identifying the various roles that users of the system will play.

Identifying human actors is a matter of asking some simple questions. If the work is being done in a workshop setting with a group, the facilitator can actually ask these questions and collect answers, drawing them on a flip chart with simple stick figures. In Chapter 7, we'll describe how these stick figures will become the first part of a model based systems engineering approach.

Here are some of the questions to ask to help identify Actors, as listed in the Rational Unified Process:

- Which user groups require help from the system to perform their tasks?
- Which user groups are needed to execute the system's most obvious main functions?
- Which user groups are required to perform secondary functions, such as system maintenance and administration?
- Will the system interact with any external hardware or software system?

Some actors are obvious and come to mind quickly when considering the system. Others are not so quick to recognize. Considering actors drive us to consider or reconsider the boundary of the system – a big part of the value of doing the work to find the right set of actors for the system being considered. A centralized defense communications system called GrandComm, for example, would seem to interact with a soldier in the field, so the soldier should be noted as an actor, right? What if, however, the centralized system really interacts with a smaller communications system, called SolComm, carried in the soldier's backpack, and the soldier interacts only with SolComm? In identifying actors for GrandComm, we might choose to include only SolComm as an actor, but not the soldier. Even though the solider is an important stakeholder for the system, and one whose input we should include in our requirements and design process, that soldier is not technically an actor, because he or she does not interact directly with GrandComm.

There is a judgment call to make here. If we change the example a bit and replace SolComm with the everyday wireless cellular phone network, then

GrandComm interacts with the soldier using his or her smart phone. In this case, it probably makes more sense to ignore the cellular network, and just include the soldier as an actor. There is no right or wrong decision here, but how actors are chosen determines what system interfaces are exposed. In the example here, if the soldier is chosen as the actor, then the use cases will describe interaction between the GrandComm system and the soldier; if cellular phone network is chosen as the actor, then use cases will describe the interaction between GrandComm and the cellular network. There is, of course, also interaction between the cellular network and the soldier, but such interaction won't be described in the use cases of the GrandComm system, since they do not involve interaction with GrandComm. In this case, it almost certainly makes more sense to call the soldier an actor and ignore the cellular network, since it is the interaction between the soldier and GrandComm that is important.

In the previous example, where GrandComm interacts with SolComm, it probably makes more sense to call SolComm the actor, since, as the designers of GrandComm, we need to focus more on how our system, GrandComm, interacts with SolComm. It is the designers of SolComm that must focus on the interaction between SolComm and the soldier. If we are designing both GrandComm and SolComm it might make more sense to consider them as one system and define our use cases from that vantage point.

6.8.5   Identifying Nonhuman Actors

We've been mentioning nonhuman actors in the examples given so far, but let's take a closer look. Most systems communicate with other systems, in addition to their human users. The dramatic expansion of computer networks, the Internet, and the Internet of Things has made truly standalone systems rare indeed. Systems communicate with each other by direction electronic or mechanical connection, or through a communication medium such as Ethernet, Wi-Fi, or the Internet. As discussed already, it's usually more useful to consider the other system as the actor rather than the communication network itself, even though literally the system might be communicating with dozens of Internet routers and servers before it ever reaches the system with which it is intended to communicate. In plain terms, when my PC gets a page of news from CNN and displays it to me, I imagine my PC talking with a CNN computer, and I just ignore all the Internet infrastructure in between. When describing systems at the level needed for systems engineering, it makes sense to take this approach so we can focus on the important interactions of the system. We choose actors to represent the systems that our system ultimately interacts with, and we tend to ignore the systems in between that provide the communication path. We pretend that our system is communicating directly with the other system.

In addition to outside systems with which it is clear there is an interaction, there are other kinds of actors we might want to add to our use case model. Sometimes it is unclear whether something outside our system needs to be included as an actor. If our system is a car, should we show the road as an actor? There is nothing wrong with doing that, but the question that should be asked is, does the car actually *interact* with the road? In a way it does, since there is contact between the tires and the road surface, but it does seem like that interaction is of a different kind from when the car interacts with the navigation satellites or the emergency services request system. The latter are clearly data interactions, and software must be developed for both the car and the remote system to make the interaction possible. But the road runs no software. It doesn't do anything. Too, the car doesn't do anything to respond to the road – it just turns the wheels which roll over the road surface. Does the road need to be an actor? How the systems engineers decide depends on whether they will need to describe the interaction between the car and the road in the use cases. We haven't yet talked about how the use cases will be described, so the usual approach is to just include actors, even when there is doubt and then see if they are needed later. The purpose of use cases is to describe the interactions between the system we are designing and building, users, and other systems, for the ultimate purpose of designing and building the system. If we won't be building functionality into the system to interact in some way with the road, other than just rolling over it, then perhaps the road doesn't need to be an actor. It won't appear in any of the use cases.

My IBM colleagues used to use a different example when explaining use cases. Rain falls on the car – is rain an actor? If all that happens is the rain falls on the car and the wipers wipe it away, then perhaps not. After all, the wipers operate whether there is rain falling or not. But what if we are designing the system to have rain-sensing wipers? In that case, the rain must be an actor since its action is what starts the use case. The system is interacting with the rain by detecting it, responding to it, and then later detecting its absence and responding to that by turning the wipers off.

Another "special case" nonhuman actor is time. Systems often have functions that happen on a regular schedule, and there can be uses cases which start, based neither on an interaction from a human user nor on a signal from another system, but based on time itself. It may be tempting to think of these use cases as internal to the system, or as breaking the rule that use cases must be initiated by an actor, but a better way is to simply represent time as an actor. After all, time is outside the system – the system does not own and control the passage of time. Time can be represented as an actor called *Time*, leaving the specification of exact times to be given in the description of each use case's flow of events, or the actor could represent a single time like "Midnight US Eastern Time on Thursdays."

6.8.6 Do We Have ALL the Actors?

Try as we might, it is nearly impossible to identify all the actors for a system at first glance, so it's often best to identify the actors we can see immediately, and then proceed to identifying use cases. It is normal that new actors will be discovered or previously identified actors modified as use cases are identified. It is also normal for some actors to be removed if we find that there are no uses cases that turn out to involve them, indicating that the system doesn't interact with the proposed actor after all. Actors represent the roles of the human users and other systems that interact with the system we are describing and later building.

6.9 Identifying Use Cases

Once we have a beginning set of actors for our system, we can proceed to identify the use cases. Use cases describe how a user achieves a result of value from the system, so we can begin by considering each actor we've identified, and simply ask how they will use the system. For the ATM, it's easy to see that the main user is the customer, who will use the ATM to withdraw cash, to make a deposit, and to check a bank balance. These become the first use cases we will list. For each use case we identify, it's best to note a few important pieces of information: the name of the use case, the actors involved in it, and a brief description of the use case. Even though it takes a bit of extra time, it's best to document these as the use cases are identified. Experience has shown that when a use case is identified, at that moment there is clear thinking about what the use case involves and how the story will go. Failing to document the brief description right then, can result in the need to remember and rethink what the use case is all about later, and lead to misunderstandings within the development team. The brief description, just a few sentences in length, removes all possible ambiguity about the use case, which actors are involved and in general how the action of the use case will proceed. A good brief description also gives a kick start when it comes time to write the full flow of events of the use case later. For the "withdraw cash" use case, the brief description might read,

> Customer inserts bank card and enters PIN to authenticate, chooses amount to withdraw, and receives cash, a receipt and the card back.

In Chapter 7, we'll show how to draw a use case diagram, which graphically depicts the use cases and their relationships with actors, but for now, just making a list will do.

In addition to the fundamental question, what does this actor use the system for, there are other prompting questions that will help identify use cases. Here's a list, as listed in the Rational Unified Process from IBM:

- What are the primary tasks the actor wants the system to perform?
- Will the actor create, store, change, remove, or read data in the system?
- Will the actor need to inform the system about sudden, external changes?
- Does the actor need to be informed about certain occurrences in the system?
- Will the actor perform a system start-up or shutdown?

Each of these questions leads to the identification of one or more use cases. In this early stage of identifying use cases, it may be unclear whether a particular usage of the system should be represented as one use case or more than one. For example, should we have a *withdraw cash* use case and also a *check balance* use case, or should checking the balance simply be part of the withdraw cash use case? If the user should be able to check a balance without making a withdrawal, then *check balance* must be a separate use case.

Switching back to the example of a car, should *commute to work* and *take vacation trip* be separate use cases? In many ways, they are similar usages of the car, and if they are similar enough, perhaps they could be combined into one use case, called something like *take driving trip*. But recall that the primary purpose of developing use cases is to communicate requirements told in stories. The question becomes, would there be different requirements for the car exposed by commuting vs. taking a vacation trip? The fuel capacity, range, cargo capacity, seating capacity, and even passenger entertainment options may be quite different when we consider each use case separately. If two similar use cases suggest no different requirements for functionality for the system, then it may make sense to combine them, however, if they should be analyzed separately and are likely to produce new requirements or a different system design, then it probably makes sense to keep them separate. Naturally, if two use cases aim at accomplishing different goals for the user, they should be separate use cases.

When naming use cases, it's best to use a verb action phrase, rather than a noun phrase. *Commute to Work*, is better as a use case name than *Commuting* or *Workday Process*. The aim is clarity and since use cases describe action sequences, an action-oriented name that expresses the goal of the use case is best. As a finer point, the use case should be named from the perspective of the actor who initiates it. So if the use case is about a student registering for a course, *Register for Course* is a better name than *Assign Student to Course, Fill Course,* or *Course Registration*.

A common trap in identifying use cases is to mistake functions for use cases. It could be tempting to claim that *start car* is a use case for a car, since of course someone needs to start the car in order to use it, but the owner didn't buy the car in order to start it. That is not the user's goal. The goal is bigger – to commute to work or go on vacation. Use cases should express complete usages – usages that produce valuable results and achieve the user's goals.

It's important that the terminology used to name a use case is familiar to the stakeholders so that even if stakeholders are not trained in use cases or model based systems engineering, they can still look at the use case diagram, read the use case descriptions, and understand them. All stakeholders involved in creating the system, whether technical or not, should recognize that the use cases are describing the things that the system will do and can verify that the team is getting it right. In a way, they should feel that what is represented in the use cases is almost obvious. At this high level, the specific goals represented in use cases should be well understood by everyone involved. Surprisingly often, however, new things come to light in this simple exercise of identifying use cases. Not everyone may have a common concept of what use cases will be, or how they will be carried out with the system. Identifying use cases, as simple as it may appear, is an opportunity to do what's necessary to bring everyone to a common understanding and agreement of the usages of the system. For a system that will be breaking new ground, serving a new kind of user or being made more intelligent, thinking through use cases is an excellent way to work out how the system can work in new ways. For instance, if in the past the user was required to enter some information into the system, we might ask, what if the system could already know that information? What if the system already knew what the user needed, and could simply present it? Or what if the system could carry out this activity without the user's involvement at all?

How many use cases should a system have? In any class I've taught on use cases, this question is always asked. The answer is, there is no answer. Or to put it in a more Zen-like fashion, a system has the number of use cases it has. Some systems have many, many use cases and some have just a few. Years ago, I worked with a military equipment manufacturer who was designing a new military aircraft. They had identified and described only three use cases for the aircraft. They reasoned that the aircraft had three types of missions that it could carry out, and they represented each as a use case and described each use case with many steps – a mission can be a long, involved story. On the other hand, an online store website like amazon.com, would have many use cases, since there are many separate things a user can do with the site, and most of them are independent of each other. Websites are like that in general – they present a large number of options through menus, buttons, and links and many of these are best represented as separate use cases. There is a certain art to the design of software user interfaces (or user experiences) and identifying use cases is a great first step, as it guides the user experience design team to better understand exactly how users will use the system. The most important use cases should be more prominent in the user experience and easier to carry out than more detailed, specialized uses. I'll single out Zoom as an example, both positive and negative. For the most part, the Zoom user interface does a great job of letting casual computer users participate in, and even create and conduct online meetings. At the same time, Zoom has many more advanced features, but they are a little bit hidden, so as to avoid confusing the casual user, but be discoverable by users with a bit more experience. On the other hand, Zoom continues to be a bit confusing as it has both a Windows App and also a web interface, and they offer different sets of options, and a somewhat different user experience. But on the whole, Zoom has struck the right combination of being both easy to use and feature rich.

Going back for a moment to the military aircraft, one might wonder how there can be so few use cases when there are hundreds of buttons and switches and a complex array of functionality. The answer goes back to how we conceptualized the system in the first place. Use cases represent usages of a system by an actor who, like all actors, is outside of the system. If we include the crew in the system of the aircraft, then the actor for each mission is likely the force or fleet commander who issued the order to the aircraft to carry out the mission. Since use cases always treat the system as a black box and avoid referring to elements (including crew members), the use case for a mission, with the commander as an initiating actor will tell the story of the aircraft's interaction with the commander, the environment, the refueling station, the enemy, etc., but without referring to the details of the parts (or crew) of the aircraft. Interactions between the crew and the subsystems of the aircraft would be described at the next level of abstraction, where the *systems* are the subsystems of the aircraft and the *actors* are the crew members. That use cases can be described at any level of system abstraction is an important feature of use cases and is central to the use of model based systems engineering, described in Chapter 7.

6.10 Use Case Flows of Events

Once use cases are identified and briefly described, the flow of events can be written. Use cases have become popular in the last few decades, so there are several good books devoted mainly to the writing of use case flows of events, including Schneider and Winters (2001) and Cockburn (2001).[1] I think that it's best to start with a simple approach, plain language, and few specialized or complex features of use cases. There are two good reasons for a simple approach: it's easier to use for those new to writing use cases, but even more importantly, simply written use cases are easier to understand by all stakeholders involved. Use cases are written to

---

1 See for example Schneider and Winters' *Applying Use Cases: A Practical Guide* and Cockburn's *Writing Effective Use Cases*.

be read and understood by everyone involved in the project. Those writing the use cases, who may have been doing it for years, might think that the perfect way to describe a particular aspect of the system is by using an extension point, a feature of use cases, but unless everyone involved in the project understands what an extension point is, clear communication will not occur. It's best to start out simply, and then add more features as the organization develops comfort and understanding with use cases.

When describing the flow of events of a use case, we are telling a story. It's a very specific story, not just any story about the system. Each use case is a story of a single, complete usage – the story of how a particular type of user, represented by an actor, proceeds to use the system to accomplish a goal. If there are various types of users playing various roles and using the system to accomplish various goals, there will be use cases to describe each.

The most important flow of events in a use case is what is called the basic flow, happy day scenario, sunny day scenario, or happy path – for some reason, it goes by many names. The happy day scenario is just that – it's the path through the use case when everything "goes right" and the user successfully completes the intended goal using the use case. No mistakes are made by the user, and the system produces no errors or malfunctions as the use case unfolds. The happy day scenario is the first flow of events to be written and the most important since it expresses the main purpose of the use case, which is of course to be completed successfully and achieve the goal, is expressed and described directly in the happy day scenario.

Just as stories may start with "once upon a time ..." the happy day scenario of a use case begins with "this use case begins when ..." Starting the use case this way makes it very clear to the reader that this is the beginning of the use case, and also states the action taken by an actor to begin the use case. It can take a bit of thinking to decide on what action starts the use case. It can be tempting to imagine that the use case begins with the user deciding something, and write, "this use case begins when the user decides to take a trip in the car." Unfortunately, with current technology, the system cannot determine when the user makes a decision so the system cannot start a use case based on a mental decision event. So what is the action that starts the use case of taking a trip in the car? With most cars, the trip begins with unlocking the car, so we might write, "This use case begins when the user unlocks the car." Most trips begin that way, but we must decide whether the unlocking of the car is part of the happy day scenario or a variation, with the happy day scenario being that the car starts out unlocked. We can go either way on this, but I suspect on balance, the normal case is that the car starts out locked, and so the event that begins the use case is that the user unlocks the car. The steps of the use case would then proceed with the user applying the brake, starting the car, placing the car in gear, and proceeding from there.

If it sounds like this example scenario is obvious and trivial, consider the innovations that have emerged from simply considering the scenario of a person taking a trip. Even in the first step, the thought must have occurred to someone, what if the user didn't have to unlock the car with a key – what if it unlocked when the owner approached? which led to the development of the electronic key fob. If adjusting the seat and mirrors is part of taking the trip, what if this happened automatically based on the driver? That idea could lead to thinking about automatic driver detection (a variety of possible technologies come to mind – face recognition, a weight sensor in the seat, personal key fobs), and how that could make using the car easier. The point is that reasoning within a use case is more likely to produce innovations that are effective for users since they are considered within the flow of events tailored to the systems users.

As we describe the happy day scenario of a use case, it is necessary to avoid any conditional statements. There should be no "if-then," or alternative branching described in the happy day scenario. Why? Because it's a single scenario – the single path through the use case when everything goes right – the normal case. Sometimes, two possible paths seem equally "happy," and it's hard to choose which alternative should be part of the happy day scenario. Above, it was hard to decide whether the happy day scenario was for the user to unlock the car, or for the car to already be unlocked and the user simply enters the car. In cases like this, we just choose one alternative, perhaps the most often occurring one, to be part of the happy day scenario, and leave the other alternative to be covered in one of the alternate flows for the use case, which we'll describe shortly. So, for every possible alternative or conditional path in the use case, we simply choose the most common or most successful path and use that in the happy day scenario. The happy day scenario will then be both the simplest and the most successful way that the use case can happen. It may be a bit frustrating to leave out important details about alternative ways that things can happen during the use case, but the happy day scenario has an important role in the engineering of intelligent systems.

It can take a significant effort to fully describe all the use cases for a system. Identifying actors and use cases, writing down the brief descriptions, and then completing the flows of events is not the time-consuming part. What really takes the effort and time is working with all important stakeholders to agree on the use cases, and how they should work; how the users will use the system. In organizations unaccustomed to thinking in use cases, important stakeholders may be surprised to find they are being asked to think about exactly how users will use the system. The investment in coordinating stakeholders, working with them to collect and synthesize their ideas and thoughts, and weaving them into a set of uses cases with which all can agree is well worth it for the ultimate success of the system. Anyone involved in systems development of any scale knows that the most expensive part of systems development is when changes must be made well into the engineering process, especially when these changes are because key people changed their minds about how the system should work. At least some of these changes can be avoided by asking these key people to work together to create use cases up front.

6.10.1 Balancing Work Up-Front with Speed

At the same time, some change is inevitable as the team learns more about the system and application while building it. The goal is to achieve the correct balance between up-front work and getting into the designing and building of a system as soon as possible. Both extremes have proven to be unsuccessful, and they have both been tried many times. Spending too much time up-front writing detailed requirements, or even writing fully detailed flows of events for all use cases, might unnecessarily delay the entire project for several reasons. Designers in the other engineering disciplines (electrical, mechanical, software, and others) are made to wait until the systems engineering group is done with this detailed work. Almost inevitably, the systems engineering team will have uncertainties, and will need information and input from the customer and from other engineering disciplines to make key decisions. The far inferior alternative is for the team to make assumptions, or guess. The final decision may not be possible until some exploratory or preliminary work is done by the systems engineers or by other engineering disciplines. In complex systems, especially those with ambitious goals compared to past systems, not everything is known about the best approaches and choices in the early phases, before any detailed engineering work is done. Doing too much work based on assumptions and guesswork can lead to the project being locked into unfortunate decisions early on, unable to change later due to the amount of rework that would be necessary.

On the other hand, avoiding all up-front systems engineering work is not wise, as it will lead to a varied understanding of the system across the teams, requiring significant coordination time and effort, and likely forcing the rework of components found later to be incompatible. The two extremes – do as much as possible up front, hoping to avoid uncertainty, wasted time, and rework later, or do nearly nothing up front, counting on being able to adjust designs as the system is developed and as more input and guidance is received from stakeholders, both have disadvantages. Admittedly, systems that involve interactive software only, such as web-based applications or embedded software applications intended to run on hardware that already exists, lend themselves much more to fully agile development approaches, which start with general ideas about the desired application and refine it throughout the development process. However, this approach too can result in rework, and may benefit from some up-front work to establish a general architecture and framework for the desired application, including the definition of use cases.

The solution would seem to be a middle ground, where the team does just enough work up front to provide the right amount of guidance to the engineers, so that they can do productive work consistent with an overall design and plan. We suggest that the middle ground utilize use cases and happy day scenarios. Most systems would benefit by developing some aspects of the full use case model up front, ensuring complete and widespread understanding among the stakeholders, and then allow more detailed aspects to be developed as the development proceeds. The up-front work should include:

- Identification of Actors
- Identification of Use Cases
- Brief descriptions of all use cases (1–3 sentences each)
- Happy day scenarios for all the main use cases (the use cases central to the system's main purpose)

With this work in hand, teams can proceed with confidence that they are aligned with the main purposes and goals of the system, without waiting for a great deal of detailed requirements work to be completed. Describing only the happy day scenario is much less work than writing descriptions of all possible alternate and exception flows, work which we'll cover in a moment. The happy day scenario is special, in that it captures the primary action of the use case. Thus, taken together the happy day scenarios of all the use cases describe all of the primary functionality of the system in a compact, easy to understand form. All of the important functionalities would be described, and what is not described can probably be safely left to describe later when it's time for the team to design and develop that part of the system.

Let's take the use of the happy day scenario one step further. In some systems, primarily software applications, we can actually implement the happy day scenario by itself in software code, without implementing other aspects of the use case. Doing so would result in a very fragile system that would only work if everything went exactly the right – the user could make no mistakes, no resources could be unavailable to the system, and no errors could occur, since the alternate and exception flows are not implemented yet. But implementing only the happy day scenario would allow the demonstration of the system to users and stakeholders and validation that the system will meet their needs. Implementing only the happy day scenario is also a great way to prove out the main concepts of how the system will work. It likely means tackling the hardest or most uncertain aspects of the system, but that's a good thing: we always learn more by doing the hard things and it enables us to adjust and make changes before we have gone too far in an ultimately fruitless direction.

6.10.2   Use Case Flows and Scenarios

The happy day scenario describes only what happens in a use case when every-thing goes right, and everything happens normally. To fully describe a use case though, we must include descriptions of everything that can happen in the use case, which means including alternate and exception flows. Unlike the happy day scenario, these flows do not give a complete path through the use case, but only describe what happens under certain conditions. Say a use case involves a driver entering the vehicle. The happy day scenario may simply state, "the driver enters the vehicle," which assumes that the vehicle is unlocked. An obvious alternative flow would be that if the car is locked, the user must unlock it with the remote before entering it; that becomes an alternate flow. It is possible that the car bat-tery is discharged, and the user cannot unlock it using the remote. We'll describe an exception flow that probably includes opening the car with a key. We think through all of these possible things that can happen or can go wrong and create little alternate flows to handle them.

We've been using the terms *flow* and *scenario*, so let's take a moment to define them. A *flow* is any set of steps in a use case. The happy day scenario is a flow, and so is the alternate flow where the user must unlock the car before entering. A *scenario* is a complete path through the use case, from beginning to end. The happy day scenario is a scenario, in addition to being a flow, because it represents a complete path through the use case. *Unlocking the car* is a flow, but is not a scenario since it doesn't describe a complete path through the use case. Only one flow is a scenario all by itself – the happy day scenario. All other scenarios are made of a combination of flows. A scenario might start out like the happy day scenario, but then take an alternate path, then come back to the happy day, then encounter an error and do an exception flow, and then the use case ends. That's a scenario. If a use case had five flows, then theoretically it would have 120 (5! – five factorial) scenarios, since there are that many ways to combine the five flows (actually even more, since not all flows might be used in each scenario, and some flows may occur more than once in a scenario). But not all of these combinations are valid scenarios that are possible for the use case. A use case with five flows might end up with about ten scenarios, meaning that those five flows can be combined in ten different, valid ways to make a complete path through the use case.

There is a special reason to pay attention to the identification of scenarios, as they become the basis for designing test cases. Each scenario becomes a test case, or a set of test cases with varying user input, that must be tested as the system is developed and when it is complete. For this reason, it's a best practice for the ver-ification and testing team to be involved in the writing of the use cases so that the use cases will meet their needs too and become the basis for the test cases they will develop. A recent trend in software development is to require the development of the test cases for a piece of the software before the software is developed. That way, the software developers can keep the test in mind as they are programming, hopefully resulting in software that is more likely to pass the test. The same principle can be useful in hardware and mechanical implementation as well; knowing how the circuit or part will be tested can be an important guide to designing and building it correctly the first time. A great way to accomplish this "test-first" approach is by creating use cases, which can simultaneously guide development and can also be the basis for the test cases.

It is an irony that in projects where use cases are not used, the test and verification group must develop test cases from scratch (based on the requirements) and those test cases will often resemble the use cases that would have been developed from the start. When bringing the idea of use cases into a project that is well underway, such as an enhancement to an existing system, the test cases can be used as a starting point for the use cases, with the awareness that the testers who developed the test cases may not have had the background knowledge of the users and stakeholders about the needs for the system, nor the exposure to the stakeholders to have a comprehensive view. So it's still best to develop the use cases as a team, including representation from the test team, so that a complete set of use cases is developed.

6.10.3 Writing Alternate Flows

Alternate flows in a use case are described in four parts: the starting point, the condition, the action, and the continuation. As long as these four are included in that order, the alternate flow can be phrased or worded in any convenient way, keeping in mind that use cases should be understood by all involved in the project. Returning to our car trip example, if an alternate flow should be that when the driver approaches the car, if the car is locked, the driver unlocks it and then then the use case continues with the next step, we can write the alternate flow in just that way. The starting point here is "when the driver approaches the car." The condition will almost always begin with "if," as in "if the car is locked…" Next comes the actions the system and actors will take if the condition happens, followed by where in the use case the action should continue after those actions are taken.

As with all the writing in a use case, the most important priorities in describing alternate flows are clarity and effective communication. Write simply so that all can understand what is being said. No special syntax is needed – use cases are not software pseudo-code – they are descriptions of how the system is intended to work from the users' perspective, so simple language is best. But simple does not mean simplistic. Use cases should be specific and detailed, but still stated in simple language, while maintaining a black box perspective of the system. As a story to illustrate, if in a use case workshop, someone proposes a step like, "the user enters their personal details into the system," then I tend to ask, is the term "personal details" well defined in this organization so that everyone knows exactly what that includes. If not, then let's just write it here in the use case, for example, "the user enters their full name, address, email, home, work, and mobile phone number." It takes little more time to be specific than to leave it vague. If left unspecified, who is expected to interpret what is meant by "personal details"? Likely, it will come to the software developer that way, and it is up to that lone developer to either stop writing software in order to research with all the needed stakeholders, or more likely, to just guess at what it means. It's better to invest the time when writing the use case and be specific.

With alternate flows, it's important to remember that an alternate flow will always contain a conditional statement of some kind, usually an *if* statement, while the sunny day scenario, also known as the main flow of a use case, will not contain any conditional or *if* statements at all. It can be tempting to write a conditional statement right in the main flow, but this creates an implied alternate flow and as a result the main flow is no longer the happy day scenario. Worse, since there is now an alternate implied in the main flow, it is impossible to determine the happy day scenario. Keep the sunny day scenario described in the main flow clear, with no conditions, and put all the conditional, alternate behavior in a separate section of the use case document.

Exception flows are also written for a use case. These are described in the same way as alternate flows, and the only difference is that exception flows refer to things going wrong, errors, and failures rather than just alternate, valid ways of the use case happening. The conditional statement that begins an exception flow will reference an error condition of some kind, such as, "if the user enters an invalid password…," "if the system cannot contact the banking system…," or "if power is suddenly removed from the system…"

6.10.4 Include and Extend with Use Cases

Include and extend are relationships between two use cases that may be included to describe desired system behavior more efficiently. They are useful, but if a practitioner or organization is new to use cases, it may be best to avoid their use initially. Remember that since use cases are intended to be read and understood by *all* stakeholders, including managers, users, testers, and developers, any nonobvious features or notations in the use case representations must be explained to all stakeholders. If the features are not explained, we risk some stakeholders ignoring It's best to start simple – the system behavior can always be described without the use of include and extend, so their use is not mandatory.

Showing that one use case *includes* another simply means that the user and system behavior described in the included use case is included in the user and system behavior of the including use case. For example, if we have a use case called *make breakfast*, a use case of the kitchen system, we might say that this use case includes the use case *make coffee*. This would mean that anytime we use the kitchen to make breakfast, we also, as part of making breakfast, make coffee. There is no making breakfast without making coffee. Of course, we could simply add the steps for making coffee in the *make breakfast* use case, so describing *make coffee* as a separate use case is not mandatory. But if we intend to make coffee as part of some other meals, or at nonmealtimes, it may make sense to pull the steps for making coffee out into a separate use case, so that we only need to write them once, and then simply include them in the other use cases. Using an include relationship between *make breakfast* (and perhaps *make lunch* and *make dinner*) and *make coffee* simply allows us to describe coffee-making once and reuse it. The reuse saves time and prevents errors if we ever need to change how we make coffee – we need only change it in one place.

Extend is another kind of relationship between two use cases and indicates that one use case is an extension of another, that the behavior in the extension only happens sometimes during the main use case. That's the important difference between an extension use case and an included use case – an extension is only performed sometimes. Like an included use case, an extension is used to describe a set of steps that can be used with more than one use case, avoiding the need to describe that set of steps in more than one place (for those with a software background, extend is a little bit like a function or subroutine call). The use case being extended does not depend on the extension, that is, the system could be implemented without the extension and still operate under at least some conditions. The extended use case defines only the point at which it can be extended (called an *extension point*), and the definition of the extension and its behavior are described only in the extension. This means there is no dependency created between the use case and the extension, a useful feature when extensions are used to express future functionality or variant versions of the system.

To pull together what we've said in this chapter on how to develop use cases for a system, here's a brief recipe:

1. Identify the system.
2. Identify the actors, human and nonhuman, that will interact with the system.
3. Identify and name the use cases the system will perform and the actors involved in each.
4. Write a brief description of each use case as it is identified, capturing the action in just a few sentences.
5. Write the flow of events for each use case, beginning with the happy day scenario (main flow) and then adding alternate and exception flows.

6.11 Examples of Use Cases

The following examples of use cases are intended to show what real use cases look like in practice. They are not perfect, but are real examples, with key terms changed to disguise their origin. There is always the danger than *examples* can be taken as *ideals*, and these are not, but in each case, these were a useful work product created in the course of a system's design. In some cases, they are very high level, almost just an outline, and details may have been added later.

6.11.1 Example Use Case 1: Request Customer Service from Acme Library Support

System: Lakeside University Coordinated Knowledge Electronic Entry (LUCKEE)
Actors: Customer
Brief Description: The Customer has logged into LUCKEE, and requests Acme customer service, which links the customer to Acme's Help system, preloading appropriate information.
Preconditions: Customer has successfully logged in to LUCKEE account.
Primary Actor: LUCKEE Customer
Basic Flow Scenario:

1. System prompts for Acme login information
2. Customer logs into Acme account using account ID and password.
3. Customer Service requests prior to log in are routed to Acme customer service.
4. Customer selects Customer Service.
5. Customer selects from Partner Company list.
6. Customer selects Link to Company.
7. Customer sees overall account information displayed
8. Customer selects Contact Your Partner Company.
9. Customer is taken to Acme Library customer assistance screen, with information transferred from LUCKEE:
   o Name
   o Address, City, State, Zip, Country
   o Account number
   o Phone
10. Customer completes customer assistance information request information including description of request and urgency, and the use case ends.

6.11.2 Example Use Case 2: Ensure Network Stability

System: Network Management System (NEMS)
Brief Description: From analyzing the current and projected network element status to deriving and implementing solutions to actual and potential transmission security risks.
Primary Actor: Time
Preconditions:

- Network is in a steady-state and stable
- Network monitoring systems operating normally

Main Success Scenario:

1. This use case begins when the network monitoring system is started and operates continuously until system shutdown.
2. System gathers network and traffic data from network providers and main hub user nodes.
3. System determines that there are no network elements over normal capacity
4. System determines that no main hubs are unreachable.
5. System determines no Level 6 or above security alerts exist.
6. Use case ends.

Alternate flows

1. At step 3 in the Main Success Scenario, if the System can't determine whether network traffic level is over normal capacity:
   1. System requests main hub operators to report overload conditions
   2. Main hubs report no traffic overload conditions
   3. Use case ends
2. At step 3 in the Main Success Scenario, if the System determines that there are network overloads:
   1. System determines corrective actions to resolve transmission overload or under voltage condition.
   2. System directs Main Hub operators or sub-unit consolidators to implement corrective actions.
   3. The use case continues at Main Flow, Step 1.

6.11.3 Example Use Case 3: Search for Boat in Inventory

System: Boats4U online boat shopping site (B4U)
Actors: Consumer, Sales Rep
Precondition: User has built vehicle using "Find boat by Payment", "Create Deal", or has entries in "My Boat Short List".

Happy Day Scenario:

1. The use case begins when the Consumer selects to search for a boat in dealer inventory.
2. System loads boat specifications from current (unsaved) boat specification list, or a boat is selected from "My Boat Short List"
3. System searches for exact year, make, model, style, and option matches in dealer inventories
4. System finds all matching boats and displays a table showing: year, make, model, style, stock #, color, icon linking to photo (if available), link to boat details
5. User selects a specific boat from inventory list displayed
6. User saves boat to "My Boat Short List"
7. Dealer is notified of the user's interest in the chosen boat(s)
8. User chooses to find financing and payment for selected boat and the use case ends

Alternate 1:

At step 2 in the Happy Day Scenario, if nothing is found using specified criteria:

1. System broadens search criteria removing criteria as necessary until search results are found. Removed criteria, in order are year, model, make, and style.
2. The use case continues at Step 3.

6.12 Use Cases with Human Activity Systems

In this chapter, we've strongly implied that use cases are intended to express the behavior of an engineered system interacting with its human and nonhuman users, and this is the most common use of use cases. But recall that in an earlier chapter we identified three types of systems – engineered, natural (biological), and human activity systems. Use cases can also be used to describe these systems. For example, imagine a university as a system. Inside the system are not only buildings, computers, and classrooms but also the university staff, faculty, and maintenance personnel. A university is a human activity system, a combination of engineered systems, physical items, and people. If the university is the system, then what are the actors for this system? Students, certainly, but also food suppliers, book publishers, scholarship providers, and student loan lenders – anyone or anything that interacts with the university system, but which is not a part of it. Taking the student for example, use cases might include *apply for admission*, *register for courses*, and *purchase books*. These use cases utilize various parts of the university system, including the people inside that system, to accomplish the user' goals.

6.13 Use Cases as a Superpower

Use cases are close to a superpower when it comes to dealing with the complexity of systems in the early stages of conceptualization and design. They cut through all of the confusion, differing perspectives, shifting priorities, and multiple user perspectives to provide a clear, straightforward and universally understandable view of the most important perspective of the system – the user's. They describe what the system is for and how it delivers benefits to its users. It's not an overstatement to say that the entire purpose of a system is to accomplish what is expressed in its use cases. Even the most complex system can be understood by reading its use cases. Use cases provide a useful abstraction by treating the system as a black box, with all of the internal subsystems and parts hidden, enabling a simple description of the interaction between the system as a whole, its human users and its external system actors. Uses cases are a powerful tool on their own, but also form the basis for the practice of model based systems engineering, described in Chapter 7.

References

Cockburn, A. (2001). *Writing Effective Use Cases*. Pearson Education India.

Jacobson, I. (1993). *Object-Oriented Software Engineering: a Use Case Driven Approach*. Pearson Education India.

Jacobson, I., Ericsson, M., and Jacobson, A. (1994). *The Object Advantage: Business Process Reengineering with Object Technology*. ACM Press/Addison-Wesley Publishing Co.

Schneider, G. and Winters, J.P. (2001). *Applying Use Cases: A Practical Guide*. Pearson Education.

7

Picturing Systems with Model Based Systems Engineering

Writing is nature's way of letting you know how sloppy your thinking is.

(Cartoonist Richard Guindon)

Mathematics is nature's way of letting you know how sloppy your writing is…
Formal mathematics is nature's way of letting you know how sloppy your mathematics is.

(Leslie Lamport (inventor of LaTex))

Use Cases are nature's way of letting you know how sloppy your requirements are.

(The Author)

7.1   How Humans Build Things

Human beings have been building things for thousands of years. In the beginning, simple objects were made from simple materials. A stone was sharpened on another stone and then used for cutting wood. Pieces of wood were tied together to form shelters and furniture. A sword was forged from a single piece of metal. For simple items, the purpose, intended use, design, and implementation all arose together in the mind of the creator. An ox cart is more complicated. There are several parts, and they must fit together for it to work correctly, but again, the usage and the design are born as twins of the same mother. Usage and design are always there in any conception of a needed product or system, no matter how simple or complex.

Sometimes, the usage is simple, and the design is complicated. When mechanical clocks were first invented in the thirteenth century, their usage was simple

Engineering Intelligent Systems: Systems Engineering and Design with Artificial Intelligence, Visual Modeling, and Systems Thinking, First Edition. Barclay R. Brown.

and straightforward – display the time – while the design required much effort to develop. Since then, many different designs and technologies have been developed to perform this same simple function. Today, atomic clocks use advanced technology and radio signals to keep the time accurate. The technology progresses, but the usage stays the same, "telling" us the time at a glance. In recent years, a second kind of usage for a clock has appeared in the form of voice assistants like Amazon Echo. They literally tell the time, but only if someone asks. So, after centuries, there are still only two usages for a clock – look at it to see the time or ask for the time to be spoken or displayed.

The example of the usages for a home voice assistant is intriguing in other ways. When you first heard of them, did you wonder what they would be good for? If you bought one, what did you expect to use it for? You may have had high hopes for clever conversation – like Tony Stark's banter with his AI Jarvis in the Ironman films. Once you discovered that such conversation is well beyond the product's capabilities, at least for now, you may have moved on to other possible uses. You found you could ask it to play music, so you asked for each of your favorite songs by name, moving on to your favorite artists and then to whole genres. Great, but what else could it do, you wondered? Perhaps surprisingly, one of the most common applications used on voice assistants is the timer – "Alexa, set a three-minute timer for the eggs." Your relationship to this product is based solely on the uses to which you can put it. You need no awareness of how the product works. You don't know how much of what it does is done locally, in the device on your table, and how much is done on a distant computer somewhere in the cloud. If you're of an engineering mindset, you may be curious about how the product is designed internally and how it works, but you don't need to know any of that in order to use it. What matters to you, and to any user or purchaser of a system – are the specific ways the system can be used.

These system usages, or *use cases,* as we introduced them in the Chapter 6, seem obvious, trivial, and even uninteresting at first glance. After all, isn't it immediately clear how a system will be used? The usages are the whole reason why the system is being built or bought in the first place. I buy a drill in order to drill holes. Drilling holes is the obvious usage of the drill – what's interesting about that? It turns out that drilling holes is not so simple, as any machinist can tell you. A handheld power drill with a conventional twist-style drill bit is fine for some ordinary kinds of hole drilling, but the careful study of the different specific usages of a drill has led to the development of many kinds of drill bits and variable speed drills. Spade bits, Forstner bits, and hole saws all have their uses. The Irwin Speedbor bit uses a screw tip to pull the rest of the bit through the wood, requiring substantially less force from the user holding the drill, making it ideal for awkward locations or overhead drilling. To tease this example a bit more, at some point someone noticed that it was difficult to drill holes in large pieces of steel. They were too big to be brought to a drill press and might already be part of a road or bridge and thus immovable. Drilling thick steel with a handheld power drill hand was a long, slow, and even dangerous process. So, in 1954 a patent was issued for a new kind of drill – a small drill press that attached itself to the steel being drilled by magnets. Which came first – the idea to make a little drill press with a magnetic base, or the use case of drilling large steel pieces in place? Of course, the use case came first – necessity is the mother of invention.

Starting from identifying use cases is an important first step in the design of any system. Identifying and describing the usages of the system and communicating them in a way that all system stakeholders can understand and agree has been found to be an effective way to avoid big problems later. A key concern for systems engineers is to be sure that the system being built is going to function properly and meet the needs of all of the stakeholders and users. In countless cases where a system is built and then rejected by the users or stakeholders, the most common cause cited in US government studies is poor requirements, indicating that the correct user requirements were not determined. It is likely that the root cause was not widespread apathy or incompetence on the part of either the users or the systems engineers. Most likely, many requirements were gathered from the available sources, documented, and then sent to some group of stakeholders for approval. By the time the requirements set is finalized, it might contain several thousand requirements statements. It's hard to imagine busy, overworked stakeholders *reading* all of those requirement statements much less fully understanding the system that would result from systems engineers attempting to meet all of them. Use cases put all those requirements into context, connecting them and explaining how the system works to fulfill them. Use cases tell the stories of how a system will be used, by those who will use it. The deceptively simple use case diagram, explained more in the Chapter 6, gives a succinct overview of all the ways the system can be used, and by whom, along with any needed interfaces to outside systems.

Use cases are a key aspect of picturing a new system in a way that all users and stakeholders can understand, reason about, and ultimately approve – and have an excellent chance of getting what they expect when the system is finished. The use case view is not the only view needed to understand a system's design. In model based systems engineering views are created to show different aspects of a system's high-level design. These views are not created in isolation and are not simply drawn using a drawing tool like Microsoft Visio, PowerPoint, or ordinary pen and paper. Instead, a model is created in a modeling software tool using a modeling language such as the Systems Modeling Language (SysML). The model contains symbolic elements that represent various parts and behaviors of the system. Views are then created from elements in the model, so that an element appearing on multiple views, or diagrams, is the same element in the model. If a change is made to that element, it automatically changes on every diagram on which it appears.

In what follows, we describe seven views that are commonly used to describe a system. These views are used in most common methods of performing model based systems engineering, so understanding them will go a long way toward developing an intuition for what it is like to design a system using MBSE. The views can be remembered using the acronym CUSTARD. These views and the concepts they represent are a useful way to think about any new system, especially early in the conceptualization and design process. To illustrate each type of view, we'll use two examples.

The first example is a military fighting vehicle. The stakeholder group includes the military leaders and Department of Defense (DoD) civilians who are specifying the vehicle, and the team from the defense contractor doing the more detailed design and then manufacturing the vehicle. In general, the vehicle is intended to be small and maneuverable for use in populated cities or towns, including residential areas. It could have both military and police applications.

The second example is an ordinary residential home. The stakeholders include the family members specifying and purchasing the home, the residents who will live there, the architect, and the builder.

7.2   C: Context

Context, the C in CUSTARD, is a representation of the system, shown in the context of its users and interfaces to other systems. In a context view, we represent the system as a single block, without showing any of its internal features or structures. This is known as a "black box" view of the system. The typical, but not required, arrangement of the diagram is to place the system in the center and then surround it with symbols representing both the users of the system, and other systems with which our system will interact. There is a key difference between users and stakeholders of a system. Users will actually touch, use, and interact directly with the system when it is built and delivered. Stakeholders are those who have a vested interest in the system, but who will not necessarily use the system. Stakeholders include those who are specifying the requirements for the system, paying for it, or affected by it in some way.

Users of a system can be either human users or other systems that will interact with the system we are building. It may sound strange to think of an external system as a user, but both human users and external systems interact with the system in the same way – they send instructions, commands, and data to the system and receive back results. Humans accomplish this by pushing buttons, flipping a switch, typing on a keyboard, using a touch screen, or even just walking past a camera or sensor. External systems accomplish the same things by sending information over a wireless connection like Wi-Fi, or a wired connection like an High Definition Multimedia Interface (HDMI) cable connecting to a television.

From the vantage point of a context view, both human users and external systems just look like things our system will interact with, so we give them a single name in MBSE models: Actor. We described actors in depth in the Chapter 6. Actors can be depicted on a context diagram as a human stick figure or human figure drawing, a simple box, or a small piece of artwork. It may seem strange to use a human stick figure to represent an external system, but this is sometimes done to emphasize that the external system is like a user of the system, even though it may not be a human being.

7.2.1   Actors for the VX

Let's illustrate further with our two example systems. First, the military vehicle, which for brevity we will refer to as the VX. We draw a simple box in the middle labeled VX. Then we ask ourselves, who are the users of the system, and what are the external systems with which our system (the VX) will interact – in other words, who are the actors for our system?

A context diagram for the VX includes all the users of the vehicle. The first one that comes to mind is probably the driver of the vehicle, because certainly the driver interacts with the vehicle. Because it's a military vehicle, there are other warfighters in the vehicle too, who will be interacting with the vehicle. It's not like a passenger car where passengers are just relaxing enjoying the scenery. Most of the people in the military vehicle will be warfighters, and will be doing things, like operating instruments that are part of the vehicle system. They could be handling weapons or other controls. There could also be passengers and since their interaction with the VX will be substantially different than the warfighters, we'll add an actor representing them as well.

Our context diagram will include the warfighters inside the vehicle and the driver. We might further specialize the warfighters inside the vehicle; for example, there could be a weapons officer and a navigation officer, and we can show that on the context diagram as well. What we're looking for is the different ways that someone would interact with the vehicle, because these differences define different roles, depending on how each individual interacts with the vehicle. Let's look further – what are some other users of the vehicle other than the crew that are in the vehicle when it's going out on a mission? Is there anybody back at a control base who is interacting with the vehicle, for example tracking its location, receiving telemetry sent back from the vehicle, or analyzing images from cameras on the vehicle, and giving interpretation of those images to the crew? Those people, let's call them remote analysts, are actors as well.

We also need to consider non-human users, external systems that interact with our system. One is the Global Positioning System (GPS) network, a set of satellites used to determine location. The vehicle is receiving signals from those satellites just as a smartphone or civilian vehicle does, though with higher precision. We can include GPS as an actor for the vehicle. Is the vehicle interacting with other communication networks – does it connect it to command and control systems, either at the base from which the vehicle departed, or other military systems perhaps anywhere in the world? Can the vehicle interact with the Internet and interact with other resources on the Internet? If so, then the Internet should be shown as an actor.

Does the vehicle interact with anything locally in its environment – can it receive radio signals, or track radar signatures? It may seem odd to include "enemy radar" as an actor for our system, but there is an interaction, so we do. Can the vehicle listen for sounds from the environment, like gunfire or voices? All those things could be depicted as actors. Does the vehicle detect weather conditions? If so, the weather itself or the environment might be an actor to the vehicle because the vehicle interacts with it in some way. Perhaps this vehicle has a way to interact with other similar vehicles in the environment using peer-to-peer communication as part of supporting a mission, in which case the other vehicles themselves would be actors to our vehicle. Can this vehicle communicate with infantry or other warfighters in the area patrolling on foot? A soldier who's not on the vehicle or in the vehicle, but is interacting with the vehicle from a distance, would also be an actor. What about drones? Can the vehicle interact with drones to see imagery of the battle space where it's operating? A drone would be an actor to our vehicle and have a certain kind of interaction with it. In addition to drones, what about other aircraft? Is the vehicle able to communicate directly with military aircraft in the area, or not? The VX may go through a command and control network which then communicates with the aircraft, in which case the aircraft doesn't have to be included as an actor to the vehicle.

What we're trying to capture here are all the entities outside our system that are interacting with our system. If we continue in this way, eventually we'll identify all the actors for our military vehicle system. Remember, these can be human users, or they can be other systems, other machines, or technologies that interact with the VX. Notice here, we're not paying any attention to what's inside our vehicle, like any of the radios or systems that are inside it, because we're looking at our system from a black box perspective. For a context view, we don't look at any of the internal subsystems in our vehicle. Later we will, as we design the details of the VX, but not right now.

As we create a context diagram, we can represent actors using stick figures, small cartoon drawings, or even photos. For example, for the driver, we might have a little photo or drawing of a driver sitting behind a wheel. For a foot-mounted soldier a mile away who's communicating with the vehicle, we can use a little drawing of that soldier. The easy alternative is to just use a stick figure, which is a symbol that's built into the modeling tool. For external systems, we have the same choices.

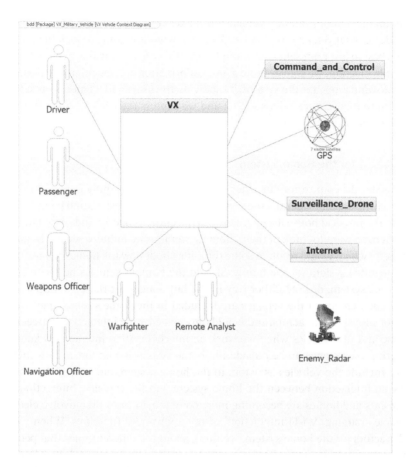

Figure 7.1 Context diagram for VX military vehicle. Source: GPS Photo, https://commons .wikimedia.org/wiki/File:GPS24goldenSML.gif. Radar Photo, https://en.wikipedia.org/ wiki/File:RES%E2%80%931.jpg.

The built-in symbol, a box, can be used, or we can take the time to find or create drawings or photos. If we are depicting the weather, perhaps a little rain cloud would be the most expressive. The VX context diagram example shown in Figure 7.1 uses a combination of drawings and boxes. Drawings take more time to find or create but may communicate better and can make the diagram more attractive to stakeholders. Diagrams that use only symbols may be clear to those familiar with the modeling language, in this case SysML, but may overwhelm more casual readers of the diagram through the repetition of the simple symbols.

In the context diagram, we put our system, the VX, in the middle of the diagram. Of course, we can represent that too with just a box, or we can have a sort of artist's conception of what our military vehicle looks like. That can be risky, because we haven't designed it yet, and we don't really know what it is going to look like. It's probably best to just represent it as a simple box. It's worth noting that even in this very simple diagram, we convey quite a lot of information and meaning. We imply a boundary and scope for the system, by showing the users and elements outside the system that have interactions with it and we express the need for all of those external interfaces.

7.2.2 Actors for the Home System

Let's consider the context for our other example system, the family home. In this case, it's a little harder to identify the system boundary. The most natural boundary would be the physical house itself, or perhaps the entire property, and some families might even want to think of their "home system" as including a second home somewhere, which would result in some thinking about how the homes might act as an integrated system for the family. Should the family's vehicles be included in the home system, or not? Either way is fine but notice that there are implications for each choice. If the vehicles are included in the home system, then we would not show them as actors, and would only consider the interaction between the house and the vehicles when we work on internal system interactions later. Considering the vehicles to be outside the home system is fine too, and in this case, we include the vehicles as actors to the home system, but only if there is actually an interaction between the home system and the vehicles. Interactions between cars and homes are becoming more common, in cases that involve electric vehicle charging, Wi-Fi interaction, or other advanced functions. When we consider actors for the home system, we think about the different roles that people might have in relation to the home system, so the actors are not, Mom, Dad, Jason, and Erika, the actors represent the roles they play. We might start with an actor called *homeowner*, which could represent one or more people who act in the role of homeowner. To this we'll add an actor called *resident*, which would represent anyone living in the house on a permanent or nearly permanent basis. Since actors define roles and not individuals, Mom might function as homeowner role sometimes, and the resident role sometimes. We might add actors for overnight guests and for visitors, which might have somewhat different interactions with the home system. Actors also represent other systems that interact with the home system, and these might include the power utility company, cable company, gas company, an Internet service provider, and perhaps others. Delivery services like UPS, Fedex, or the growing number of independent delivery drivers could be an additional actor, as we might consider how the home system will receive deliveries when someone is or is not home. An example of a context diagram for the home system is shown in Figure 7.2.

Figure 7.2 Context diagram
for home system.

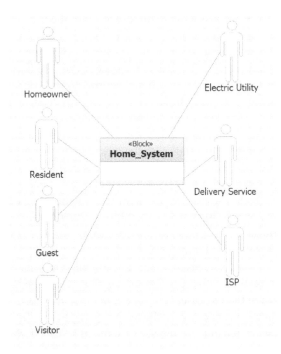

7.3 U: Usage

The U in CUSTARD stands for usage. Usages can be represented in a number of ways, but the best and most common way is by specifying use cases. We introduced use cases in the Chapter 6 as representing complete usages of a system. They are sequences of events performed with a system that produce a result of value. Use cases are initiated by an actor and are complete usage of the system. In other words, a use case represents a goal that the user wants to achieve by using the system. A use case is not just a function that the user can perform with the system. We have seen how to describe use cases in detail, how each is fleshed out in terms of a story, a set of steps, and a small narrative to describe the events of the use case.

For the usage view in CUSTARD, we simply identify the use cases, and draw them on a use case diagram connecting the use cases to the actors with lines. Here we are using the SysML language notation. There are more subtleties to what these lines and symbols mean, but we don't need to go into that here. Fundamentally, a use case diagram is quite simple. It shows the use cases, using an oval-shaped symbol. Each use case is connected to one or more actors, which are symbolized by human stick figures, a labeled box, or a little picture of the external system that our system is talking to. As you recall, the context diagram shows the same actors; we just bring them forward into the use case diagram, add the use cases, and connect the actors. On the use case diagram, we enclose the use cases in a box called the *system boundary box* which represents the system. Since use cases always represent usages of the system, they are shown inside the system boundary box, while actors, which are external to the system, are shown outside the box. Since we're going to describe use cases in detail later, we'll proceed to look at the two example systems and see what the usage view would show.

For our military vehicle, the VX, what are the usages that we should identify? It's typically best to start with the most common or primary usage of the system – the use case that's at the heart of the reason that the system exists. The fundamental purpose of this military vehicle is to carry out a mission. It might be that there are different use cases for different types of missions; each type of mission could take place in a different way. Recall from the Chapter 6 that a use case is a complete usage of a system, so the use case would identify how someone uses the vehicle to carry out the entire mission.

If the VX needs to be able to carry out a reconnaissance mission, that would be a use case. A use case should be named as an action phrase, so the team would name the use case, *Perform Reconnaissance Mission*. Another use case might be an assault raid, where the team is going to use the vehicle to attack a building in enemy territory, so the use case might be called, *Perform Raid* or *Carry Out Raid* or *Prosecute Raid*, whatever terminology makes sense to the stakeholders of that system. There are probably other use cases for that vehicle. Perhaps there is a use case for loading the vehicle with supplies, putting ammunition into the weapons, and preparing it for a mission. It might be that different people prepare the vehicle, so different actors might be associated with this *Prepare VX for Mission* use case.

There are likely some other use cases for the VX. There could be a use case for when the VX is going to be part of a convoy to go from one place to another. There could be enough difference between this use case and *Perform Reconnaissance Mission* that it should be shown as a separate use case. In the *Drive in Convoy* use case, perhaps the VX must maintain a certain kind of communication with other vehicles in this convoy. Or perhaps the VX is under control or monitoring from a remote station during the trip, or the VX may use different sensors when it's in a convoy. If any of these differences exist, then *Drive in Convoy* should be a different use case from *Perform Reconnaissance Mission*. There might also be use cases like *Perform Maintenance* if there is active involvement of the VX in the maintenance process. Maintenance can be of two kinds. One kind of maintenance doesn't really involve the vehicle at all; the maintainer just uses tools to work on the vehicle and the vehicle doesn't have to do anything – it is just sitting there passively as the maintenance is performed. That kind of maintenance may not need to be represented as a use case, since there is no functionality of the VX being described – it is not really a use case of the vehicle. But sometimes there is functionality built into the vehicle, say, a port for diagnostics, like we have all had in our cars since the

Figure 7.3 Use case diagram for VX.

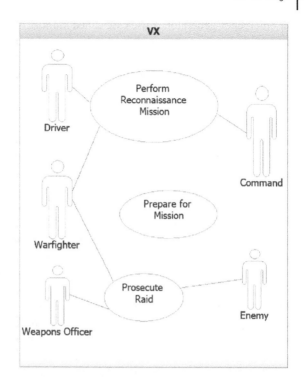

1990s, where the maintainer connects a device to the vehicle to diagnose what's going on. Since that's clearly an interaction with the vehicle, there should be a use case called something like, *Diagnose Vehicle*, which would describe what happens in the diagnosis process. If we continue in this way, we'll soon identify all of the usages of the VX, and place them on a use case diagram. An example with some of the VX use cases is shown in Figure 7.3.

For the home system, it may be quite fun to apply use case thinking to the use of a system that is so familiar to all of us. Here, we ask what the actors shown on the context diagram use the home system for, or what the home system uses them for – both ways indicate use cases. Perhaps the first thing to come to mind, is that the residents and guests use the home to *stay comfortable*. We name the use case from the perspective of the actor who initiates the use case. The *stay comfortable* use case is initiated when someone turns the HVAC (heating, ventilation, and air conditioning) system on. The home system is also used to provide power to many other systems. The *power up systems* (where power is used as a verb) use case would involve providing power to all the systems and devices in the home, from the large kitchen and laundry appliances to the charging of a guest's mobile device. It would involve distributing power from the outside utility and from any sources inside the home such as a solar array, wind turbine, or emergency generator. A use case called *manage access* would include how the home admits the right people and prevents entry by unauthorized ones, likely using a combination of mechanical locks, garage door openers, or Wi-Fi enabled smart locks. While we are talking about security, we may want to include some so-called negative use cases – use cases that we don't want to succeed. A would-be burglar would try to *break in*, and the home system would resist or detect the intrusion in the described ways. Sophisticated security systems are designed by considering stories and scenarios, and with multiple levels of detection and reaction. A neighbor walking her dog by the front yard might trigger only some automatic lights, but an intruder walking up the driveway, might trigger more lights plus an alert inside the home and a camera capture. The forced opening of a door would trigger an alarm and a call to the police via a professional monitoring service. The home system could also be used to play music throughout the home, and there are a variety of ways to implement that function ranging from a set of Amazon Echo devices, to a professionally installed whole house audio speaker system. Many other use cases are possible for the home system.

I often find myself thinking in terms of use cases when considering home improvements in areas like entertainment, bicycle storage, and computing. As an example, when moving into a new home, I've found it useful to live in it a while, perhaps months, before deciding how to automate the lightning. Once the patterns are known, it becomes clear which lights are used every day and which are used only occasionally and need not be automated. The beginning of a use case diagram for the home system is shown in Figure 7.4.

Figure 7.4 Use case diagram for home.

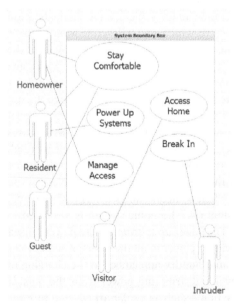

7.4   S: States and Modes

The S in CUSTARD is for states and modes. States and modes, like context, are a very familiar concept to engineers and to systems engineers in particular, from long before model based systems engineering was developed in the early 2000s. Thinking about states and modes in systems is a natural, intuitive way to understand important aspects of how systems operate. A home alarm system has obvious states that are apparent even to the user of the system. The alarm system can be in the *off* state, during which it may still monitor for fire or smoke conditions. It can be in the *disarmed* state, during which it may chime when doors and windows are opened, but will not set off alarms. The *armed* state normally has two sub-states. The system can be armed in *home* mode or armed in *away* mode. In home mode, the alarm goes off instantly, but only when a door or window is opened, and not when motion detectors are triggered. In away mode, a door being opened starts an entry delay to allow residents of the home to enter and disarm the alarm, while a window being opened would trigger the alarm immediately. The system also has a state we could call *alarmed,* where the alarm is currently activated, triggering a siren, lights, and a potential call to the homeowner or a professional monitoring center who will alert the police.

Understanding the states and modes in an alarm system is the key to understanding its operation. States are connected to each other by state transitions. Generally, a system cannot simply move from any state to any other state. Our home alarm system, once it's in the alarmed state, can probably only move from that state to the off state. From the off state, it could then be put into the armed/home or armed/away states. A car in the *driving* state cannot usually be put directly in the *parked* state, without first going through the *stopped* state. Systems engineers use state transition diagrams to think about of the system's intended states and modes, the possible transitions between them, and the triggers that cause the system to move from one state to another.

The terms *state* and *mode* have no universal definitions that would differentiate them from each other, so strictly speaking, the terms are interchangeable. In a particular context, a project might define them more specifically, for example, by deciding that *states* will be the highest-level states of the overall system, and *modes* will be the sub-states within those high-level states. Subsystems may also have their own states. A car might have overall states of *on, started, off, parked,* or *locked,* while the transmission of the car (a subsystem) would have its own set of states, including *park, drive, neutral,* and *reverse.* Additionally, there is a relationship between the states of the car and the states of the transmission. Some transmission states are only possible in some states of the car. If the car is locked, the transmission must be in park for example – some cars enforce this by not allowing the key to be removed if the transmission is not in park. Putting the transmission in park while the car is driving is an invalid combination of states.

States and modes of a system often come up naturally, when engineers are talking with stakeholders. Even stakeholders who have no particular engineering background may naturally give you the states and modes of a newly imagined system as they talk about it. A stakeholder might say something like, "OK, first, somebody is going to unlock it, and then the car is unlocked, and it stays unlocked until the car is moving and then it locks again automatically." They are naturally referencing states and modes of the system, and it's very useful to capture that information as input to the identified states and modes of the system. Of course, the systems engineers will add other more detailed states and modes they flesh out the design in depth.

Let's take our two example systems, the VX military vehicle and the home system, and identify some states and modes in those systems. Because the VX is a vehicle, some of the states and modes will be easy to identify. One state might be *parked*. If the vehicle is parked, it means it's stopped and people can get in and get out of it safely. Another state might be *on* or *started* meaning the engine is running, the power is on, power is applied to the various subsystems of the vehicle, and it's ready to move. Once it's been put in gear, and it's moving, then it's probably in another state called *driving* or *moving*. When the VX stops at a stop sign, assuming military vehicles still need to stop at stop signs, it is still in the driving mode, it just happens to be motionless at the moment; therefore, we might call that state *stopped*, which is different than *parked*. The vehicle does not park at a stop sign; it just stops. There are likely other modes that are particular to the VX's role as a military vehicle, including perhaps a state called *weapons hot*, which means that all the weapons are activated, staffed, and ready to fire at a target. There's no universal set of states and modes – they are specific to each system and its purpose and function.

Some of the examples given suggest the notion of sub-states. The VX might be in a state, and then there might be some states within that state, which we call sub-states. If the vehicle is in the state of *driving*, it may be either *accelerating* or *braking*. We can represent these as sub-states to better describe the overall behavior of the vehicle. Transitions can be shown both between the overall states of the vehicle and between the sub-states. If it is only possible for the VX to go from the overall state of *driving* to the overall state of *stopped* only when it is in the *braking* sub-state, but not when it is in the *accelerating* sub-state, this can be shown.

In identifying states and modes, we only identify the ones that make sense and that are important for our system. In our house example we find that the states may seem more abstract, because the house is not one machine. While the house is a system made up of other systems, it is still useful to consider the states of the house as a whole. One important state of the house might be *vacant*, which means that nobody's home. When nobody's home, we may want things to happen differently in the house, such as lights operating automatically to simulate people being home

Figure 7.5 States for the home.

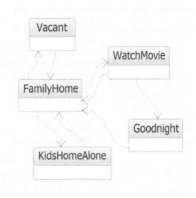

or certain appliances like stoves or curling irons being turned off automatically. *Vacant* is a particular state of the house. We may want to differentiate between the state of *no one is home* and the state of *no humans are home*. If there are pets home alone, it might make a difference in how the air conditioning and heating systems are set. An example of a diagram showing states and modes for the home system is shown in Figure 7.5.

We might also imagine that there are different states depending on *who* is home. If the teenage son is home alone, he might want his favorite music blasting through all the rooms in the house, but if Dad's home also, then the music is more localized and everyone is listening to something different. The ultimate might be having states that represent who is in each room of the house.

States and modes are identified not only for control of electronic or automated systems – states can also be used to help guide physical design. One state of the house might be called workday; in the workday state, one or more people are working at their jobs from home, while the kids are at school or away from the house. How should the house be designed to accommodate that state? Perhaps there would be two separate offices, and perhaps they should not be right next to each other, so that noise doesn't interfere with the other person. Maybe it makes sense for mom and dad's offices to be at opposite ends of the house, so that they have quiet spaces to work from during the day when they're both working at home. Thinking about states can be illuminating regarding the home's design.

There might be another state that happens on a weeknight, where dad is doing a little bit of extra work. Maybe mom is not working at that moment, and the kids are home, and they're enjoying some kind of entertainment, or working on home-work. Again, analyzing such states might cause us to think about designing in soundproofing, or noise isolation between mom and dad's offices, the family room, and the entertainment center of the house, because we want to make sure different activities don't interfere with each other. It might lead us to the currently popular open design of homes, where for example the person preparing food in the kitchen can still be together with others who are socializing or watching TV.

I was once in a house that was designed by a wonderful architect, and the house was beautiful. It was designed around a very open kind of plan, perhaps too open. It had a dramatic two-story open living room space, and the bedrooms were to the sides and above this large open area, with an open balcony on the second floor. Perhaps accidentally, it was designed so that if you set a fork on the counter in the kitchen, you would hear it in every room of the house. There was just no sound isolation because things were so open. If you have one, or maybe two people living in that house, that might be fine, but if you start filling up those bedrooms with children, that wide open sound environment might not be so beautiful after all. Thinking through usages, states, and modes can reveal problems like that before the house is built.

7.5 T: Timing

Continuing to the next aspect of MBSE and working through the acronym CUS-TARD, we arrive at T, which stands for *timing*. Here we describe the timing that occurs in the system. Some kinds of systems have very critical timing aspects, and these aspects are important to reason about in the early stages of the systems engineering process. Other systems might not have timing concerns, so the area of timing may deserve more or less attention depending on the particular kind of system. Timing involves several aspects of the system, and the main aspect of timing is sequence – the sequence of events that must happen in the system. Do the events in this sequence require very precise timing? Maybe certain things in the system can't take too long, or they'll make other things wait, and the whole performance of the system suffers. The system may need to be optimized to prioritize the speed of certain operations over other operations so that the whole system works smoothly. For example, if in driving a car, the driver selects a gear change and it takes 10 seconds to make the shift, that's probably unacceptable, no matter how smoothly the shift occurs. There may be a certain acceptable range in the timing of some system operations. Let's look to our two systems to see some examples of this.

First, for the VX military vehicle, what kind of timing aspects are important to consider? The time it takes to go from the *secure* state to the *weapons hot* state might need to be below a certain limit, so it can be accomplished in a very short amount of time, which could influence certain kinds of designs choices. If getting the weapons ready to use involves getting out wrenches removing some bolts, that might take too long. Designers might consider a different way to release the weapons so that they can be ready to use quickly. Another aspect of timing might be the sequence of interactions between a sensor or radar device on the vehicle and the tracking of a target. Does the VX need to be able to track an aircraft flying overhead? Or does it only need to be able to track other vehicles? That's a big difference. Vehicles can go at most 100 mph, but aircraft can go many hundreds of miles per hour. If the VX needs to track aircraft, it might need a faster computer to meet the faster timing requirements.

In our house example, we first notice that houses operate on a slow time schedule, compared to military equipment, so we may not have many timing concerns. There's not much that has to happen instantly or in milliseconds, so there may not be a lot of critical timing aspects to the design of the house. The systems engineer could consider how long it takes to get from the house to some other destination, as an aspect of the house. And if we're shopping for a house, its location relative to traffic patterns and the travel time to work, stores, or the beach could be an important concern.

What kinds of timing things might be important inside the house? Well, one aspect of timing is in the entertainment system. When trying to stream 4k ultra high-definition content to your TV screen, you need sufficient Internet bandwidth, but also adequate speed in all the connections between the devices in the path from Walt Disney's servers to your TV screen. The house might require a current version HDMI cable to be able to stream the latest 4k content. What networks, wired and wireless, are needed in the house to accommodate all the use cases we've identified? Thinking through requirements of that kind might cause us to run fast network cables through the walls when we build the house, so that we can hook up various kinds of devices in the future and have faster and more reliable access. Or we might do what I did in the last house I built: I ran large plastic conduit pipe in the walls between rooms and between floors, allowing me to fish new cables anytime I want. These days, what I primarily fish through those pipes are Ethernet cables, but sometimes other cables are needed. I ran speaker wires in the walls since there is still no good substitute for hardwiring speakers, and it's hard to fish speaker wires to the right locations later on. By thinking through what we need the subsystems in the house to do, we think systematically about these requirements and design elements, and timing can be an important aspect.

7.6 A: Architecture

The A in CUSTARD stands for architecture. The systems-level architecture reflects the overall structure of the system. Many books have been written about system architecture and systems architecting, and we're not trying to cover that subject in depth here, but let's try to get a fundamental intuition. When we think about the overall system, we first picture it as a black box, the way it was shown on context diagram we created earlier. The single box in the middle of the context diagram represents the entire system, no matter how complex it may be. In the architecture stage, we think about decomposing that single box into the subsystems that make up the overall system. All systems are made up of subsystems, which of course are systems in their own right. Most systems are also subsystems of other larger systems. Therefore, all systems can be considered systems of systems. To paraphrase the Yogi, it's all systems (or turtles), all the way down.

When looking at a familiar type of system, one that already exists, like a passenger vehicle or personal computer, it's easy to see the subsystems, but when faced with the creation of a new kind of system, deciding how to break the system down into subsystems may not be as obvious. It takes a combination of experience with that kind of system along with good knowledge of the principles of system architecting to arrive at the best decisions.

Two of the central concepts in architecture are cohesion and coupling. An ideal architecture has high cohesion, that is, each subsystem has a tightly related set of functions or services for which it is responsible. At the same time, a good architecture has low coupling, that is, subsystems talk to other subsystems as little as possible. With a high-cohesion, low-coupling architecture, subsystems carry out their work largely independent of each other, communicating with other subsystems only to respond to requests or return results and updates. The architecture of a corporation typically follows this pattern. Finance, Marketing, Sales, and Human Resources are all highly cohesive divisions, carrying out their specific functions and communicating frequently within themselves, but interacting much less frequently with the other divisions.

To put it another way, if two subsystems have a lot of interaction going on between them, it makes for a lot of communication overhead and added complexity, and it could make us question whether those subsystems should be combined into one subsystem. On the other hand, if we have a function like power, we probably want that to be a single subsystem supplying power to all the other subsystems, rather than giving all the other subsystems their own power supply capability. Let's make this even more clear by considering our example systems.

With our VX military vehicle, we consider what subsystems should be identified within the overall system of the vehicle. There's no single, correct, or obvious answer. If the system is being designed from scratch, then the designer is free to choose the subsystems as the architecture is designed. But military vehicles have been created before, so there could be a kind of normal or even standard architecture. Sometimes there are reference architectures that can be used for some categories of devices and systems. Organizations such as a national defense agency even standardize architectures (or try to) so that new products will follow consistent architectural patterns. But fundamentally, it's up to the systems engineers to come up with the architecture for the system that they're designing.

For the VX, we imagine we would have some subsystems that are common to any vehicle, including the one parked in your driveway. In addition, there would be subsystems specific to this vehicle and its purpose. Most vehicles will have some kind of engine or motor to propel the vehicle, as well as a suspension system. Typically, there is some kind of body, and there is likely a chassis or a frame. The architecture might treat these all as second-level subsystems, or might take an alternate approach by combining the suspension subsystem with the chassis or frame subsystem in an attempt to maximize cohesion and minimize coupling.

As a military vehicle, the VX would have sensor systems, including perhaps radar, LIDAR, Radio Frequency (RF) sensors, or optical sensors. With so many kinds of sensors, it might make sense to combine the sensors into one sensor subsystem, and then allow other subsystems to request sensor data from it. Then at the third level of architecture, there could be sub-subsystems for each different kind of sensor.

Continuing with the second level, the VX will probably have some weapon systems and communication systems that provide radio, satellite, or other kinds of communication from the vehicle to other vehicles, bases, or aircraft. That's just one possibility for the architecture, though. Take communications, which we treated as a subsystem above. Another possible architecture would be to give each appropriate subsystem a communications capability of its own, rather than one integrated communications subsystem that serves all of them. Which is better? That trade-off is exactly what systems engineers struggle with when designing the architecture of a system. It's not always clear which trade-off option is better. One engineer might argue that a single common communication subsystem could be vulnerable to failure or attack – a single point of failure. Another might argue that it's easier to secure a single communications subsystem than to secure multiple little communications capabilities scattered throughout the subsystems of the vehicle. On the other hand, the communication needs of a particular subsystem might be so specialized that it makes more sense for that subsystem to handle its own communications instead of throwing in with the common communication subsystem.

Another common example of the trade-offs that must be considered when architecting a system is power. It might be intuitive to think that there should be one power system for the vehicle, distributing power to all subsystems. That could make sense, but another option would be for the power to come into the vehicle and be immediately distributed to all the subsystems, with each of them having their own power supplies that create power in the form needed for that subsystem. For example, if AC power comes into the vehicle, say from an on-board generator, we could have either one big DC power supply that distributes DC to every subsystem, or the AC could be distributed to all the subsystems, and they could each have their own DC power supply. Again, that's a trade-off that the designers will make as they design the architecture of the system.

An important recent trend and goal in many kinds of complex systems is to make architectures more modular. For example, if a new kind of communication capability becomes available, and the architecture of a system is modular enough, it should be feasible to take the old communication system out and put a new communication system in. A modular architecture would just require that interfaces between the new communication subsystem and the other subsystems are the same as they were for the old communications subsystem. If communication was distributed in little bits and pieces all throughout all subsystems, it could be much more difficult to pull out the old communication communications and put in a new one. Modularity and standardized interfaces have become very important in architectures of all kinds due to the rapid pace of technological progress.

For the home system example, we must consider the architecture of the house. The architecture of a house is more familiar to most people than that of a military vehicle. To a systems engineer, the architecture of the house as a system concerns much more than the style and appearance of the home and its floor plan. We might start with the foundation and consider it to be a subsystem of the house, since it is typically made of different materials, is installed differently, and serves a different purpose and function than other parts of the house. Similarly, the framing, including the walls, floors, and ceilings, might be considered a subsystem of the architecture (though it could be argued that house framing is not a system, since it consists only of wood and nails, and has no interactive functionality). It probably makes sense to consider the roof to be a separate subsystem because it serves different purposes and is designed separately from the foundation and framing.

It's important to realize that designing an architecture based on historical knowledge of how similar systems have been built in the past can limit innovation. Remember that Connolly's law suggests that the architecture of systems will mirror the architecture of the organization that designed them. If in the past our company has built homes by having three separate departments (or contractors) to handle the foundation, framing, and roof, it will be no surprise that these three parts of the house will always be separate and distinct, connected only when the home is built. An innovation that suggests a way to build, say, the roof and walls as an integrated single subsystem (perhaps a dome design) would have a hard time being accepted since it would mean radically changing the structure of the organization. We could combine the roof and wall divisions and direct them to create an integrated roof-and-wall approach, or eliminate both of these groups altogether and hire some new innovative designers. Either of these options would likely be highly destructive to productivity, at least in the short term so organizational inertia acts to keep designs fairly consistent over time.

Connolly's law works both ways, however, so if we design the architecture based on a new innovation that may allow roof and walls to be engineered as one, we could set up our new organization in just that way – with a roof-and-framing division, alongside the foundation division. In either case, Connolly is always right – the system architecture and the organization architecture will tend to match, for the good or otherwise. The lesson here is to be careful of assumptions in the early phases of designing the architecture that might limit innovation later. If the architecture shows that the roof and framing are separate subsystems, it is likely that these two subsystems will be assigned to different divisions (or contractors) and thus it is highly unlikely, even impossible, for a single integrated design approach to be taken for these two subsystems. It's another trade-off. To some extent systems engineers must lean on the knowledge of how past systems have been built – they can't start with a clean sheet of paper for every new house or car model. But every now and then, it makes sense to challenge past assumptions and consider that, to take a real example, we don't need an engine in the car at all – just individual electric motors at the wheels and a large battery system built as an integrated part of the chassis. An electric car design that followed the traditional architecture would probably result in one large electric motor, replacing the engine under the hood, and a large battery bank where the fuel tank used to be – workable perhaps, but neither optimal nor innovative. In sum, examine early assumptions about the architecture of the system to be sure they are justified and not innovation-killing.

Back to our house, we note that in this case that the assumption of the roof being a different subsystem from the framing is justified, since the roof will be designed using different principles and materials, such as trusses and plates instead of wood and nails. Identifying other subsystems in the house architecture, we probably want a plumbing subsystem, and an electrical or power subsystem that takes in electricity from a utility, generator, or solar panel system and distributes it throughout the house. Then there will be a heating and cooling subsystem (or HVAC). We consider this one subsystem, even if eventually it may consist of multiple furnaces, heat pumps, wood stoves, or other kinds of devices to do heating and cooling. This subsystem also includes ventilation, such as fresh air exchangers, fans, and ducts. From there we might imagine more specialized subsystems that will depend on the homeowner's particular requirements. We could sketch in a subsystem called communications that would include the home's connection(s) to the Internet and other kinds of communication within the house, such as voice assistants, distribution of Internet to computing and streaming devices, and intercom systems. The communications subsystem would also include managing the ways that mobile devices operate inside the house and how they connect to the cellular network and home Wi-Fi. The communications subsystem might also include computer networks within the house. Today, that's probably a combination of Ethernet and Wi-Fi. Moving from technology into the hobby domain, the house might have some kind of workshop, art studio, or workout facility, and depending on the complexity and needs of each, it could be useful to consider each of these as separate subsystems, rather than just ordinary rooms. A workshop might have some different power requirements, or different structural requirements for the floor or walls, or special ventilation requirements, depending on what kind of work will be done there.

As is clear from these examples, the process of designing the system architecture is far from a straightforward, fill-in-the-blanks kind of exercise. In the CUST part of CUSTARD, we focus mainly on the needs and requirements for the system. When we reach the A for architecture, we must start wrestling with the design of the system. English fails us here, because design can refer to several aspects of the systems engineering process. Some may refer to the creation of use cases as an aspect of design, while others think of design as only applying to the physical construction details, drawings, blueprints, and machining instructions. The important thing is to understand the concepts of how a system comes together from its earliest vague ideas to its concrete implementation. It's in the next stage of CUSTARD that we bring together the *needs* for the system and the *implementation* of the system.

7.7 R: Realization

The purpose of the realization stage is to show how the system will be realized, that is, how the behavior described in the usage, states, and timing views will be carried out by the system when it's eventually built. Obviously, all functions of the system are carried out by the system in some way. In order to build that system, however, we need to break the system down into parts, which we did in the architecture stage just a moment ago. We now need to break down the needed behavior of the system into functions that can each be carried out by a subsystem of the overall system. The sequence of steps describing a use case must map to a subsystem that carries out each step. Or, since use case steps might not have a one-to-one correspondence with functions in subsystems, it may take the actions of several subsystems to carry out some use case steps.

Mapping out the realization of a system using model based systems engineering is a two-step process. First, we draw out the sequence of events for each use case, treating the system as a black box performing the functions of the system, but not showing any of the subsystems inside. Then second, we draw the same sequence again, showing how each of these system functions is handled by a subsystem, or a combination of subsystems. To accomplish the first step, we create a diagram in our model that shows the interactions between the system and its actors for each use case. Usually this is done using an activity diagram in the SysML modeling language, which resembles a traditional flow chart. We add swim lanes to this diagram to show who or what is responsible for each action. In this black box activity diagram, there will be swim lanes only for the system itself, and each of the actors who are involved in this particular use case – not all actors are involved in every use case. We walk through each step in the use case and model that step as the set of actions taken by the system and the actors. Let's use our example systems to make this clear.

For the VX military vehicle, one of the use cases we identified back in the *Usage* stage in CUSTARD was *Perform Reconnaissance Mission* so let's walk through a few steps of that use case and see how we translate it into the actions of the actors and the system. The first step reads,

This use case begins when the user accesses the VX for carrying out a reconnaissance mission.

Though this line sounds a bit awkward, it makes sense as the description of the beginning of the use case. Remember that the use case is about carrying out a sequence of actions using the system. We can't say that the use case begins when someone decides to use the vehicle for the mission, because making a decision is not something that involves the system. We also don't want to say that the use case begins when we unlock the door of the vehicle, since we want that unlocking to be *part* of the use case, and too, the vehicle might not even be locked.

To show the realization of this step, we show the actor performing the action *Access VX* and then the action *Start VX*. After the user performs Start VX, the VX performs the action *Power Up*. This action may involve many different actions within the vehicle, but remember that right now we are considering the VX as a black box, and from that perspective all we can see is that the vehicle powers up. In the next step, the user initiates a data communications check, and then the VX performs a sequence of data signal tests with both the command and control system and the Internet gateway system, which are represented as actors since they are external to the VX.

As we walk through the story of this use case, we break down each step into one or more actions taken by a user or by the system, including any interactions between the system and other outside systems like the command and control system or GPS satellites. In creating the black box version of the activity diagram, we are simply retelling the story of the use case in terms of the interactions between actors and the system – we aren't inventing anything new yet. Doing this retelling may not be a trivial task, however. Recall that we placed no significant restrictions on how the use case should be described in the *Usage* stage of CUSTARD, only that it should be expressed in terms familiar to all stakeholders so that it can be understood by all. That means that the descriptions may leave out specific descriptions of what each actor and the system actually *do* as the use case proceeds. In *Realization*, we must think a bit more precisely and describe each action in detail. The original use case may have simply read, "The user starts the vehicle." That's fine as a step in the use case, but in order to accomplish the goals of the realization stage, we must think about what this step means in terms of the action of the user and the action of the system. Doing this should not be terribly difficult, but it requires some focus and diligence. In the present example, the user takes some action to request the vehicle to start and the vehicle responds by trying to start, and either succeeding or failing at this attempt, and then giving some indication of the outcome to the user.

With a little thought, it becomes apparent that the user does not actually start the vehicle. The user only requests that the vehicle start itself. It can be challenging to express this without falling into mentioning components of the system, which we must not do when using a black box perspective of the system. We could be tempted to write that the user turns a key or presses a button, but keys and buttons are low-level components of the system and should not be mentioned here. To do so risks dictating design decisions that should be made later, after considering alternatives. But if we think of what the user does as making a request, we can simply assign the action of *request vehicle start* to the user and the action of *start vehicle* (essentially, *start yourself*) or just *start* to the vehicle.

So far, we have taken only the first of the two steps in *Realization* by creating the black box description of the use case, in terms of the actions of the actors and the system. Once we complete this for all use cases, we will have produced some important and useful results, even though we have only taken the first of the two steps. First, we have a succinct description of everything the system must be able to do. Collecting all of the actions shown in the system swim lane of the activity diagram, we have a nice list of all actions to be taken by the system. In all likelihood, there are duplications in this list, which is to be expected since it is quite likely that the system will need to take some actions during more than one use case. What's also likely is that the systems engineers doing this work didn't name some identical actions in exactly the same way across different use cases. Even if just one systems engineer did the work, naming variations are likely. To keep the descriptive system model correct and compact, it is important to check the list of actions and give identical actions identical names across all the use cases in which they appear.

In addition to a nice list of the system actions (functions) needed to accomplish all use cases, our work so far in *Realization* has also produced a description of all actions required of our actors. In the case of our human actors or users, this list could guide the training we offer to actors. In the case of the external systems this list is the beginning of an external interface description, a key part of the full system description. Even at this early stage of systems engineering, we have identified all the occasions where our system will need to interact with an external system, and more importantly, the precise context of each interaction. It's one thing for systems engineers to make a general list of the systems with which a new system will interact, and this is often done, but it's far more useful to describe when and why each external system interaction must take place, and this is what we are doing by starting from use cases, and identifying *where* interactions with external system actors are needed in each use case flow.

Let's move on now to describe the second step of the *Realization* stage of CUS-TARD. As before, the purpose here is not to describe exactly how to perform this aspect of MBSE modeling, but to give you a good intuition of what needs to be accomplished in an MBSE approach overall. In this book we focus mainly on the early stages and of how to describe the needed behavior of the system, for two reasons. First, this is the aspect of MBSE that is most needed for intelligent systems, and second, this is the aspect that is often given limited treatment in descriptions of MBSE methods.

The second step of the *Realization* stage is the creation of a white box representation of each use case. The purpose is to show how the use case is carried out or realized by a combination of the subsystems we identified in the *Architecture* stage. Before, in the black box representation we showed how the use case is carried out by the system and the actors. Now, in the white box representation, we will show how the use case is carried out by the subsystems of that system. The good news is that we don't have to start over with each use case – we can begin with the black box representation and transform it into the white box representation. To perform this transformation, we simply make a copy of the black box activity diagram we created above, and then add swim lanes for the subsystems. We then have a diagram that has swim lanes for the actors, the system, and the subsystems. It doesn't make sense to have *both* the system and the subsystems on the same diagram in this way, but this is just a temporary form of the diagram.

We then proceed to examine each action that is in the system swim lane, and ask the fundamental question, what subsystem of the overall system carries out this action? There are two possible answers: either the action is carried out by a single subsystem, or it is carried out by some combination of more than one subsystem. If the action is carried out by a single subsystem, then we can simply move the action to that subsystem's swim lane and go on to the next action. If the action is carried out by a combination of subsystems, then we have a little more work to do. We'll need to split this action up into multiple actions, each of which can be carried out by just one subsystem. In this way, each action in the system swim lane becomes one or more actions in the subsystem swim lanes. The result is that we have allocated all actions in this use case to the subsystems.

As we do this work, we maintain the connections to the actors. If in the black box representation, an actor performed an action A followed by the system performing an action B, followed by another actor performing action C, and in the white box representation action B became actions B1, B2 and B3, performed by various subsystems, then we would see a connection between action A and action B1, and then from B1 to B2, from B2 to B3, and then from B3 to C. The original black box action B is replaced in the sequence by actions B1–B3. When all the replacements, called *white box expansions,* are completed, there will be nothing left in the original system swim lane and it can be removed from the diagram, leaving only the actor and subsystem swim lanes.

Once we have completed the white box expansions for all use cases, we can collect the actions for each subsystem, and rationalize them as we did for the system functions, renaming any identical actions with differing names to use the same name. The result is a complete list of actions for each subsystem, with those actions shown in the context of the use cases in which they are used. Identifying the needed actions or functions of subsystems is one of the primary purposes of systems engineering, and MBSE provides a systematic way to accomplish it, representing the result in a widely understood graphical language like SysML.

7.8   D: Decomposition

The final stage of MBSE and the last letter in CUSTARD indicates that much of the process we've described is performed repeatedly as we decompose the system into multiple levels. Above, we saw in the *Architecture* and *Realization* stages how the system is decomposed into subsystems and the needed behavior expressed in the use cases is allocated to those subsystems. In most system development efforts, this doesn't go far enough to complete the work of the systems engineers. The subsystems of the system are still too complex to be handed off to the discipline-specific engineering teams, which include electrical engineering, mechanical engineering, software engineering, and others. A subsystem like the engine system in a vehicle still requires a combination of these engineering disciplines to design and build.

Systems engineers must decompose the system further, into more detailed levels. How many levels are needed depends on the complexity of the system. Generally, it is best for systems to have approximately 7–12 subsystems at a single level. This guideline comes from the awareness of the ability of human beings to fully understand and hold in mind only a limited number of elements at once. A system with 50 subsystems at the second level would be hard to keep in mind, so almost inevitably in such a case, engineers would start grouping the 50 into a few subgroups, creating a level in between the system and the 50 subsystems. By starting with the assumption that we should limit the number of subsystems at each level to a smaller number from the start, we can avoid backtracking.

The process of decomposing the system is commonly illustrated by systems engineers in the *Vee diagram* shown in Figure 2.1. Systems engineers have used the Vee to explain and understand the development of complex systems. Some will argue that the Vee model suggests or even requires a waterfall development process and that agile is a better approach. I tend to think this is a misunderstanding of the Vee. The Vee is not meant to express a development process at all, but to express the idea of decomposition. Complex systems are not created the way one would knit a scarf – beginning at one end and knitting until the scarf is completed. Complex systems are inevitably created by breaking the system down into subsystems, the subsystems into sub-subsystems, and so on until a level of decomposition is reached where the components can be built. For example, to build a bicycle, we think through the subsystems of the bicycle – frame, wheels, gearing, brakes, steering – and then proceed to build each one of these, and then integrate them together. A system as complex as the space shuttle is built the same way, decomposing the overall system far enough to be able to build each component, and then integrating the components into sub-subsystems, the sub-subsystems into subsystems, and finally the subsystems into the final spacecraft. The Vee model makes this clear and shows the inevitable phases of system development where early phases deal with the system overall – its requirements and overall architecture, followed by phases that deal with decomposition of the system and allocation of needed behavior to subsystems, sub-subsystems, and components. Then at the bottom of the Vee, components are built in their own engineering specialties including electrical, mechanical, and software. Components are tested and verified within in their own disciplines before being integrated into sub-subsystems. Each sub-subsystem is tested, before being integrated into subsystems, which are tested and finally integrated into the whole system, which is tested as a whole, before being put into actual use.

Now, let's apply *Decomposition* to our example systems. For the VX, we've already hinted at the next level of architectural decomposition when we referred to the sensor subsystem and the different types of sensors that could be included within it, such as radar, LIDAR, RF sensors, or optical sensors. We could call these sub-subsystems, but many projects will give a more sensible general name to each level of decomposition. How about if we call sub-subsystems *elements* in our examples? For a particular model of the VX, we have elements within the sensor subsystem for radar and optical sensors. It's the VX sport model – the luxury model would have more. In our MBSE process, we do more than just decompose the architecture into the system, subsystem, and element levels, through – we need to allocate behavior to the items at each of these levels.

For the VX, let's say we've already identified the action of *detect target* and assigned it to the sensor subsystem. When we decompose the sensor subsystem into its sub-subsystems, called elements (the subsystem of a subsystem is an element), we must figure out how *detect target* is going to be accomplished by the sensor subsystem, which is to say, how the sensor subsystem's elements will collaborate together to accomplish this action or function. We'll do this in the same way we described in *Realization,* though we only need to apply the second step of white box expansion. We already have the black box description of the actions that the sensor subsystem needs to take from the white box expansion we did at the level above. So now all we need to do is another white box expansion based on that previous white box expansion. Exactly how this is shown in diagrams is beyond the scope of this MBSE overview, but the main idea is that we take all the actions allocated to the sensor subsystem in the initial white box expansion, and then do a white box expansion on each – it's all white box expansions, all the way down. For the VX, our *detect target* action, allocated to the sensor subsystem, must be allocated to some combination of the elements (sub-subsystems) of the radar element. In this example, we might design that a *detect target* action is performed simultaneously by both the radar element and the optical sensor element (camera), and then the results are compared and presented back to the operator. In this case we would have two actions, both named *detect target* – one on the radar element and one on the optical sensor element. While it might seem confusing to have three different behaviors with the same name (and they could be named differently if desired), it makes a kind of sense. These two sensors can both detect a target, even though they do it in completely different ways, so they each have a function named *detect target*. In addition, the system overall has an ability to detect a target, meaning that the system can be used, if desired, to perform only the detection of a target, without doing anything else like firing at it, or navigating toward it. This idea that a complete usage of the system could consist of only detecting a target is clear from the existence of the *detect target* use case. To reiterate, the system overall and the two sensor subsystems can all perform a behavior called *detect target,* though the meaning of this behavior is different for each.

Moving on to our home system example, let's apply decomposition to the home heating, ventilation, and air conditioning system (HVAC). In the *Architecture* phase, we began the decomposition of the home system and identified HVAC as a subsystem. At the next level of decomposition, we must identify sub-subsystems (or elements). There are several approaches we could take here, and the right choice depends on how we want to think about the architecture of the HVAC system. It may seem curious to those not familiar with engineering and systems that there could be multiple valid ways of representing the architecture of the same physical system, but this is indeed the case. Put another way, there are various architectures that might result in the same system being engineered and built. In the case of the HVAC system (and by the way, it's completely valid to refer to a subsystem as a *system* in the general sense, since subsystems are in themselves systems) one way to represent the architecture is the obvious one – draw out three sub-subsystems for heating, ventilation, and air conditioning. If we are designing an HVAC system from scratch, and with no prior knowledge of how these systems have been designed in the past, this might be a logical way to start, but this approach could lead us to a very non-optimal design. Taken literally, this three-element architecture would result in three independent systems for heating, cooling, and ventilation despite the potential efficiencies of making some aspects of these three operations common. All three typically involve the movement of air, so architecting the system in this three-part way could result in three separate air-moving systems, one for each of the sub-subsystems. This is not necessarily impractical – a home with an oil-based heating system and mini-split air conditioners is an example of this architecture used in many homes. The point here is that architectural decisions made early on can sometimes exert unintended influence on the later design of the system. Here, by assuming that heating, ventilation, and air conditioning are three separate sub-subsystems, we lead designers to make them completely independent, each with their own air-moving capability (assuming they even need such a capability).

An alternative architecture would be to designate the sub-subsystems as forced air, heating, and cooling. This subtle difference in terminology could have big impacts on how other engineers approach the design. The forced air sub-subsystem could be designed to move the air for the entire system, while heating adds heat to this moving air flow, and cooling adds cooling (actually, subtracts heat). This is a workable architecture and is exactly how things are done in many homes. The issue of what's common, and what's separate also comes up in the control of the HVAC system. Should we assume there is one unified control system, with one or more thermostats that control the entire HVAC system, or do we assume that each of the three sub-subsystems have their own control system?

In practical terms, the architecting of a system at this stage is governed by two competing influences, and the systems engineer must strike a balance between them. One influence is the pull to design the system following the architecture of existing systems – to do it *the way we've always done things*. Following historical patterns of design is not necessarily a bad thing; often those designs have been proved and refined over years or decades of iteration. The competing influence is innovation. The more we avoid assumptions about how a system should be architected based on past systems, the more we allow for innovation and new ways of doing things. For HVAC systems, the question of whether one large system, perhaps with multiple zones and duct dampers, is better or worse than several independent systems, does not have a clear answer, and many homes are heated and cooled in each of these ways. The systems engineer must, paradoxically, consider allowing for innovations that have not yet been made. The sources of heat and cooling in HVAC systems are quite separate in systems installed in cold climates, with heat being generated by burning gas or oil, but in milder climates the innovation of the heat pump generates both heat and cold from the same system components by simply reversing the cycle, switching the cold side from outside to inside in summer and the reverse in winter. Allowing for an innovation is often more a matter of avoiding making architectural decisions too early in the design process. Keeping heating and cooling in the same sub-subsystem enables design teams to consider a single system that does both, while separating them early on may result in two teams, each focusing on one subsystem, with little chance of a single, integrated system being the design choice. If the systems engineer knows that the house is in Boston or Minneapolis, then the separation makes sense; if it's in Miami or San Diego, not so much.

We'll continue the HVAC example to illustrate another aspect of architectures – what systems engineers call *multiplicity*. In an architecture, some elements may be used in the system more than once. It's a simple notion, but profound in implication. In the HVAC system for the house, instead of implying that the house is heated by a single system, we can designate that the architecture should contain one or more heating units. Doing so would allow the designers to consider having multiple heating units, perhaps one per floor, or even one per room of the home. There are practical HVAC systems used today with each of these variations. Having multiple units has the advantage of redundancy – if the upstairs heating system goes out, the downstairs unit can work harder and provide backup heat for upstairs. On the other hand, multiple systems can cost more than one larger system. An interesting design used in some large buses and recreational vehicles is to have multiple air conditioning units on the roof, all flowing into the same ducting system. On mild days, only one unit is activated, but on hot days multiple units power up. The units are inexpensive and easily replaceable, compared to a large system built permanently into the vehicle.

The point of our rather lengthy treatment of HVAC systems architecture is that decomposition of a system is no mere formality or detail – it is an important part of the overall design process, setting in motion and guiding the design activities that come afterward. The architecture of a system is not so much the way the system will be built, but more the way systems engineers and designers think about the system, and how it carries out its needed functions and use cases. Good architecture results in a system that is often simpler, less prone to failure, more reliable, and easier to extend and modify in the future.

7.9 Conclusion

In this chapter, we have reviewed the main stages and concepts in the development of a model of the system when a model-based systems engineering (MBSE) approach is used. MBSE came into focus as the most effective way to do systems engineering in the early 2000s and has steadily gained popularity among systems engineers in the last 20 years. MBSE is now the primary approach taken in most new complex systems development projects and is increasingly required by government and defense organizations when they buy new systems. There are various methods and modeling languages used for MBSE, but in recent years the SysML language has become prominent, with several modeling software packages available. The main purpose of this chapter has been to provide an overall view of the MBSE process. In Chapter 6, we focused on the early stages of the process, *Context* and *Usage,* since these are at the heart of the purpose of this book.

8

A Time for Timeboxes and the Use of Usage Processes

Time is an illusion

(Albert Einstein)

Despite Einstein's assertion, we seem to live in time. We imagine life to be the passage of time, and time is essential for nearly all our activities. Time is also essential to our systems and their design. To use a system is to use it in time. No use case takes place outside of time or in zero time. Time is also important inside of a use case. Response time of a system to a human user's button press or screen tap can make the difference between usability and a failed product.

To design intelligent systems, we need better ways of describing how a system works in time, what happens when, in what sequence, and what behavior can happen concurrently. Despite the fact that nothing could be more intuitive than dealing with the past, present, and future of time, behavioral system models can cause unnecessary confusion due to an inattention to the representation of time. Analyzing how human beings think about and relate to time in everyday life sheds new light on how the complex behavior of systems can be represented. For instance, the past is easily represented on a timeline, because past events are seen – in hindsight – as occurring linearly. Processes represented as flow charts or activity diagrams usually appear to be happening in the present, as if all actions were taking place in the infinitesimally short period of time we know as the present moment. The future, in contrast, is seen as a landscape full of potential events or tasks to be done – each may be assigned a start time and an intended finishing time, but there is often tremendous uncertainty about when the events will actually occur, how long they will take and in what order they will happen. Time itself, however, knows nothing of our distinctions of past, present, and future – these are human mental concepts – time is an illusion.

In this chapter, we take a closer look at how we human beings deal with time from the perspective of system design. We will see that some of our current ways

Engineering Intelligent Systems: Systems Engineering and Design with Artificial Intelligence, Visual Modeling, and Systems Thinking, First Edition. Barclay R. Brown.
© 2023 John Wiley & Sons, Inc. Published 2023 by John Wiley & Sons, Inc.

of representing time in descriptions of a system we are attempting to design may be inadequate. In many cases time may simply be ignored considering the needed behavior of a system. We have a rich set of concepts and language to talk about time and events occurring in time – past, present, and future, but our ability to express these in system models is very limited. In what follows, we'll suggest some new ways of representing time – ways that allow us to express what we know about systems and time with a new level of clarity. We can use these new techniques to capture more intuitive views of the system behavior, processes, and use cases we model as part of our system design process.

8.1 Problems in Time Modeling: Concurrency, False Precision, and Uncertainty

There are three main problems in the typical ways of expressing time in system models and in descriptions of business processes and activities – concurrency, precision, and uncertainty: We will examine each in turn.

8.1.1 Concurrency

Mark Twain said, "time is what keeps everything from happening at once." It's good that not everything happens at once, but many things do happen at the same time, and we do have some ways to represent this concurrency. Activity diagrams, introduced in the previous chapter, allow for forks and joins, which enable the modeler to show behavior happening concurrently in time on a single diagram. However, what if there is concurrency between behavior represented on one diagram and behavior shown on another? Subsystems within a system often send signals among them, triggering behavior that may happen in one subsystem as other behavior happens in another. Unless we can put the behavior of the entire system on a single diagram, there is no way to depict this concurrency. As we read the descriptions of the system behavior, we tend to build mental models of what is happening in various parts of the system, but it's hard to represent these on a diagram. If the system we're considering is a human activity system, say a university, we can imagine many things happening at once – classes taking place, food deliveries to dining halls, sports games, parties, administration meetings, and more, but imagining how those might be shown in an integrated way bogles the mind, especially when we realize that each of these is not a single process, but is itself a series of interconnected processes, with their own concurrencies.

8.1.2 False Precision

Commonly used representations of time in systems suffer from a false precision. Gantt charts showing tasks in a project show each beginning on a specific date, sometimes even a specific time, most realize that there is a great deal of uncertainty. A task scheduled for a date next year, might occur on that date, but could easily occur days or weeks before or after the planned date (usually after). Two events appearing on the chart may look like one will start after the other, because the scheduled start dates are a week apart, but it's very likely that it would be fine for the project if they started in the reverse sequence. The chart expresses a false precision regarding the sequencing of these two events – they don't really mean what they seem to show. Gantt charts also allow us to show that one task must complete before another, but is the meaning that the previous task must complete, or just that the second task should start around the time scheduled for the previous task to complete?

We have no problem speaking of time in approximate terms:

"Let's get this project started in Q1 of next year."
"We should start the new project when the previous one is almost done."
"Graduation is planned for June of 2025; let's plan the trip for the following Fall."

While we routinely talk this way, computer-based scheduling and calendar tools cannot usually handle approximate dates, like "Q1" or uncertain dates, like "sometime in June." When we try to schedule long-term projects with uncertain dates, we tend to be forced into a false precision, choosing exact dates for tasks that are 18 months in the future, though we know it is highly unlikely the task will begin on the planned date.

8.1.3 Uncertainty

We often build uncertainty into our thinking and speaking about time. It's easy enough for us to say things like, "The project should take between three and six weeks, but likely closer to six," but it's difficult or impossible to express this in scheduling and modeling tools. For example, say a project has a set of six sequential tasks and each of them is estimated to take between three and six weeks. What is the best estimate of the completion time of the project? Given the well-documented planning fallacy, perhaps the most accurate estimate would be 36 weeks or even more. But the more accurate answer would be between 18 and 36 weeks. We might even go further and add that the most likely duration is 30 weeks, with a 5% change of it being less than 20 or over 35. What we are expressing here is a distribution of completion times. Again, commonly used modeling methods do not allow us to express this kind of uncertainly.

The passage of time is, of course quite familiar to all human experience. In systems modeling, however, time presents some particular difficulties, resulting in a tendency to ignore time in models of the complex behavior of systems, especially at higher levels of abstraction. There are two primary paradigms for the expression of complex behavior. Here, we will examine each with respect to its treatment of time and then proceed to describe an approach to the modeling of time using timeboxes.

8.2 Processes and Use Cases

System behavior can be expressed either in a series of processes, or in a set of use cases. We've introduced use cases in the Chapter 6. A process is series of related events, actions, or other processes, tied together in a sequential flow. Since processes can include other processes (sometimes called sub-processes), they can be used to express behavior at any level of abstraction. A national election of a new president is a process, as is the way in which an omelet is prepared. Processes are expressed in flows using tools such as flow charts, SysML activity diagrams, or business process modelling notation (BPMN) diagrams. All of these show processes in a similar way, in a flow of events, with a start, finish, and optional conditional branching and simultaneous events shown in parallel flows. The passage of time is implied, since events follow one another in sequence, but it is not possible to see the time-based relationships between events appearing on different diagrams, nor can it be seen which events occur simultaneously, when they are part of different process flows. Even within a single process flow, business process models do not strictly require that the previous event be completed before the next can begin, making it even more difficult to relate events in time to each other. This is not necessarily a shortcoming of these modeling methods since as it is possible that timing may not be relevant to some applications.

To illustrate with an example, imagine that getting the family ready for the day involves three processes: making and serving breakfast, getting dressed, and dressing the kids. Each of these three is a separate process, which could fairly easily be drawn out as a process flow, in a flow chart or SysML activity diagram. But what is the relationship between these three processes. We can say that they all occur at the same time, starting around 6:30 am and ending around 8:00 am, but that doesn't give very much information. We may want to convey that the coffee maker must be started before getting in the shower. One approach would be to simply combine all three processes into one process, and then every activity in each process can be shown in sequence, with the appropriate finish-to-start relationships, but it may be hard to do this. Does starting the toaster come before or after approving the kids' clothing choices for the day? What seems to be needed is a way to express activities and processes in time, but with the flexibility we have as we speak about them informally.

As covered in the Chapters 6 and 7, use case models express system behavior as a set of interactions or dialogs with users (people or other systems). They express the complete usages of a system to accomplish specific goals. Use case models are made up of two main parts. A use case diagram shows all (or a subset) of the use cases of a system along with their relationships to actors. For each use case there is also an expression of the flow of events for that use case, either in text form as a narrative or set of steps, or in a graphical form, typically using an activity diagram. The expression of the flow of events of a use case uses the same format and technique as the description of a process, and thus has the same limitations as to the expression of time. The use case diagram, on the other hand, eschews the notion of time altogether. No sequence among the use cases appearing on the diagram is expressed or implied. All use cases are processes, but not all processes are use cases.

In practice, the relationship between processes and use cases is an interesting one, with some teams and project employing only one of the two, and others applying both. Of the two, use cases are the more direct and specific way to express the functionality of a system and we've spent considerable time on their use in Chapters 6 and 7. In what follows, we'll show how these two ways of describing the behavior of a system and its users rest on different paradigms for describing behavior in general, and it happens in time. We'll also show how the fundamental assumption that describing the business processes within which a system operates enables the team to proceed directly to the description of the needed system functionality is false, and how this assumption causes wasted effort in large projects. Finally, we'll show two innovations in the modeling of behavior that enable us to integrate business process modeling with use case modeling and produce a single model that achieves the objectives of both kinds of models. Business process modeling is often employed as a precursor to the development of a system to be used in a business organization, while use case modeling follows to specify the needed system functionality, again spending significant time with stakeholders to gather the needed input.

The two innovations described in this chapter are the *usage process* and the *timebox*. Usage processes allow use cases to be identified and integrated with business process models as they are developed. Timeboxes allow processes to be positioned in time-relation to each other without the need to combine processes into higher level processes. The combination of usage processes and timeboxes allows any level of complex behavior to be modeled in one pass, without the redundancy and waste of separate business process and use case modeling work.

8.3 Modeling: Two Paradigms

Business process modeling teams and stakeholders may spend months or years developing detailed business process models, expecting that these models will provide a useful base of information for system designers. Unfortunately, as the business process model is analyzed by the system designers, it is found that information needed to specify the functionality of the system does not exist in the business process model. System designers may then employ use case modeling to specify the needed system functionality, again spending significant time with stakeholders to gather the needed input.

Stakeholders find this two-pass process redundant and wasteful of time and money since the input they provide to both modeling teams is largely identical, with each team capturing only the aspects relevant to their form of modeling. As we will see, an integrated approach that achieves the objectives of both business process modeling *and* use case modeling in an integrated form, in one analysis pass, should result in time savings, increased accuracy, and improved communication among all participants in the systems development process.

While there are a number of legitimate objectives that justify business process modeling, the intent in the some of the large-scale development projects we've observed over the years was to provide an understanding of how the business operates so that a proper system could be designed to support and automate it. For example, an insurance claims processing organization would model the business process of receiving, evaluating, and paying claims to policy holders in order to better develop a claims processing system (Kajdan 2008; Hammer and Champy 1993; Bergener 2015). Understanding the business processes at work in an organization can be a worthwhile and important goal (Li et al. 2014), and can be helpful in designing a system to support those processes. This logic is what usually justifies the investment in the business process modeling work. Such work is not inexpensive, due to the extraordinary amount of time it requires to interview the necessary people in an organization, synthesize and reduce their input, formalize it in a business process modeling notation and perhaps most significantly of all, gain widespread agreement on the resulting models.

Following the business process modeling work, which may require months of concentrated effort by a team of analysts or consultants, the system design team would begin its work. Their first step is to determine exactly how the system to be developed is required to work to support the business, thus they are asking a different question from the business process analysts whose work preceded theirs. To determine the requirements for the system, a system use case model is developed in the manner described in Chapters 6 and 7.

8.3.1 The Key Observation

It is generally assumed that in this systems development process, the business process models developed in the first phase will be useful to the designers of the system in the second phase (Krogstie 2013 in Glykas 2013). However, as business process models are analyzed by system designers, they observed that the information they need to describe the use cases and design the system did not exist in the business process model (Mili et al. 2010). To create the use cases, the team had to retrace the steps of the business process modelers, capturing different information as they talk with many of the same subject matter experts involved in the business itself.

The costly part of developing a business process model is not the conceptualizing or drawing of the model, but the time spent eliciting information from those familiar with the business. In most cases, multiple people in the business organization are interviewed, often for many hours or days each, to find out what they know about the business processes being employed. Invariably, people have different views of the processes, and these views must be reduced to a common understanding. Later, when the development team needs to develop the system use case model, a very similar process is conducted, walking through the same business processes but gathering different input. Business process modeling asks the question, "How do the business processes happen?" and they describe the processes, step by step. While occasionally they might capture the use of a system, for the most part of the business process, modeling ignores the role of systems in the process. The origin of ignoring technology when doing business process modeling goes back to a time when most business processes were carried out manually, with people writing and printing documents on stand-alone word processors, copying documents on copiers, filing documents in physical file cabinets, making phone calls, writing entries in paper logbooks, and so on. In those days, it made sense to capture the business process as it was being done, without the use of technology, so that it could be analyzed and an automated system developed. Today, most business processes are already automated to some degree, possibly with various online systems and the task is more likely one of integration or replacement of these systems, or the enhancement of currently automated processes to implement new requirements and changes.

When technology is already an essential part of a business process, it makes no sense to ignore it when modeling the business process, just as it would make no sense to design a new city while ignoring the presence of cars, as we did in the days before cars existed.

In some system development projects we've observed, the same subject matter experts are involved in both business process modeling and the later system use case modeling, and worse, they are asked to give much the same information. This suggests that the information needed by the system designers to create use cases may have been given the first time, when the business processes were defined, but this information was not captured since it was not needed for the business process models. In addition to being a potential waste of time, such as double interviewing risks frustrating and fatiguing the subject matter experts, reducing the quality of the information.

It is not the fault of the business process modelers – rather, it is a limitation of the business process modeling paradigm. Business process models are not designed to capture the usage scenarios of systems used during the business processes. Thus, any information about system usage given by the business stakeholder and subject matter experts during the business process modeling activity is lost, simply because there is no systematic way to capture it in the business process model. The same ground must therefore be trod again by the systems modelers to create use cases.

When business process modeling is employed, the result may be a better understanding of the business or mission, but despite expectations, system use cases cannot be determined from these business process models. Both business process modeling and system use case practitioners tend to assume that business process models do provide a foundation from which to determine system use cases, but despite these expectations and claims in the literature, there is no clear method for deriving system use cases from business process models (Alotaibi and Liu 2014; Mohapatra 2013; Deutch and Milo 2012; Sinha and Paradkar 2010).

Two-pass system modeling, with system use case modeling following business process modeling, has some other disadvantages as well. Even though the information put into the business process and system use case models is distinct in form and content, contradictions between the two may be introduced. Technology suggested or implied in the business process may differ from the technology specified during the use case development phase, or the technology may fail to meet the real needs of the business process. It is a common complaint about systems that they don't fully meet the needs of the users and the lack of connection between business process and use case models may be one of the causes.

8.3.2 Source of the Problem

The lack of a clear connection between business process models and use case models is neither a deficiency of business process modeling nor of system use case modeling – it is mainly a difference in intent. A business process model is intended to capture the business process, not to specify how some technology will enable, or already enables, the process. Business process models generally take one of two approaches to technology, ignoring it completely, leaving out any mention of computer systems or other technology used in carrying out the business process, or simply including the work carried out by the technology in the process descriptions. In either case, the distinct role of the technology system is generally invisible when looking at the business process model (Gerth 2013).

As mentioned, the clear separation of business processes and technology may have been appropriate when addressing a business process that is currently carried out manually, and for which automation is being planned, but an integrated approach is a better match when automation exists, and technology improvement or replacement is being designed. Since the automation of business processes using computers and software has been underway for more than 40 years, it is likely that technology is involved or even essential to most business processes today.

As an example, consider an airline's business processes. When describing the passenger reservations *process*, likely the business process model would include activities such as "passenger makes reservation" with the responsible business entity being the reservation agent. If further detailed, this activity would describe the information given by the customer, the actions taken by the agent and the response back to the customer. Most often, there is no description of how technologies, such as the telephone network, computers, and software support this activity. When the system designers then proceed to design a new reservation system, oddly enough, they will want to write a "passenger makes reservation" *use case*, in which the roles of the passenger and agent, represented by actors, interact with the software system to accomplish the reservation. Separate use cases would be developed for a passenger making a reservation via the worldwide web, a passenger approaching an agent at the airport, and a passenger phoning the airline. In a way, each of these use cases follows a flow of events similar to the business process, but these use cases are not derivable from the business process flow, so the analysis process, including interviewing subject matter experts, reduction, analysis, and synthesis, is repeated under the system use case paradigm in order to create the use case model (McSheffrey 2001; Hruby 1998; Dhammaraksa and Intakosum 2009).

8.4 Process and System Paradigms

The fundamental reason that business process models and use case models cannot be integrated is that they derive from two incompatible paradigms: the *process* paradigm and the *systems* paradigm. In the process paradigm, upon which business process modeling is based, behavior is described in processes, which consist of sequences of activities. Processes may also consist of other processes, which may be called sub-processes. The process paradigm is evident in approaches such as business process modeling, flow charts, and task procedures. The process paradigm is also intuitively familiar in ordinary life in the form of instructions for say replacing the brake pads on an automobile, a recipe for making lasagna, or instructions for filling out a tax form.

In the systems paradigm, upon which system use case modeling is based, behavior is described as sequences of activities and interactions between systems, sub-systems, and users, in order to achieve a specific goal. Each activity in the systems paradigm is performed by a system, sub-system, or user, and the overall sequence describes the usage of a system to accomplish some desired goal – a use case. The systems paradigm is evident in approaches such as use case modeling, concept of operations (CONOPS), model-based systems engineering, operational scenario modeling and in instructions for using a system, such as user guides.

Once the need for a new system behavior modeling paradigm, enabling the integration of business process modeling and use case modeling has been identified, several approaches can be considered. One alternative considered was to simply eliminate one of the two modeling approaches and use only the other by itself. But, business process modeling on its own has been shown to be inadequate mainly because it does not produce adequate software and system requirements. Use case modeling on its own may be inadequate because it provides no stakeholder-friendly way to represent the business or mission context within which the use cases of the system are carried out. Use case modeling on its own will be successful if the business processes involved are well known and understood among the stakeholder, and the task is only to design a new system. In general however, since both methods provide value, and neither is reducible to the other, the need arises for a new paradigm integrating both (Sinha and Paradkar 2010; Bahill 2012; Knauss and Lubke 2008; Lübke 2006).

In past work (Brown 2013), a set of goals was proposed for a new paradigm of behavior modeling. Among the goals were the following.

Relative Temporality. The ability to show how various processes and activities correspond in absolute or relative time, much as tasks appear on a GANTT chart aligned in time, but overcoming the false precision problem described earlier.

Complex Behavior. Be able to represent any scope of complex behavior (e.g. a machine, an organization, a city), over any time scale, or mix of time scales (e.g. microsecond weapon timing, insurance claims processing, national energy strategy development).

Easily Readable. Models should be understandable and readable by untrained readers using only the aid of legends, labels, and the like.

Express Uncertainty. Allow for "fuzzy" definitions of responsible entities or actors, time scales, and interactions, as may be appropriate to express limited or evolving knowledge levels.

8.5 A Closer Examination of Time

It is intuitive to think of time as a combination of three components: the past, the present, and the future. The difficulty in representing time in models of complex behavior stems from the conflation of these three aspects of time. Taking each one in turn and examining our mental models may shed light on how we might express time in behavior models.

History, whether personal and recent, or grand and ancient, lends itself to a purely linear representation. This is because we know when each event happened in time – there is no doubt about exactly *when* anything happened (assuming complete knowledge). Two baseball games between the same pair of teams, no matter how similar, happened at two different times. No matter how many times someone eats a steak at a favorite restaurant, these are clearly distinct and separate events, if only in time. In the past, a space shuttle launch is not a complex arrangement of actions that may happen at various times – it is a specific series of events in time.

As these examples illustrate, viewing time in the past is easily done using the notion of a timeline. Since the absolute dates and times are known, it is a simple matter to arrange these events in a linear form. By using different time scales, spans of minutes, days, years, or centuries can be conveniently represented. Virtually all historical descriptions include timelines illustrating past events.

Examining how people perceive the *present* and what is happening is akin to asking the question, "what are you doing now?" For the most part, the answer will be given in the form of a process that is being carried out. Of course, the actual present moment is an infinitesimally small point in time during which nothing can actually happen, but in common colloquial language, the question of "what are you doing now?" seems to be taken as "in what process or processes are you participating now?" For example, if the answer is, "I am baking a cake," we understand that the respondent is carrying out the *bake a cake* process and is somewhere in the time interval required for that process. In our common usage of language, there is no need to distinguish the exact point in the process, only that we have begun it and have not yet finished it. Though the entire process is not actually occurring in the present instant, we understand what is meant by, "Now I am baking a cake."

What is relevant for our purposes here is to note that both process models and use case flows of events are normally specified as if they occur completely in the present. Though we know nothing actually happens without the passage of time, the passage of time is not indicated in process models; processes all appear to be happening "now." Except for an instruction such as, "Bake at 350° for 30 minutes," any mention of time is omitted from the process, and may not even be necessary for the process model to accomplish its intended purpose.

We also note that the use of a timeline to express this present-oriented process is not very helpful. We might put the events needed to bake the cake on a timeline, but immediately we are faced with uncertainty about exactly how long each action will take, which actions might precede others and what choices need to be made along the way. Timelines, in their traditional form, are unsuited to expressing events that have not happened yet.

We also note in passing that the present is actually quite full. Many processes are underway. Giving a complete answer to the question, "What are you doing now?" would include not only baking the cake, but also working at a job. While the respondent may not be at work that very moment, mapping out the *working at a job* process began when he or she was hired, and will continue until that job is left behind. If asked, "where are you working now?" the respondent is not likely to answer, "well, I am working in the kitchen, but tomorrow I'll be working at my job across town." The question is understood to refer to the long-term process of employment in the job – the person is working at that job now, even though her or she is at home baking the cake. The same goes for living in the house. The process of living in a house begins when the house is purchased or rented and ends when the house is vacated. So in the present moment, in addition to baking a cake and working at the job, our busy respondent is also living in a house. Of course, we all could go on and on, listing a great number of things we are "doing now." All of the processes began in the past, days or years ago, all converge into the infinitesimally small point we call the present moment. Even though the present has no duration, there is a great deal happening – a paradox perhaps.

When people think about events that will happen in the future, there is always uncertainty and a lack of precision. A high school student planning a college degree program may know that it will involve four years, a number of semesters, courses, exams, and social activities, but the lack of specific knowledge about when and even in what sequence any of these activities will happen would make it difficult to represent the future in either of the two ways described above for past and present events – something else is needed.

Conventionally, the future is specified as a plan or schedule. A plan, such as a Gantt chart, shows a series of planned events on what appears to be a timeline, but it is not a timeline in the same sense as a historical timeline, specifically, the planned activities and events can move around as time passes. Something placed on the timeline for the tenth of next month is understood only to be planned for this date, and might move earlier or later dozens of times between now and then, depending on how it is related to other events and activities in the schedule. For example, some events are indeed fixed to a specific date and time, while others must wait for a previous activity to finish before they can begin.

If, instead of considering the baking of a cake to be a present activity, for instance, we consider baking a cake in the future, we'll tend to imagine the process of baking the cake beginning at some point in the future, perhaps with a deadline in mind for its completion, and in between these two points in time, the processing being carried out. The process could be decomposed into specific activities each of which might be planned out with time dependency relationships to the others (we must mix the batter before pouring it into the pan).

If we accept that the natural and intuitive expressions of past, present, and future events and activities are different and incompatible in their most commonly used forms, then we can consider what it might take to model the past, present, and future in an integrated way. As a precursor, we'll consider an illustration of how the intuitive notions of past, present, and future might be expressed in one view.

A patient in a cardiac care hospital is connected to an ECG (electrocardiogram) machine. The patient's heart is intended to follow a planned process which describes the heart's intended (future) behavior, beating in a healthy, repeating fashion and responding to a stimulus, such as climbing a flight of stairs, by increasing its rate. This is the intended *future* of the heart. The ECG monitor is a window into the *present* where we see the process happening, heartbeat by heartbeat. The chart recorder on the ECG is producing an ongoing record that is receding into the *past*, like a historical timeline. Taken together, the intended future beating of the heart, the current display window and the chart recorder provide a representation of the past, present, and future behavior of the patient's heart.

8.6   The Need for a New Approach

Addressing business process modeling and system use case modeling in an integrated form, with a flexible approach to the modeling of time should yield benefits in time saving, accuracy, and communication. The integrated approach described in the rest of this chapter is intended to accomplish the following key objectives:

1. Capture both business processes and system use cases, in an integrated form, achieving the objectives of both activities in one pass of stakeholder interactions.
2. Represent any degree of complex behavior (e.g. machine, organization, city), over any time scale, or mix of time scales (e.g. microsecond weapon timing, insurance claims processing, national energy strategy development).
3. Avoid specifying duplicate elements that represent the same behavior
4. Eliminate the need for unnecessary or unnatural paradigms, e.g. force an organization to think of itself as a "system" rather than an organization.
5. Allow for all normal forms of behavioral patterns including simultaneous action, asynchronous and synchronous behavior, invocation, return, event triggering, and continuous action.
6. Be understandable and readable by untrained readers using only the aid of legends, labels, and the like.
7. Use familiar modeling semantics, syntax, notations, etc. such as SysML/UML, IDEF, BPMN, etc. as far as is possible.
8. Use familiar conventional notations such as timelines, flowcharts, block diagrams, etc. as far as is possible.

The approach described here resulted from an examination of the fundamental concepts underlying both business process and use case modeling, noting where the fundamentals are similar and where they diverge. Supporting goals (7) and (8) above, the following existing modeling notations were also examined so that the new paradigm builds on existing concepts and notations where possible, rather than inventing wholly new formulations:

☐ Business Process Modeling Notation (BPMN)
☐ Systems Modeling Language (SysML, focusing on Activity Diagrams)
☐ Unified Modeling Language (UML, focus on Activity Diagrams)
☐ Task modeling approaches used in Human Systems Integration (HSI)
☐ Enhanced Functional Flow Block Diagrams (EFFBD)

The new approach is based on two key innovations – the usage process and the timebox. A usage process is a use case of a system inserted into a business or mission process flow, allowing the role of the system or systems involved in the process to be described once, unambiguously, and fully integrated into the process flow. Adopting this method requires only that business process modelers understand the concept of a use case, be able to recognize where one is needed, identify it, and include it as the business models are developed. Usage processes within process models are able to express any combination of manual (ordinary) processes and system usages, whether sequential or concurrent, using only familiar flow notation already common to process modeling.

Timeboxes and timelines are introduced as a way to integrate and synchronize business processes and use cases at any level of abstraction. Timeboxes allow a process or set of processes to be shown as happening during a period of time, without requiring precise scheduling or the strict hierarchical decomposition of each process into sub-processes. Timeboxes may contain other timeboxes and timeboxes may overlap each other, allowing the flexible grouping and decomposition of complex processes. Timeboxes may also optionally contain timelines which enables timeboxes to be shown in temporal relationships to each other.

The combination of usage processes and timeboxes with timelines is able to display a wide variety of complex behavior, at multiple levels of abstraction, and allow flexible representations of the relationships between business processes and the usage of systems. These two key innovations are described in more detail in Sections 8.6 and 8.8.

8.7 The Timebox

To represent the aspects of past, present, and future in an integrated fashion, the notion of a timebox is introduced. The idea of a timebox was originally intended to represent a fixed interval of time within which certain activities were planned to occur. The usage here is does not conflict with that idea, since, when seen in the past, all timeboxes do indeed have a fixed duration – the duration that actually occurred. A timebox in the present has a fixed beginning time which is the time that the timebox actually began, but will have an uncertain (though planned) ending time and thus duration. A timebox containing activities planned to begin and end in the future, will have both an uncertain beginning and an uncertain ending point in time.

The timebox provides a way of integrating loosely connected, multiple processes into a cohesive diagram or model. A timebox can be thought of as a container for a process which is made up of other processes. In its simplest form, a timebox simply indicates a collection of activities or processes located in time. Timeboxes can be nested in various ways to represent the composition of processes (Figure 8.1). Since timeboxes are connected only in time relation to one another, and not in a causal process flow, they allow a much greater flexibility than traditional process decomposition.

When placed on a timeline, this distribution is indicated using a trapezoidal timebox representation, as shown in Figure 8.5. This representation allows for the expression of both estimation uncertainty and approximate or vague time specification. For example, a project may be scheduled to begin in the month of March and finish sometime between July and August, indicating 30- and 60-day distributions for the starting and ending time, respectively.

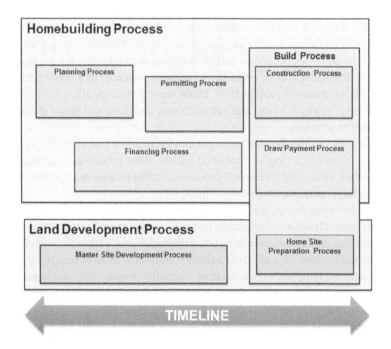

Figure 8.1 Nested timeboxes illustrating home construction processes.

A timebox, then, will have a beginning point, an endpoint, and a duration, though any or all of these may be unknown until the timebox has moved into the past, at which point all three are known. In the method proposed here, timeboxes contain activities and events, expressed as processes, and may also contain other timeboxes. Any set of activities from the small to the large may be contained in a timebox. The process of getting a master's degree might be shown in a timebox that would start sometime after the completion of an undergraduate degree and continue for a planned duration of between two and three years. Showing this process as contained in a timebox allows it to be shown in relation to processes in other timeboxes, say the purchase of a home or a move to another city.

Processes can maintain an independence from each other but at the same time be positioned relative to each other in time. This is the normal and familiar way of thinking about activities and processes in everyday life. A person's career process is distinct from that person's marriage and family process, but they are connected in time, since they are both aspects of the same person's life, and likely overlap in time. It's the same way with systems, subsystems, and components that are connected to each other in various ways during their operation. The activities of an air traffic controller are distinct from those going on in an airplane overhead, but the two processes are related in time beginning when the tower accepts responsibility for the aircraft's route and ending as it lands or is handed off to another controller.

The notation for a timebox is quite simple – a rectangular box containing a representation of the activities occurring within that time interval (Figure 8.1). The activities inside a timebox may be omitted, or can be represented as a flow chart, activity diagram, or business process flow. These representations allow for conditional branching, iterative loops, and simultaneous activities and flows of the behavior inside the timebox.

The simplest form of a timebox is a rectangle labeled with the process contained in the timebox. Timeboxes may be contained within other timeboxes, and may overlap each other, indicating concurrent processes, or that processes are a part of more than one larger process, as shown in Figure 8.1.

In a complex behavior model, it is often desirable to show behavior at multiple levels of abstraction. This can be done easily using timeboxes to enclose other timeboxes and timelines. A timebox containing other timeboxes, but without a timeline (Figure 8.1) would indicate that all of the behavior contained in all of the smaller, contained timeboxes, happens within the time interval represented by the larger, containing timebox. In this way, timeboxes can be nested as deeply as required to model and complexity of behavior. It should also be noted that the nesting of timeboxes is not the same as the decomposition of the processes inside the timeboxes. Process decomposition breaks process elements into small process elements, while maintaining the same place in the causal chain represented by the process. Strictly speaking, timeboxes are not decomposed into constituent timeboxes, but since timeboxes simply represent time intervals, smaller time intervals may be contained within larger intervals.

If a timebox's interval is completely in the past, and the beginning and end of the timebox are known, the left and ride sides of the timebox rectangle are shown as a single vertical line, making the timebox a simple rectangle. Any timebox occurring in the past has a definite and specific beginning time, ending time and duration, but these are not always known. Events occurring in the distant past or events that were not recorded exactly may have uncertainty, even though we know that they did occur at some specific point in time. The date and time of one's birth is carefully recorded, but the exact moment when the person first took the wheel of an automobile may not be. It may be known that this point was sometime early in the person's fifteenth year, since we know that learner's permits can be issued, for example, at age 15 years and 8 months.

Unlike strictly specified tasks on a Gantt chart, timeboxes may express uncertainty through a distribution of possible starting and ending times. Timeboxes that represent future events have uncertainty as to their starting times, ending times, and durations. Next year's summer vacation may start sometime in June and end sometime in June or July, with a duration of two to three weeks. Timeboxes can express this kind of uncertainty changing from the rectangular shape to a trapezoidal one and turning the left and right edges into triangles expressing the range of possible starting and ending times, as shown in Figure 8.2. Each triangle expresses a statistical distribution of possible starting times and ending times. The distribution may be as simple as a triangular distribution, which indicates a most likely starting time along with a best case (earliest) starting time and a worst case (latest) starting time, but with no other indication of the likelihood of the actual time.

As the work of Murray Cantor, an early mentor of mine at IBM showed, using even a simple triangular distribution when estimating times is an improvement

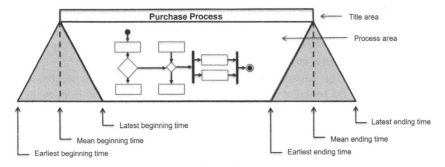

Figure 8.2 Trapezoidal timebox showing a distribution for starting and ending times.

over the usual single point estimate. He would tell a story to illustrate. Suppose a supplier tells you that the project will be finished on March 31 of next year. You intend to show the results at your annual conference in mid-April. Should you feel comfortable with the supplier's estimate? On the one hand, you feel that this is no problem. You have the supplier's promised date and a couple of weeks to spare. On the other hand, what if you asked the supplier how likely it is that the date will be met? If the supplier responds that the date is 90% certain, are you now comfortable? Suppose you ask the supplier what the chances of completion by April 15 would be and the answer is 95%. Is this good enough? It would mean that there is a one in twenty chance that the completion will not happen by the time of the conference. Now instead imagine that the supplier had a bit more of a mathematical and probabilistic bent, and stated that the most likely completion date is March 31, with a best case completion of March 1, and a worst case of April 30. From this you might conclude that there is only around a 50% chance of completion by April 15. Based on this more complete information you might make changes in the project or even the supplier, based on your needs.

The idea is that it is always better to express an estimated time or date as a distribution, rather than a single point in time. Cantor (2008) makes the point that a single estimate without an expressed uncertainty is meaningless. Cantor goes on to point out the variance in the estimated time – the relationship between the most likely, best case, and worst case may be a good indicator of successful progress in the project. There's a big difference between two estimates of completion that both show March 31 as the mostly likely date, but one shows a range of March 1 to April 30, while the other shows a range of March 15 to April 15. One is justified having more confidence in the latter estimate.

The triangular distribution is the easiest to conceptualize and to request from the estimator. It takes no statistical understanding of distributions, but only the answers to three questions: what is the most likely time, the best-case time and the worst-case time for the estimate? A normal distribution can also be used to express the uncertainty of a time, but give the well-documented planning fallacy, the completion date is much more likely to be late than early, so a better distribution might an exponential or other kind of Gamma distribution where the area to the right of the mean is larger than that to the left.

Uncertainty in the starting time, ending time, and duration of a timebox can also come from the inherent uncertainty when a time is specified. If an event is intended to occur at 4:00 pm then there is an implied interval of 60 seconds during which the event may occur and still fulfil the intention. If an event is intended to start next week, the implied interval is seven (or five) days. All specifications of time, no matter how precise, have this "specification uncertainty." Making the uncertainty explicit allows for the representation of a wide variety of time-based behavior, at a variety of timescales. Overlaps in the distributions between two timeboxes shows that there is a chance, but not a certainty that the two timeboxes may overlap in time, as shown in Figure 8.1.

8.8 Timeboxes with Timelines

To locate timeboxes more specifically in time, they may be positioned on a timeline in order to show the planned or actual beginning, ending, and duration of the time interval represented by the timebox. A timeline placed inside a timebox indicates the explicit timing and temporal relationships of the contained timeboxes. Of course, all timelines must be synchronized and aligned within a model since in non-relativistic timeframes there is only one timeline.

As shown in Figure 8.1 above, timeboxes can convey quite a lot of information even without the addition of a timeline. Without a timeline, the arrangement of timeboxes within a larger, enclosing timebox is not significant. The Homebuilding Process contains seven other processes, one of which is shared with the land development process. Without a timeline, the diagram conveys that these seven processes may occur in any order. This non-specificity can be useful to convey that a set of activities will occur within a larger timebox, but the sequence is not known or is not important. An example mentioned earlier is the set of activities a university student will do – taking classes, joining a fraternity or sorority, working at a part-time job, student teaching, etc. – will all occur within the timebox *going to college,* but may occur in any other and in various combinations at the same time.

8.8.1 Thinking in Timeboxes

One of the key objectives of the timebox approach is the representations of multiple levels of behavior, from general high-level behavior to low-level detailed behavior. Such flexibility allows a modeler to "start anywhere" and construct a model with both higher and lower levels than the chosen starting level. Thus, a model may have no fixed top or bottom level – new levels may be added in either direction. In order to depict the relationships between timeboxes in a more concrete manner, timeboxes may be placed on timelines, indicating the starting and ending times, with optional uncertainty distributions, of the processes contained in the timebox. The collection of timeboxes placed on a timeline may itself be placed inside a larger timebox, depicting a higher-level process (Figure 8.3). The two timelines shown cannot be completely independent of one another, because, ignoring relativistic effects and science fiction-like time travel, time is a single continuum. Therefore, a point on the upper, smaller timeline must be the same point in time

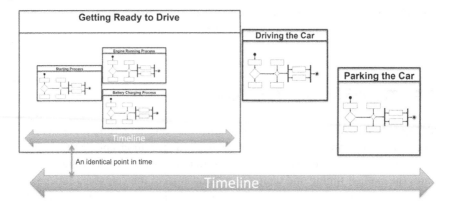

Figure 8.3 Timeboxes placed on a timeline.

as the corresponding point on the lower timeline. The correspondence of multiple timelines gives the entire model continuity and integrity, not matter how large the model grows or how many nested timeboxes and timelines it contains.

The notion of timeboxes when used in conjunction with timelines, allows a great flexibility in representing complex and interrelated behavior. To summarize what has been said so far, the main benefits of using timeboxes and timelines include:

1. Processes can be shown in relation to the passage of time, without committing individual process elements to specific points in time.
2. Iterative and conditional process flows can be shown where they are planned to occur in time, without reducing them to a series of specific events and activities.
3. Processes can be shown as related to each other in time without implying causal relationships and direct connections.
4. The difference between past, present, and future events and activities, is incorporated in a single model. Planned (future) activities flow seamlessly into the past as uncertainty is eliminated.
5. Enough information about the uncertainty of timing is included that overall behavioral models can be constructed and simulated using Monte Carlo and other simulation techniques.
6. Variable precision is accommodated using uncertainty so that start, end, and duration times can be expressed in scales of seconds, minutes, days, months, or years. Zooming in and out in time can be automated in graphical tools.

8.9 The Usage Process

Any series of activities or actions, when put into an ordered sequence is a process. A corporation may have a *new hire* process, that includes numerous activities performed when someone joins the company. A family may have a morning process that includes all the activities that make the family ready to begin the day. Brushing one's teeth is a process as is writing a book, as is someone buying that book online. Processes may or may not include the use of machines, computers, or other systems, and they may or may not mention their use. A *purchase book* process might or might not describe whether the book is being purchased at a store, online, or at a yard sale. A more complex process, like new hire, might use multiple systems to record the employee's information, but in a pure process description, these systems would not be mentioned, or might be mentioned in passing with a note indicating that this process involves the mentioned systems.

The concept of a usage process derives its usefulness from the observation that, based on the accepted definition of a process, all use cases are also processes. Processes are generalized sequences of activities. The definition of a process requires no reference to the use (or lack or use) of a system or systems so it follows that a process must be one of three types:

1. **Manual Process.** An activity that is carried out without the use of any technological system – a completely manual activity.
2. **Usage Process.** An activity that consists of the use of a system to achieve a goal or purpose.
3. **Combination Process.** A combination of activities of the above two types.

It is easy to see that a process of type (2) above is a use case. A process of type (1) contains no use of systems and thus includes no use cases, and a process of type (3) may contain one or more use cases. It is therefore straightforward to decompose a process of type (3) into a combination of type (1) and type (2) processes. When this is done, the use cases become obvious since they are precisely the type (2) processes. After factoring the type (3) processes into manual processes and use cases, what remains are only type (1) and (2) processes – manual processes and use cases.

Since use cases are also processes they can be combined into sequences with other processes. In business process modeling all processes are "compatible" that is, they may all be placed in sequences with each other. This is an obvious but subtle point. While it may not be useful to place the process "make a sandwich" in sequence with the process "negotiate the peace treaty" it is possible and allowable within the rules and method of process modeling. Similarly, it violates no rule of process modeling to include these use cases-as-processes from multiple systems in the same process. Since a process sequence is not owned by or bound to any particular scope or system, use cases representing usages of various systems can simply be included in any process flow.

Processes that represent and are identical with use cases, the type (2) processes above, are called *usage processes*. Processes that are not usage processes will be referred to as ordinary processes. Ordinary processes may be manual processes that do not involve the usage of a system (type (1) above) or they may be combined processes that involve system usages and manual processes together (type (3) above). A combined process could later be decomposed and the usage processes separated from the manual processes, or it may be left combined as it is because the system being used is not of interest in the modeling effort at hand.

The most frequent reason for leaving a type (3) process as it is, and refraining from decomposing it, is that we simply don't care about the usage of the systems occurring in it. As an example, consider if we are designing an avionics system to assist pilot communications with an air traffic control tower. Part of the process we want to describe is say the process of the pilot establishing contact with the tower before and during takeoff. As we describe that process, we are particularly interested in the usage processes that involve the system we intend to design and build, so we'll be sure to separate those out and describe them as usage processes. We are not so interested in how the air traffic controllers interact with their system during the process we are describing, so we may simply leave those parts of the process expressed as combination or ordinary processes.

Since usage processes are distinct from ordinary processes, a different symbol is used to depict them in a process flow. In this way it is immediately obvious to the reader of a process flow which processes are usage processes and indicate the usage of a system, and which processes are ordinary processes which indicate manual or combination processes. The symbol used for a usage process is the same as the symbol for a use case, when shown on a use case diagram – an oval with the name of the usage process either inside it or under it. The usual symbol for a process is either a rectangle or a round-shouldered rectangle (sometimes called a round-angle), so the oval is visually different, making usage processes obvious. Color can be used to further differentiate usage processes.

In the new paradigm of behavioral modeling described in this chapter, all processes are either usage processes or ordinary processes. What must not be allowed is a process that includes the usage of a system of interest but does not make this usage plain in describing how the process proceeds. Doing so conceals important information about what the system of interest is required to do – information which is needed to completely identify use cases. The obvious benefit of this integrated business process and use case modeling approach is that the use cases can be directly read from the process model, making a separate use case modeling process unnecessary. The systems engineering team can proceed directly to describing the flows of events of the use cases identified during the integrated process modeling work. Before considering some illustrative examples, two additional aspects of process modeling must be addressed: simultaneous behavior and responsibility swim lanes.

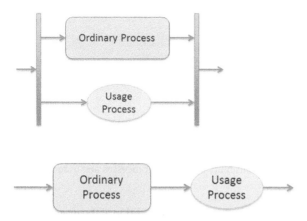

Figure 8.4 Concurrent and sequential execution of ordinary process and usage process.

Processes may occur sequentially or concurrently. All popular process notations including Business Process Modeling Notation (BPMN) and UML/SysML Activity Diagrams, allow processes to be shown as occurring before or after other processes, or simultaneously with them. Borrowing this notation, ordinary processes and usage processes can be shown in sequential or concurrent ordering as shown in Figure 8.4. An ordinary process may continue, as a system is used to perform some function as part of the overall process. The human resources specialists might use a system to create the employee's record, while simultaneously the onboarding process for the employee continues. Alternatively, a usage process and an ordinary process may be sequential, if say the creation of the employee's record must be completed before the onboarding process can continue.

Swim lanes, which are already used in both business process modeling and use case modeling practice, indicate the responsibility for a particular process. With an ordinary process, it is clear that the swim lane should represent the entity responsible for carrying out the process – typically an individual or organizational unit such as a department or organization.

With a usage process, we have two choices. The usage process can be placed in a swim lane representing either the actor who will initiate the use case, or the system for that use case. In either case the other piece of information must be indicated with the usage process separately. In other words, if the usage process is placed in a swim lane representing the actor, then the system must be indicated with the usage process. If the usage process is placed in a swim lane representing the system, then the actor must be indicated with the usage process. Figure 8.5 shows the former approach, which is preferable in terms of clarity of the diagram. One reason is that there is probably already a swim lane for the actor, since the actor is probably responsible for other processes in the diagram, so there is no need to add additional

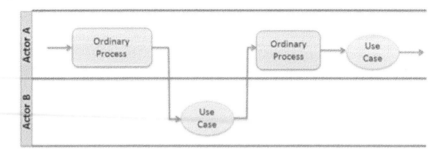

Figure 8.5 Swim lanes showing responsibility for ordinary processes and usage processes.

swim lanes for each system involved in the process. In addition, if the usage process is placed in a swim lane representing a system, it will probably be necessary to include an additional process in the actor swim lane representing only the fact that the actor initiates the use case. Without this additional process, the flow may be unclear as the usage process would immediately follow some unrelated ordinary or usage process, implying that is those that initiate the use case.

8.10 Pilot Project Examples

During the development of usage processes and timeboxes in 2015, three pilot projects were conducted in order to validate the usefulness, flexibility, and ease of creating models using the new paradigm concepts.

8.10.1 Pilot Project: The Hunt for Red October

The first pilot project utilized a familiar set of complex behavior, based on the popular novel and movie, *The Hunt for Red October*, which is a complex story line concerning a realistic military engagement. Creating a model of the overall storyline as well as a more detailed model of the final battle sequence was found to be easy and intuitive using the new concepts. The result was then shown to two kinds of respondents – some experienced in business process and use case modeling and some with no such background. Importantly, no explanation of timeboxes or usage processes was given to the respondents prior to seeing the finished diagrams, in order to assess whether the diagrams were intuitive and understandable, even without special explanation or the modeling concepts.

In response to questions, respondents indicated that the models were easy to understand and follow. In fact, respondents tended to ask about content-related

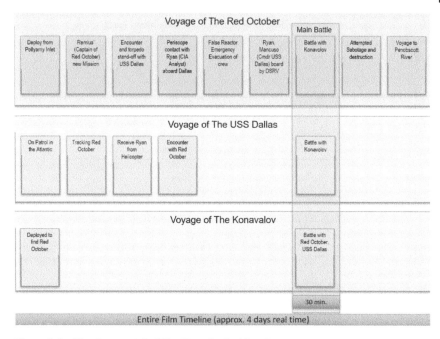

Figure 8.6 Timebox model of The Hunt for Red October.

topics, indicating that they had completely understood the models and thus were able to focus on their content. The timebox model of the film's overall timeline is shown in Figure 8.6. An application of usage processes is shown in Figure 8.7a,b focusing on the torpedo system of the submarine Konavolov. In this pilot project, the usage processes, representing use cases of the torpedo system were located in the swim lane for the torpedo system. During this pilot project, the conclusion was reached that a better way is to place the usage processes in the swim lane for the initiating actor of each use case. In this example, the diagram would be simplified, since each use case would be placed in the swim lane for the Konavolov and activities that simply indicate the firing of the torpedo would be superfluous and could be eliminated.

The Red October pilot project showed that an arbitrary, complex behavioral description can be reduced to simple diagrams using timeboxes and usage processes. Participants in the project were able to understand diagrams intuitively, likely because the diagrams come closer to the way behavioral sequences are represented in human mental models. We cannot hold dozens of scenes, actors, and locations in our minds at once, so we naturally break things down into "chunks" of time, represented by timeboxes, with rough sequencing and concurrency

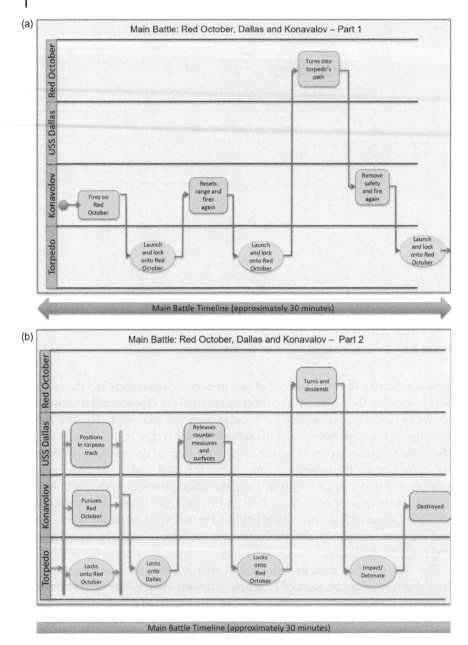

Figure 8.7 Usage process model of Main Battle Timeline, (a) Part I and (b) Part II.

shown in the diagram. Of course, we do not suggest that these simple models replace more detailed models, but they are very beneficial in representing complex behavior at a high level of abstraction, gaining widespread understanding, and forming a framework for more detailed models.

8.10.2   Pilot Project: FAA

The second pilot project involved a group of systems engineers at the Federal Aviation Administration (FAA) who contributed samples of previously developed business process and use case models. The business process models had been intended to form a basis for describing new functionality needed in a system, but as we've described, in many cases they did not, and the systems engineers who came along after the business process models had to start mostly from scratch to develop the use case models. The group was interested in how these two activities could be accomplished in a more efficient and less redundant way.

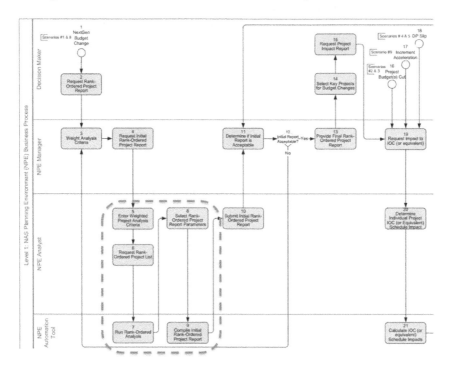

Figure 8.8   FAA example business process before use of new modeling approach.
Source: Used with permission from FAA.

With input from the FAA team, the business process models and use case models were analyzed and some examples of the processes reformulated using usage processes. FAA systems engineers responded positively to the changes, confirming that integrating business process and use case information, previously in two separate and unintegrated models, would provide significant benefits. In addition, it was noted that integrating the two teams previously assigned to create the two separate models, would bring additional efficiency since the combined work would be completed much more quickly than the two separate consecutive efforts. In addition, it was found that the new paradigm approach could be employed directly with internal client organizations, instead of waiting for the client organization to formulate its own business process model first, before requesting a new system or system enhancement. Figures 8.8 and 8.9 show an example of the before and after FAA business process models.

In Figure 8.8, the original business process diagram is shown. The large oval shows the business processes that were identified as being processes that are performed by someone using a system, and therefore could be represented as a usage process. In this example, several business processes (numbered 5, 6, 7, 8,

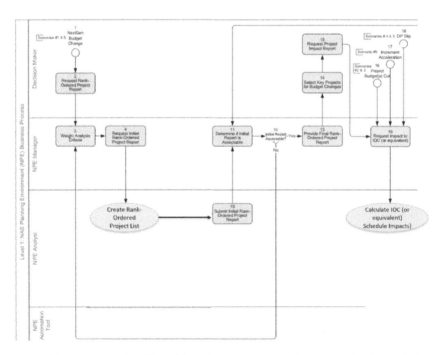

Figure 8.9 FAA example with addition of usage processes. Source: Used with permission from FAA.

and 9) were actually parts of a single use case, so that set of business processes were replaced by a single usage process, as shown in Figure 8.9. The new usage processes were placed in the swim lane representing the actor, as the swim lane already existed on the diagram and the resulting diagram is more clear, succinct, and intuitive. At the end of the pilot project, the FAA Director for Systems Engineering recommended that the new paradigm modeling method be incorporated into the FAA Systems Engineering handbook and become the recommended method for future projects.

8.10.3 Pilot Project: IBM Agile Process

The third pilot project attempted the application of new paradigm modeling concepts to an agile systems engineering process developed by IBM. Reviews of the resulting new paradigm models by the primary author of the new systems engineering process indicated that the clarity of the process had been enhanced and that the new paradigm models may be a clearer and more intuitive way to explain and illustrate his method and process for systems engineers (Figures 8.10–8.13).

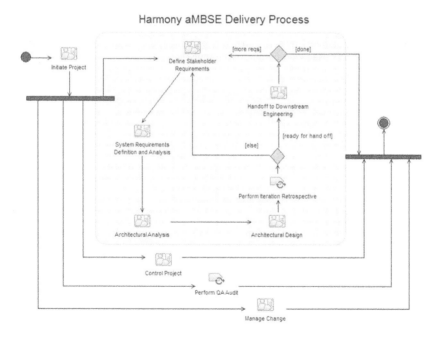

Figure 8.10 Harmony AMBSE delivery process. Source: From IBM Harmony Agile MBSE Process (2015), used with permission from IBM.

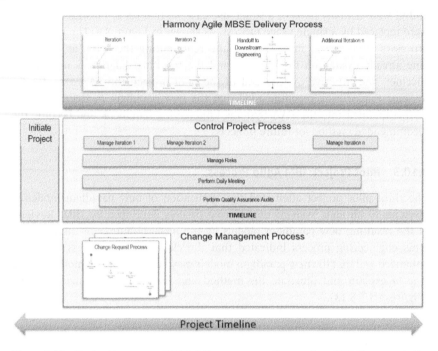

Figure 8.11 Revised harmony aMBSE delivery process. Source: from IBM Harmony Agile MBSE Process (2015), used with permission from IBM.

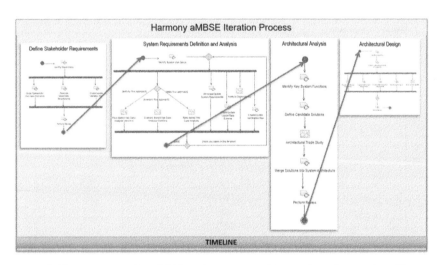

Figure 8.12 Harmony agile MBSE iteration process with notations. Source: From IBM Harmony Agile MBSE Process (2015), used with permission from IBM.

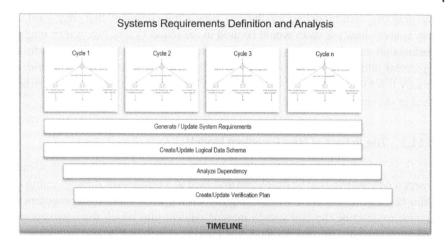

Figure 8.13   Revised harmony agile MBSE iteration process.

8.11   Summary: A New Paradigm Modeling Approach

In this chapter, we've introduced a new approach to the modeling of behavior in systems. The new approach builds on and integrates methods from business process modeling and use case modeling into a single process, able to be performed by analysts with little or no additional training or background. Use of this new modeling approach will save as much as half of the time, effort, and cost normally incurred when employing the traditional approach of business process modeling followed by use case modeling. The new modeling approach produces a more holistic and integrated model, incorporating both business process and use case information in a unified and integrated model. The two main innovations included in the new approach are timeboxes and usage processes. Timeboxes express behavior without requiring exact sequencing of activities, enabling modeling of approximate or unknown start times, end times, or durations. Timeboxes with timelines express behavior in relative sequence without requiring exact specification or prediction of absolute starting or ending times. Usage processes allow the inclusion of use cases into business process flows, producing an integrated view of both processes and use cases of multiple systems that are used in the processes.

It is no surprise to anyone that systems are increasing in complexity and becoming more intelligent. The only defense against system complexity and the resulting unwanted emergent behavior is a more complete and common understanding of the system. Use cases are a big help but apply only to a single system. New paradigm models can span the behavior of many systems, showing how they operate together as an integrated whole. In fact, systems engineers learning use case modeling for the first time often comment that they want a way to show how use cases would be used in sequence, that is, how a user might perform on complete system usage and then go on to another, and so on, in a particular situation. Business process modeling with usage processes provides exactly this kind of model, showing how not just one, but many systems may be used in the context of a larger process, task, or mission.

8.11.1 The Impact of New Paradigm Models

The new paradigm modeling concepts developed are applicable to situations where a new system is to be developed and used in a business or mission context. Prior to the development of system requirements, process models are sometimes created to describe the business or mission context into which the new system will be deployed. This process modeling effort is normally followed by a system requirements development effort, often involving the development of use cases. New paradigm models allow both types of modeling to be accomplished in one pass, saving project time and money. The main savings of the integrated approach comes from avoiding repeated conversations with stakeholders about the two overlapping and largely redundant subjects – their business processes and their desired use of systems in those processes.

There are several types of projects where the new paradigm models may be employed including large scale IT software development projects, systems engineering projects, and process modeling projects. Large-scale software development projects in the information technology field often employ both business process modeling and use case modeling. New paradigm models can be used to interact directly with business stakeholders, capturing both the business process into which the new software will be deployed, and also the use cases for the new software in one integrated process. Systems engineering projects such as those designing defense related systems can use new paradigm models to conceptualize and document concepts of operations (CONOPS) using process flows and integrating use cases for the systems that will be built to support the mission operations.

These new models may be built by the system acquirer, likely a military organization, or by the contractor or integrator designing and building the system. New paradigm models can form the basis of a model-based systems engineering approach that further elaborates the use cases identified using the process described in the Chapter 6. New paradigm models are especially applicable to innovative defense systems development where new capabilities are constantly being incorporated with an increasing emphasis on network-centric warfare and integrated systems, since they enable the expression of complex behavior occurring at the same or different times, in various places.

New paradigm models may also provide advantages in the design of advanced automobiles that require complex interactions with the driver, with other systems outside the car (telematics) and with other cars and the environment (autonomous operation). As the Internet of Things (IoT) continues to expand and consumer electronics products are increasingly network-based, the increasing complexity may also suggest the use of new paradigm models for design of these products.

Business process modeling projects, even when not intended to be used as a precursor to a software or systems engineering development project, can utilize new paradigm models to produce clearer and more complete process models than what is possible with conventional business process models alone. The use of timeboxes and timelines can make models more realistic and chronologically organized, without introducing the false precision of either Gantt-oriented project plans or hierarchically decomposed process models. New paradigm models can thus also be used for project planning and enable the depiction of complex planned activities, represented as processes, to be shown using timelines and timeboxes, without committing to a full and detailed schedule of the subordinate activities.

In sum, the new paradigm and modeling approach described in this chapter is applicable to software development projects and the systems engineering phase of large systems development projects, where the system is intended to be used by people to carry out a business purpose or mission, including defense and military systems, space systems, transportation systems, and complex medical devices and systems. The modeling approach is also applicable to large-scale software systems development. Beyond the cost and time savings, using the new modeling approach can reduce wasteful and repetitive stakeholder involvement and the resulting frustration.

8.11.2 The Future of New Paradigm Models

The pilot projects described here are a start at validating the usefulness of new paradigm models. As the approach is still relatively new, modeling tools do not have built in capabilities for timeboxes, timelines, and usage processes. When implemented in a modeling tool, more flexible models and model automation may be possible, for example, nested arrangements of timeboxes may allow zooming in and out from very high levels of abstraction to very detailed levels of behavioral descriptions within a single model. Timelines may allow executable simulations based on time, either in forward or reverse direction. Attention can also be given to the visualization of new paradigm models in other than two dimensions. For example, both BPMN and UML/SysML allow for swim lanes to be drawn vertically or horizontally, or both, which may allow for the representation of additional aspects of the behavioral model. Timelines are usually thought of as one dimensional, but Rosenberg and Grafton (2010) contains a fascinating example from 1912 (p. 182) of two parallel timelines, one representing the starting time of a train trip and the other representing the ending time, resulting in a series of diagonal lines for a group of trains. The thickness of a single diagonal might be used to represent the uncertainty of the beginning or ending time, a concept introduced herein for timeboxes.

Given the limited abilities of BPMN, UML, and SysML models to represent time, it is suggested that those standards might benefit from the inclusion of timelines and timeboxes as a way to organize process representations in this modeling languages. The UML and SysML standards could also benefit from the inclusion of the usage process as model element, possibly as an adaptation of an activity element.

References

Alotaibi, Y. and Liu, F. (2014). A novel secure business process modeling approach and its impact on business performance. *Information Sciences* 277: 375–395. https://doi.org/10.1016/j.ins.2014.02.088.

Bahill, A.T. (2012). Diogenes, a process for identifying unintended consequences. *Systems Engineering* 15: 287–306. https://doi.org/10.1002/sys.20208.

Bergener, P. (2015). Detecting potential weaknesses in business processes. *Business Process Management Journal* 21: 25–54. https://doi.org/10.1108/BPMJ-07-2013-0103.

Brown, B. (2013). Goals for a new paradigm of behavior modeling. *Paper Presented at the Proceedings of the INCOSE International Symposium 2013, Philadelphia, PA* (26 June 2013).

Cantor, M. (2008). Risk and relevance. In: *Proceedings of the 4th international workshop on Predictor models in software engineering*, 1–2.

Deutch, D. and Milo, T. (2012). *Business Processes: A Database Perspective*. [Electronic resource]. San Rafael, CA: Morgan & Claypool.

Dhammaraksa, K. and Intakosum, S. (2009). Measuring size of business process from use case description. *Paper Presented at the Computer Science and Information Technology. 2nd IEEE International Conference on Computer Science and Information Technology, Beijing, China* (8–11 August 2009).

Gerth, C. (2013). *Business Process Models. Change Management*. [Electronic resource]. Berlin: Springer.

Glykas, M. (2013). *Business Process Management*. Berlin: Springer.

Hammer, M. and Champy, J. (1993). *Reengineering the Corporation: A Manifesto for Business Revolution*, 1e. New York: Harper Business.

Hruby, P. (1998). Mapping business processes to software design artifacts. In: *European Conference on Object-Oriented Programming*, 234–236. Berlin, Heidelberg: Springer.

Kajdan, V. (2008). Bumpy road to lean enterprise. *Total Quality Management & Business Excellence* 19 (1/2): 91–99. https://doi.org/10.1080/14783360701602338.

Knauss, E. and Lubke, D. (2008). Using the friction between business processes and use cases in SOA requirements. *Paper Presented at the Computer Software and Applications. COMPSAC '08. 32nd Annual IEEE International, Beijing, China* (28 July 2008 to 1 August 2008).

Krogstie, J. (2013). Perspectives to process modeling. In: *Business Process Management*, 1–39. Berlin, Heidelberg: Springer.

Li, Y., Cao, B., Xu, L.D. et al. (2014). An efficient recommendation method for improving business process modeling. *IEEE Transactions on Industrial Informatics* 10 (1): 502–513.

Lübke, D. (2006). Transformation of use cases to EPC models. In: *EPK*, 137–156.

McSheffrey, E. (2001). Integrating business process models with UML system models. *Popkin Software*.

Mili, H., Tremblay, G., Jaoude, G.B. et al. (2010). Business process modeling languages: sorting through the alphabet soup. *ACM Computing Surveys (CSUR)* 43 (1): 1–56.

Mohapatra, S. (2013). *Business Process Reengineering. Automation Decision Points in Process Reengineering*. [Electronic resource]. New York: Springer.

Rosenberg, D. and Grafton, A. (2010). *Cartographies of Time*. New York, NY: Princeton Architectural Press.

Sinha, A. and Paradkar, A. (2010). Use cases to process specifications in business process modeling notation. *Paper Presented at the Web Services (ICWS), 2010 IEEE International Conference on Web Services*, Beijing, China (5–10 July 2010).

Part III

Systems Thinking for Intelligent Systems

9

Solving Hard Problems with Systems Thinking

As introduced in Chapter 3, the main purpose of systems thinking is the solving of problems that occur in systems. Systems thinking can be applied to natural systems, engineered systems, and human activity systems, the three types described in Chapter 3 but is most often applied to human activity systems and natural systems. This is likely because when considering a problem in an engineered system, the most logical approach is to refer to the original intended operation of the system and compare its observed, problematic behavior to its designed, intended behavior. Any discrepancy, known as a failure, defect, or bug can be traced, using the engineering designs, to a particular part of the system. Engineers dive down into subsystems, components, and finally individual parts to find the issue and fix it by repairing or replacing some part. This is the normal break-fix process used to correct problems in an engineered system. If the system ever functioned correctly but is not now functioning correctly, the problem can only be the failure of one or more parts. While it's possible for two unrelated parts in a system to fail at the same time, in most cases a newly appearing problem is caused by the failure of a single part or a set of closely related parts, and even then, it is probably just one part that was the origin of the failure. A failing part can certainly "take others with it" so engineers learn to look for the original cause, even though the fix may involve repairing or replacing several parts. In an engineered system there is a known, correct operating state and if the system had ever operated in that state, the objective is to return the system to that correct operating state.

9.1 Human Activity Systems and Systems Thinking

Human activity systems and natural systems are different. They are much more dynamic, and in most cases do not have a clearly identified, original, correct operating state. Parts of these systems are by definition, living things that grow and change on their own, so the systems of which they are parts change as a result.

Engineering Intelligent Systems: Systems Engineering and Design with Artificial Intelligence, Visual Modeling, and Systems Thinking, First Edition. Barclay R. Brown.
© 2023 John Wiley & Sons, Inc. Published 2023 by John Wiley & Sons, Inc.

To the frustration of engineers, human activity and natural systems cannot be debugged like engineering systems. An engineer parent might imagine that a son or daughter struggling with teenage relationships is a system with a defect, and simply needs fixing, but the parent will likely be disappointed with the results of applying such a fix.

It may come as a surprise that many systems engineers know very little about the theory and practice of systems thinking. They may assume that systems thinking is simply the kind of thinking that they do, since after all they are systems engineers. Or they may assume that systems thinking is simply thinking about systems. But systems thinking, as we'll describe in this chapter is a method and practice of its own, with its own concepts, methods, techniques, and even diagrams. To further differentiate it from systems engineering, we note that both the subject system and the goal or objective are different.

In systems engineering, the subject is an engineered system. We (or someone, somewhere) know (or should know) the requirements for the system, its design, and how it works. In systems thinking, the subject system is usually a human activity system, and often profoundly mysterious. Likely it grew up over time, without an original, complete design. A city, for instance, may have had some original planning, but most aspects grew organically, driven by the humans in the system, into what it is today. If there's a problem with how it is working, there is no designer to blame. It's hard to even call the problem a system defect, since that would imply that the system was supposed to work a different way, and did at one point work that way, but now something has broken, and the system no longer functions as it once did. Human activity systems aren't like that. They are changing all the time and in largely unpredictable ways. They are more akin to living beings, so human activity systems actually have more in common with natural systems than they do with engineered systems.

The goal of systems engineering is to conceive, design, build, and deploy systems that meet the needs, wants, and expectations of their intended users. The goal of systems thinking is to solve a problem within a system. Some problems are not hard to solve so quickly observed, linear solutions work best. If the problem is that the garage is too small for the new RV, then it doesn't take a lot of systems thinking to see that a larger garage is probably the best solution. Of course, we can phrase the problem and solution in system terms, speaking of the interaction between the RV system and the garage system, but perhaps we should forgo the thinking and get busy calling contractors. Systems thinking is more suited to what are sometimes referred to as *wicked* problems – problems that have many interrelated aspects, where a single, simple intervention is unlikely to solve the problem. We typically know this because by the time we get to the problem, all the single, simple interventions have already been tried. If the murder rate is too high, you can bet that someone tried simply increasing the severity of the punishment for murder, probably with limited or no effect. The murder rate in a city, society, or country is a wicked problem, with many interrelated, contributing factors. It's a good candidate for systems thinking. If a problem has persisted for a long time and has resisted most attempts at improving the situation, despite honest and sincere efforts, then systems thinking might have something to contribute.

9.2 The Central Insight of Systems Thinking

To begin our exploration of systems thinking, let's consider the central insight of systems thinking, which is, *the system is always working*. As we look at systems, and here we are thinking primarily about human activity systems, it is tempting to think that a system is somehow broken. If we look at the American justice system, we might note that it results in the incarceration of more people (by percentage of population) than any country in the world. It may be tempting to think of this system as broken because it incarcerates too many people. From a systems thinking perspective, however the American justice system is working. It is operating the way that the American people, through their representatives in government have designed it, implemented it, and are operating it. The system is doing what it is designed to do, and operating according to the way it is implemented. The system is working. Yet people still say that the incarceration rate is too high and they want to lower it. This cannot happen with the current system, but of course the system can be changed, and then it will be a different system. What if all prison sentences were capped at 10 years, or if an entire category of former crimes were declared to be lawful, as some states have done with the possession of marijuana? These changes to the system would of course affect the incarceration rate, but the systems thinking would not stop there. Would such changes have other effects on the economic and social systems involved, including potentially unintended consequences?

To take another example, consider the concern that the unemployment rate is too high. First, we need to examine what is meant by "too high," but we'll save that for the discussion later on formulating the problem. The business, economic, and employment system is working, and producing the unemployment rate that it is, given the current conditions. The unemployment rate cannot simply be changed by turning a dial to a different number. To change the unemployment rate, one or more aspects of the system must be changed. The system is working.

As a somewhat playful example of this insight that the system is always working, I have a large van that I've owned for many years. One day it began making a strange sound, much like I was driving over rumble strips in the road, when I wasn't. It didn't happen all the time, only when accelerating, and the harder the acceleration, the more the sound. Someone assuming that the system (the van) is not working, might proceed to try to correct the problem by identifying the malfunctioning part and replacing it. There are mechanics and technicians of all kinds who work this way. They assume that something in the system is broken and start replacing parts until the system works as it is supposed to. This approach is sometimes workable with engineered systems, but of course doesn't often apply to human activity systems – people cannot usually be declared defective and simply replaced. The approach of assuming the system is not working may also fail with engineered system as the van story indicates. If a mechanic, assuming that some part in the system is broken, started replacing parts, he or she would have to work all the way up to and including the transmission before a part-replacement approach would solve the problem. The sound is indeed coming from the transmission, but the problem is not a broken transmission. Some web searching on the message boards devoted to this kind of van reveals that this sound is the sound made by the transmission when the transmission fluid and filter need servicing, a very simple and inexpensive procedure. Knowledge of how the system is working enables the correct intervention to solve the problem. Without the knowledge the only approach is the replacement of parts until the problem goes away.

As a student engineer, I was taught a related lesson about finding problems in systems. Instead of assuming something is broken and replacing parts until the problem was solved, I was taught to imagine how the system might be producing the symptom or problem I'm observing. In other words, assume the system is working, and imagine the ways that the system could be producing the observed, unwanted behavior. Once it is understood how the system is working, the intervention may become obvious. When a master mechanic listens to a running engine and announces the exact problem and its solution, this is what is happening. The mechanic compares the remembered sound of a perfectly running engine with the sound actually being heard and imagines how the system could be producing the difference. With enough experience, the mechanic may appear to be a magician or psychic. The magic of course is the knowledge of how the system works. When someone fully understands a system and how it is doing what it is doing, then symptoms can lead to causes which lead to possible interventions. System knowledge allows possible interventions to be evaluated in terms of the effect they will have on the entire system. To say that the system is always working does not mean that it is producing the results that are desired. When a different result is desired, then a system is needed that is different from the current system in some small or large ways. If you want a different result, you need a different system. We now turn our attention to the area of problems in systems.

9.3   Solving Problems with Systems Thinking

Solving problems with systems thinking follows a general process, though it is possible to work on more than one of these steps in parallel or to skip around in some situations. Here's a summary of the process.

1. Identify a problem
2. Find the real problem, or the problem-behind-the-problem
3. Identify the system or systems that are producing the problem
4. Describe, model, and fully understand the system and how it works
5. Apply systems thinking concepts, diagrams, archetypes, and leverage points
6. Carefully, propose system interventions
7. Test and verify potential interventions
8. Implement the solution incrementally

Even from a quick glance, it is easy to see that this process is not always followed when addressing some important and large-scale problems in human activity systems. It is not unusual to identify a problem, jump to a possible solution, and implement it for everyone, everywhere all at once, and only then find out what happens. Lawmakers do it all the time – new laws change many systems all at once and the real impact may only be felt later. Regulation, deregulation, tax codes, immigration laws, and even the legalization of certain substances have effects that span societal, business, and economic systems. This chapter walks through each step of the systems thinking process.

9.4   Identify a Problem

Systems thinking is about solving problems in a different way. To begin there must be an identified problem. Identifying the real problem is not so easy. Often the initial statement of the problem is simply someone's solution or agenda in disguise. If someone says that the problem with our country is that the minimum wage needs to be raised, is this a good statement of the real problem? Or is it that the person simply has a commitment or agenda to try to have the minimum wage raised? If the true statement of the real problem is that the minimum wage needs to be raised, then there is only one solution to this problem – raise the minimum wage. Systems thinking asks us to look more deeply into the situation. For example, if everyone were making more than the minimum wage, raising it would make no difference. The real problem in this situation is probably something more like: some people in some jobs aren't making enough income to afford to live in some areas. Seeing the real problem this way, we open up our view to the systems that are involved in causing what we call the problem.

In my courses on systems thinking, I ask students to create a problem catalog by reviewing current news publications and thinking about what's concerning in our society or the world. The challenge is to correctly state the problem, in such a way that it leads to the possibility of understanding the systems within which the problem exists. Most students start out with headline issues, and give the agenda-based formulation of the problem. One might write that there are too few women in engineering. What's the real problem here? If the problem is actually that there are too few women in engineering, then should we force some women into engineering? Or should we somehow reduce the number of men in engineering? With more thought, it might be that the real problem is that women are somehow being blocked or discouraged from entering engineering. That's a completely different problem. Or is the problem that not enough women want to go into engineering? That's a different problem too.

When it comes to gender imbalance issues, one has to be careful to not assume that the problem is outright discrimination. Elsewhere in this book we spoke about male and female nurses. Does the fact that 93% of nurses are female mean that the problem is widespread discrimination against men entering the field of nursing? No – there could be discrimination, but perhaps it's more likely that other factors are driving that gender imbalance in nursing. Is there even a problem here? Well, maybe. If a society primes its children that nursing is a good profession for girls but not for boys, then perhaps that is the source of the imbalance. If we feel that this imbalance is a bad thing, then perhaps we have a good statement of the problem – that children are taught inappropriate occupation/gender stereotypes. At the same time, we need to ask if all imbalances are problems. There are more tall people (men and women) playing collegiate and professional basketball than short ones. Should we try to encourage more short people to play basketball? Should we have rules or laws that implement a kind of height-blindness in basketball recruiting? Or perhaps tall people are just more interested in basketball because they have advantages in the game that make them, in general, more successful. Our purpose at this stage is simply to identify the problem, and it's not so easy, is it?

9.5 Find the Real Problem

Problems usually present themselves as a difference or gap between a current situation and a desired situation. What makes stating the problem difficult is that the causes of the gap are not apparent in the statement of the problem. If I state my problem as *I don't make enough money*, it gives me no insight into the cause of my not making enough money. Whatever that cause is, is the problem. If I'm at the top of my profession but still don't make enough money, then the problem is almost certainly my profession, and should be stated as, people in my profession don't make enough money. This isn't quite right though; there might be people in my profession that feel they make plenty of money. I should really state that the problem is that people in my profession don't make as much money as I want to make. Now I have a problem I can work on.

Returning to the problem of unemployment, we need to ask several questions before we can arrive at a good problem statement. What is the best level of unemployment? Economists agree that an economy with zero unemployment is not healthy for a variety of reasons. Perhaps even before asking about the desired rate, we should define unemployment. There are several measures used for unemployment that are used to compute ratios between those who are employed, those who are unemployed but available and looking for work, and those who are not in the labor force, including disabled, retired, military, students, stay-at-home parents, and the incarcerated (https://www.bls.gov/cps/cps_htgm.htm). The most popular and often-cited measure of unemployment is that ratio of unemployed people to the entire labor force. Right away we notice that if what we have in mind is trying to get more people to work instead of relying on social programs, family or other support, the unemployment rate does not help us, since those people are not in the labor force. Interestingly, people who drop out of the labor force *lower* the unemployment rate – they reduce the denominator increasing the ratio of unemployed people to the entire labor force. Have we arrived at a solution? To lower the unemployment rate, we just need to get more people *out* of the labor force. Likely this is not what we had in mind when we started addressing the problems of unemployment.

In finding the real problem, we need to look behind the unemployment rate, to what it is we are actually trying to change or solve. Are we picturing people who have, for example, lost their jobs due to business or economic slowing, who are looking for work, but unable to find a new job – and we need to help more of them get back to work? Addressing this problem would be quite different from thinking about, say the people who leave their jobs voluntarily and for whatever reason drop out of the labor force. Perhaps they find that their skills are out of date, become discouraged about finding another job, and so decide to leave the labor force in one way or another. In 2020 the generous expansion of unemployment benefits is thought to have contributed to an increase in the number of unemployed people. Is this a part of the problem perhaps?

In formulating the problem, it can help to apply the *five whys* process, where we ask *why* as many as five times to dig down to deeper causes and problem areas. To illustrate, say we take the area of human-generated CO_2 emissions and their effect on the climate as the problem area. Instead of jumping to solutions, we try to formulate the problem. Why are human beings generating CO_2? Because they need power and transportation, and those are largely powered by combustion, which produces CO_2 as a byproduct. Why do we need transportation? To get to work, to the store, and to meet with others. Why do we need to get to work? And so on. Each why is suggestive of alternate paths of exploration for the formulation of a workable problem statement. As a companion to the five whys, consider the *five what-ifs*. Anytime a why question is answered, also ask, *what-if*. What if combustion were not necessary for the production of power? What if we didn't need to travel to get to work? What if we didn't need to work? What-ifs lead to how-coulds, and a new problem statement is born.

Naturally, during the problem formulation process, solutions will come to mind. That's fine. They can be noted, but it is wise not to explore them deeply yet, and in particular not to become enamored with a particular approach or solution. We are at steps one and two in the systems thinking process and still need to address steps three through five before we are ready to evaluate potential solutions. In practice, once you are familiar and skilled with the entire process, you will find yourself running up and down the list of steps very quickly and many times over during when applying systems thinking. Thinking of a solution leads right to thinking about the system the causal relationships, system patterns, possible interventions, various consequences, etc. then returning to a differently characterized problem, new ideas, and so on. Let's move on to the next step.

9.6 Identify the System

Once we have a workable problem statement, and are certain that it's not a kind of Trojan horse, with someone's pre-baked agenda-based solution hidden inside, we can proceed to consider the system that's producing the problem. Are all problems produced by systems? Philosophically, if everything in the universe is a system, or part of a system, then of course all problems come from systems. More practically, problems that have not been immediately solved already are very likely to be products of a system or a set of systems. If a problem is so simple and straightforward that it can be completely solved without any systems considerations, then it would probably already be solved and we would probably not be discussing it.

Our task in this step is to identify the system or set of systems producing the problem. The important question here is, how far to go in characterizing the system. A problem in the area of unemployment concerns the economy of course, but do we need to consider the local economy, the national economy, or perhaps the even the global economy to address this particular problem? If the problem concerns morale at your company, the system could be the department, the division, the company, or perhaps even the industry in which the company operates. Problems with terrorism are often addressed at the local level, that is, the problem concerns only the terrorists and the victims. Asking a few why questions could lead us to a larger system, perhaps the country or countries involved or even global value systems and their differences. How far we need to go is determined by the point at which we feel we understand the system well enough. This is certainly a subjective judgment, but the point is to work at understanding the system until a confidence level is reached sufficient to predict the result of an intervention.

One of my students' favorite problem areas concerns traffic. Most large cities have enough traffic that it is a problem to people due to delays, accidents, maintenance closures, and other annoyances, to say nothing of the bad behavior of other drivers. On first glance, the system producing the traffic is simply the roads and the vehicles – too little of one and too much of the other. Is that all there is to it? Roads are only one part of a transportation infrastructure that provides choices to commuters and other travelers. If the traffic problems center around morning and evening rush hour, the system must include the businesses that require workers to report during those hours and other systems that affect the traffic and the roads including nearby construction, tolls, and even perhaps the sun itself (ever heard of a sunshine slowdown on west-bound roadways in the late afternoon?).

In 1996, I lived in Atlanta when the Olympics came to town. Of course, everyone feared traffic Armageddon, and hyped up by media-driven fear, most drivers vowed to stay off the roads during the time of the games. Many employers allowed the (at the time) unthinkable – allowing people to work from home, and Marta, the subway system, had record ridership with accompanying crowds and delays. The amusing irony was that in the end, the best way to get around town during the Olympics was to drive on the roads, since they were largely empty. The point is that we think of traffic as inevitable, but as the Olympics, and more recently COVID-19 showed, traffic is not a fixed phenomenon. Economists will point out that the amount of driving people do goes down a bit with every increase in fuel prices, showing that demand for gasoline is to some degree elastic.

To be fair, the initial identification of the system that is producing the problem is a bit of a guess. Until it is known how the system is producing the problem, we can't be sure we have our heads around the whole system we need. There's no need to be hung up on finding the right system, or systems. In fact, it makes no real difference if we think of the problem as a problem of one big system or a set of smaller, interrelated systems. Since we are usually talking about human activity systems, it doesn't make a lot of difference where, or how big, we draw system boundaries initially.

9.7 Understanding the System

Once we can identify the system producing the problem, we begin to try to understand this system. In systems thinking, understanding a system doesn't mean simply that we can recognize it, describe it, and name its parts. Understanding a system the way we need to means understanding the causal relationships in the system. Here we need to take a brief side trip into the nature of causality. When we say that one thing causes another, that A causes B, we mean that a change in A causes a change in B. We may think of this a simple relationship, such as that an increase in A causes an increase in B, but this is not the whole story. If A is not the only cause contributing to B, an increase in A may be swamped by other causes and B may still decrease. Our statement that A is a cause of B is correct, but we may still not see B increase whenever A increases. We can state a causal relationship more accurately as, if A increases, then B will increase by more than it would have increased without the increase in A, or that B will decrease less that it would have decreased without the increase in A.

It makes sense in systems thinking to consider all causes as partial causes. Complex human activity systems are normally what systems thinking is dealing with, because that's where the big problems lie. In these systems, little if anything is caused by a single factor. Let's take an example. In the 1980s, American president Ronald Reagan said, "We must reject the idea that every time a law's broken, society is guilty rather than the lawbreaker. It is time to restore the American precept that each individual is accountable for his actions." Is this good systems thinking? Was President Reagan correct that it is *either* the individual *or* society that is responsible for the crime? Perhaps there is more to crime than just a single cause. If we ask, why did that person commit that crime at that moment? the answer is not so simple. Certainly part of the cause is that the person decided to take the action to commit the crime, but we must then ask, why did the person make that decision? If we say, it was because the person was poor and needed money, this does not seem correct, since there are many, many poor people who need money who do not commit crimes. If we try to claim that the person's race, age, or living situation is completely responsible, again, we find many people with those same factors who are not committing crimes. At the same time, however, we might find that there are patterns and partial causes in the system. If in two groups of people with all factors equal except for say, income, if we consistently find that more crime is committed by the low-income group, then we can consider that a causal factor. However, we may find that educational level is a stronger factor and more predictive of criminal behavior. On the other hand, I would bet that this relationship itself is not so simple, but will vary by type of crime. So-called white-collar crime like tax fraud, embezzlement, insider trading and the like, probably skew toward the more highly educated, while violent crime may be more prevalent among lower educational and socioeconomic levels. Some types of crime, or even some types of crime in some situations might have different causal patterns. Legal gun ownership on its own might not be predictive of crime, though I would guess that shootings at home may be more prevalent in homes that own guns, than in those that do not. If many violent crimes are committed with illegally obtained guns, then putting further controls on the legal acquisition of guns may not make any difference. The headline question of, do more guns cause more crime? is not one that is easily or simply answered.

One of the most fascinating aspects of systems thinking is how it leads into so many different fields. Systems thinking might be a misleading term, since much of the work of systems thinking feels more like research than just sitting around and thinking. The systems thinker/researcher must boldly look into fields of science, psychology, technology, logic, economics, anthropology, probability, and statistics to fully understand the most important systems in our world. Of this list, the subjects of probability and statistics are perhaps the most fundamentally necessary for clear systems thinking. I'm sympathetic to those who feel we should teach school children probability and statistics in their early years even at the expense of trigonometry, geometry, and calculus. Humans have been found to be quite lacking in good statistical and probabilistic reasoning. Here are just a few examples.

Let's say you learned that a certain behavior, say eating peanut butter, doubles your risk of cancer. You might be alarmed and swear off peanut butter forever. But when you dig deeper you find out that the chance of getting cancer in general, peanut butter or no, is one in one million. Eating peanut butter doubles it to two in one million. Now, you're not so worried about peanut butter anymore. If you flip a coin six times, what are the chances that you will flip all six heads? To make it more concrete, how many times would you need to repeat the six flips in order to have a 50% chance of hitting all six heads? The answer, 32 may surprise you. Professor Michael Starbird of the University of Texas at Austin, illustrates this with his students by asking every student do the six flips and record the results. With a class of 64 or more, he is virtually guaranteed that someone will flip six heads. He makes the point even stronger by asking that person his or her secret – how were you able to flip all six heads; was it in the wrist perhaps? To the person involved, it is a surprising and unexpected result, but seen in the large context of a roomful of coin flippers, it makes perfect probabilistic sense. How about this: in a room of fifty people, what are the chances that two of them will have the same birthday? It might surprise you to learn that the chances are over 97%. The point with these enjoyable illustrations is that we are not naturally good at probability estimation. We may never develop good intuitions about probabilistic events, but through education we can at least learn to pause, do a few simple calculations, and then have correct information with which to reason about systems.

Understanding systems also involves avoiding being fooled. I have an illustration of this I've presented to seminars called "Dr. Brown's Miracle Elixir," a magical drink that makes you feel better, no matter what kind of ailment you have. Here's how it works. The experience of most common ailments or chronic conditions varies day to day, with the person having good days and bad days. The directions for Dr. Brown's Miracle Elixir say that you should take it on bad days, and it will make you feel better by the next day or at the latest, within two days. In the seminar I show graphs of data of improvement, and sure enough, 76% of people taking it improve by the next day and over 90% improve by the second day. Before everyone in the seminar can click over to Amazon to purchase the stuff, however, I reveal that all of the data is randomly generated. The daily feeling level of each of the fictional subjects in the "clinical trial" is a random number between one and ten. The overall average is of course, five. So when the person feels below a five, they take the elixir. The magic, however, is not in the elixir, it's in a phenomenon called *reversion (or regression) to the mean*. Over time, the randomly generated series of numbers will tend to move toward the mean. So, when a value occurs below the mean, the most likely occurrence afterward is for the numbers to trend upward; when the value is high, the most likely occurrence is for the numbers to trend downward, in both cases moving toward the overall mean. Since we carefully instruct the customers to take the elixir only when they are feeling bad, that is, below their average, it is highly likely that their feelings will trend upward afterward. Of course the elixir is useless – how they feel is likely to improve after bad days, elixir or no. The placebo effect of taking the elixir contributes even more to the perceived effectiveness. Countless millions of dollars are made selling supplements of all kinds based on this idea. As you might notice, even a real clinical trial will show these bogus results. Only the addition of a control group that takes no elixir, or better yet a placebo elixir, will show that the elixir has no effect. Systems thinkers must become familiar with the ways that systems can fool people into believing in causal effect that simply aren't there.

9.7.1 Rocks Are Hard

A common trap in identifying systems and causes of problems is something I like to call "rocks are hard." If our analysis of a problem and the system that's producing it can sometimes lead us to a conclusion that blames the problem on something that is a fixed and immutable truth. We illustrate with a parable about a mountain road frequented by cyclists and motorcyclists. Recently, there have been incidents of riders being hit by rocks falling off of the mountain. Of course, this is a problem and so we investigate. We find that the problem is that rocks are injuring riders. When then ask why the rocks are injuring the riders and the answer that suggests itself is that rocks are hard – that's why they are injuring riders. If we conclude that we have fully understood the system that is causing injury, we have reached a "rocks are hard" formulation of the problem and the system. It is true that the reason for the injuries is that rocks are hard, but this is not in itself an explanation – rocks *are* hard, that's why we call them rocks. If feathers had been falling off the mountain, no injuries would result. The problem is not that rocks cause injury – that's what rocks do, because they are hard – the problem is that rocks are falling off the mountain. Of course, this is all clear in this simple parable, but we can easily find "rocks are hard" explanations of problems everywhere. When a news article explains that the not enough middle-class workers can afford homes because they don't make enough money, this is "rocks are hard" – what we mean by *can't afford it,* is that they don't make enough money. Nothing is explained. Stories that "explain" the increase in violent crime by showing how there has been an increase in murder, gun violence, and assault, mean nothing since these things are what we mean by violent crime. Stating that the cause of pollution is companies dumping waste into the river, with the fish and birds are negatively affected, doesn't get us anywhere – this is what we mean by pollution. Pollution is bad. Rocks are hard. What's the cause of obesity in America? If you are tempted to say that it is people eating too much, you're caught in the "rocks are hard" trap.

When dealing with human activity systems, it helps to have some understanding of psychology. Since generally we are only interested in the behavior of the system, we may take a more behavioral approach to psychology, such as that used in the popular cognitive behavioral therapy method, than a depth psychology view such as used in psychoanalysis. Systems thinking cares more about what people do, not so much what they think or feel, except as causal relationships can be identified between these and their actions. Say we are dealing with the problem of suicide and trying to understand the system that produces the level of suicide that we observe. We might assume that depression is causal, and it probably is to some extent, but there are lots of depressed people who are not committing suicide or even attempting it, so there must be other, "more causal" causes. We can generalize this insight by asking an important question in systems thinking – why isn't it happening more?

The natural kind of question to ask when trying to understand a system is, why is the phenomenon of interest happening – in the present example, why is suicide happening? It's important to ask this, but asking only this question can lead to a distorted understanding of the system. If we look at the area of obesity and diet, we might ask why people become obese. One answer could be that they eat too much. If we ask why they eat too much, we might answer that they enjoy food, or certain kinds of foods. If we stop there, thinking we understand why people become obese, we don't have a complete understanding. Enjoying food does not always result in obesity. We have to ask, given obesity, why aren't even more people becoming obese? Obesity rates are about 36% in the United States as of this writing, but we should ask, why isn't it higher? After all, most people enjoy food, so why isn't everyone obese? Why isn't obesity happening more than it is already? This question will lead us to other causal factors that are operating. For example, we might answer that some people place a high value on being fit, or at least on not being overweight. If that's right, then it's a cause that contributes to limiting obesity. If we forget to ask, why isn't it happening more? we may proceed with a limited or inadequate understanding of the system.

If we ask why a person committed a crime, and the answer is that the person has attribute x, y, and z, we need to immediately ask if there are other people with those same attributes who are not committing crimes. If there are, then there are other causes at work which we must uncover to have a full understanding of the system. If we ask why a person growing up in a disadvantaged environment, in a certain racial or societal group, has lower socioeconomic achievement in life, and we answer that it is their disadvantaged environment, we must also ask if there are people growing up in the same environment who have much better socioeconomic achievement. If there are, then the cause is not solely the environment, and improving the environment may or may not have a dramatic effect on each person's outcome. At the same time, we should also ask if there are those growing up in very advantaged circumstances who fail to achieve the high socioeconomic result we might expect. If so, what are the causal factors?

Let's look at an opposite case or two. Say we observe someone who has achieved remarkable success, perhaps a Nobel prize nomination, and we want to know the cause. If we reason that it is because of his or her upbringing, highly educated parents, and ability to attend the best schools, then we need to ask, does everyone with those factors receive a Nobel prize nomination? Of course not, so there must be more to the cause, and in fact the factors we've identified so far might not even be significant compared to the factors we have yet to uncover. If we look at a company who achieves great things, we can always find business books that describe in great detail all of the wonderful philosophies, strategies, and management practices utilized by the company, which together resulted in their success. The system thinker should think, well, wait a minute – aren't there other companies with the same philosophies, strategies, and practices who did not achieve success? What we see here is an example of survivor bias: the simple fact that when we go to analyze successful businesses, the only ones we have to examine are those that survived. The others are gone, and can't be examined. We also see success bias, in that we may only analyze the successful companies, so our conclusions will be biased in favor of the ones who happen to have succeeded. It's amusing to systems thinkers that a phenomenon that is known to be a function of random chance, when combined with success and survivor biases, can spawn an entire philosophy of how to succeed in that particular area. The classroom coin flip mentioned earlier falls into this trap. The "successful" six-time-head-flipper did nothing different from all the others, but is the one that "succeeded" so his or her attributes and behaviors are likely to be held up as an example of how to succeed at consistent coin flipping.

9.7.2 Heart and Soul

Since this book is about intelligent systems, let's consider one more example from the field of AI. In times long past, the heart was thought of as a mysterious organ, central to the mind, soul, and very being of a person. If the heart stopped, the person was considered dead. We retain this idea in our language when we speak of putting one's heart into something, or giving one's heart away in love. The heart's mystery disappeared however when we learned the role of the brain. Before then, we might have assumed that if we replaced someone's heart, either in a Frankenstein-like laboratory experiment or by replacement with an artificial heart, the person would awaken from the operation a different person – after all, the heart, the seat of emotions, and the soul, had been changed. Now we know the heart is just a pump, and can be replaced with an artificial pump, even one that does not beat, but pumps continuously, according to the latest developments. We have discovered that a person with such an artificial heart, and perhaps no heartbeat, would be expected to be the same person, with the same thoughts, identity, and personality. We understand the heart and thus can replace it with an artificial device, and retain the patient's personality. The brain is different. Our current science does not understand the brain well enough to build a substitute, and if we were to be able to perform a brain transplant, we would not expect the person to wake up as the same person at all. An interesting thought experiment is to imaging transplanting the healthy brain of an auto accident victim into the healthy body of a brain-dead person – which person is saved? By understanding the system called the heart we can be certain about how to repair it or even replace it, while keeping the overall system of the human being functioning as before. We do not understand the system of the brain, so we are not often able to repair it, and we cannot replace it, or build a machine that does what it does.

Imagine if we did have heart-level understanding of the brain. We could conceivably take some readings or measurements of a person's brain, and then create a machine that does what that person's brain does, creating either a replacement brain or a sort of clone of the person's mind. There are approaches to AI research that attempt, not to mimic the brain's behavior in software but to actually create a synthetic brain. These efforts are hampered by our very limited understanding of the brain and how it works. Understanding a system fully is a high bar to reach, but we must strive for this level of complete understanding if we are to solve complex problems in complex systems.

As we've said, understanding a system is centered around understanding the causal relationships in the system. Judea Pearl in his *Book of Why*, develops a new system of logic around the seemingly simple idea that one thing may appear to cause another due to its clear correlation, or because it is a reliable antecedent of it chronologically. When a child reaches six years old, they begin their formal education. Reaching the age of six always happens right before the child entering school, very reliably, but clearly it is not the reaching of age six that causes the child to enter school, but the actions of the child's parents to enroll and deliver the child to the schoolhouse. It's vital for the systems thinker to see the difference. We are tempted to draw an arrow in our causal loop diagram from *reaches age six* to *starts attending school*, but thinking that this is how the system works would be a distorted view. We might miss that the actions of the parents are necessary, along with whatever causal factors support or hinder those actions. Pearl emphasizes that actual cause requires both that the cause always produces the effect, and that removing the cause reliably removes the effect. The child reaching age six does not always produce the result of starting school, drawing suspicion that it is not causal, or that it is partially causal, but may be overcome by other stronger causes. We also note that children sometimes start school without reaching age six, making it even more unlikely that reaching age six is the actual cause at work here.

Identifying true causes in the quest to understand a system, is a noble but difficult pursuit, full of curious traps and deceptions. It's almost as if it were designed as a puzzle from the start. Another of my favorite traps is called mistaking effects for causes. On first glance, this sounds like crazy talk – who could mistake something that is an effect, for something that is a cause of the same phenomenon? The rain falls, and the lawn furniture gets wet – it's pretty clear what's the cause and what's the effect. Perhaps it's not so obvious all the time. What is obvious is that effects are correlated with causes. They must be, otherwise they could not be effects of those causes. Effects may also be correlated with noncausal factors, like the taking of Dr. Brown's Miracle Elixir and feeling better in a day or two, but certainly real effects are correlated with their causes, unless swamped by other stronger causes. What is also true is that the causes are correlated with effects. Devoted runners feel better on days that they run – which is the cause? Feeling good is correlated with running, suggesting that running is the cause and feeling good is the effect, but running is also correlated with feeling good, suggesting that feeling good is the cause and running is the effect. Does the person feel good because they run, or run on days when they feel good?

9.7.3 Confusing Cause and Effect

Correlation is not truly commutative. If A is correlated with B, it is not a mathematical certainty that B is correlated with A, but it happens often, or is perceived to be happening. If we see that a lot of fit people eat fruit, and also that a lot of people who eat fruit are fit, then it may be hard to see which is the causal factor – we can't conclusively say that eating fruit makes a person more fit, nor can we say that being fit makes you eat more fruit. To take a somewhat amusing example, let's stay with fitness, but look at resting heart rate. It is well known that fit and athletic people have a lower resting heart rate than those who are neither fit nor athletic. This means that low resting heart rate is correlated with fitness. If we study people with low heart rates, we might also find that enough of them are fit and athletic to also show a correlation between fitness and low resting heart rate. If both are true, we could be puzzled as to which is the cause and which is the effect. If we imagine that low resting heart rate is the cause, and fitness is the effect, then we may be tempted to try to make people fit by lowering their resting heart rate. We might give people Metoprolol, or another beta blocker, thereby lowering their heart rate and blood pressure. While these drugs have important and valid uses for the treatment of other conditions, it is of course unlikely, even ridiculous, to expect that lowering someone's heart rate medically will make the person more fit and athletic. We would have mistaken an effect (low heart rate) for a cause (being fit).

Determining how cause and effect work in a system is such an important part of finding out how a system works, a type of diagram was developed to conveniently show cause and effect relationships. In a complex system, which is usually the kind we find when confronting hard problems, there are many causal relationships, so keeping track of them is a task best accomplished graphically. In its simplest form, a causal loop diagram shows quantities in a system, from concrete and tangible to general and abstract, and then uses arrows to show how more of one quantity leads to more (or less) of another. Quantities can represent actually quantifiable information, such as the number of births, deaths, or hospital admissions, or can be more abstract like an amount of confidence, good reputation, the chances of being admitted to a university, or the likelihood of a successful launch. As long as one can sensibly speak of something increasing or decreasing, it can be on a causal loop diagram. What can't be on a causal loop diagram is something that can't rise or fall, like the climate, or housing. Something more specific like the 100-year average temperature in New York, or the number of houses currently on the market in Miami, can be. Factors on a causal loop diagram can be either causes, or effects, or more often, both.

Causal loop diagrams can be created by starting anywhere in the system. Begin by identifying just one factor in the system and then ask, what causes this factor to rise or fall, or what other factors does this factor cause to rise or fall. Draw arrows from the causes to the effects. By convention, we put a plus sign (+) at the head end of an arrow if the causal relationship is positive, meaning that an increase in the cause tends to result in an increase in the effect. A minus sign (−) at the head end of an arrow conveys that when the cause increases, the effect tends to decrease. Remember that we must say *tends to* because other factors may overwhelm or *swamp* any single factor. Adding a lane to a road, increasing road capacity, would tend to decrease commute times in a city, but if the city is growing fast enough, there may still be an increase in commute times, even with the road expansion. The commute time will be less than it *would otherwise be*, due to the road expansion, but still may not actually decrease. There is no reason to draw *road expansion* as a causal factor for *road capacity*. These two are the same thing in different terms; showing them as a causal relationship will just clutter up a causal loop diagram without adding any important information – an example of a *rocks are hard* explanation.

If you draw out enough causal relationships in a system, before long some patterns will become apparent. One of the most common patterns is called a balancing feedback loop. Consider drawing the causal loop diagram for a home heating system, controlled by a simple thermostat. We might start by drawing a causal arrow from *current temperature* to *heat output*. As the current temperature in the house drops, more heat output is produced. Then we must draw an additional arrow from *heat output* back to *current temperature*, indicating that as heat output increases, it causes the current temperature to increase. What we have drawn so far, is not wrong, but it doesn't tell the whole story about how the system works. If the home heating system worked the way we have drawn it so far, the heating system would be turning on and off rapidly all the time. Here's why. Say the thermostat is set to 70°. When the temperature falls to 69.9°, the heating system turns on. Then when the current temperature rises to 70° again, it turns off. This constant on-and-off is not efficient for the system. What we really want is a *dead band*, a range of temperature during which the heating system stays off. A typical deadband is around 2°. We can amend our diagram to show this by calling the first element something like *calls for heat*, which means that the system calls for heat when the current temperature falls to 69°, and stops calling when the current temperature reaches 71°. This makes sense, since it's really not the current temperature in the house that causes heat to be produced, but the difference between the current temperature and the thermostat's setting. The two arrows we have drawn form a balancing feedback loop because as one factor increases, it increases the other, but as the other increases, it decreases the first. A balancing feedback loop will always settle into a steady state as it balances the two factors that are each affecting each other. It will act like a thermostat maintaining the temperature in the house.

Balancing feedback loops are not limited to just two causal factors. The well-known relationship between price, supply, and demand is a good example. Higher prices cause lower demand, lower demand causes lower sales, lower sales causes higher supply, and higher supply causes lower prices. In a normal situation, this balancing feedback loop causes prices to stay in check. When drawn as a causal loop diagram a balancing feedback loop can be identified by observing a complete loop of factors, where all the signs are positive, except for one negative. A complete picture of this system would add additional factors that could cause lower supply, higher demand or a change in elasticity, which indicates how willing buyers are to go without or substitute something else. Gasoline prices fluctuate for a number of reasons, but one is demand. Gas prices fall after the summer vacation season, when there is less driving. At a sufficiently high price level, some people will forgo driving. Here we are using examples that almost everyone knows and understands; the point is that causal loop diagrams can help us reason about relationships of this type and communicate them to others who don't already understand them.

The other primary pattern seen in causal loop diagrams is the reinforcing feedback loop. In this kind of loop, all the signs of the causal relationships are positive (or all are negative), leading to the loop reinforcing itself, increasing (or decreasing) endlessly. In our world of systems, there are many examples of reinforcing feedback loops in operation, though often the seemingly infinite reinforcement is checked by other outside forces. A car rolling down a hill, will gain momentum, which helps it roll faster, gaining even more momentum, which makes it roll even faster, endlessly, until it reaches terminal velocity, crashes or someone hits the brakes.

It is very common for a system to have both balancing and reinforcing feedback loops in operation at the same time. In a free market, as the adoption rate for a new product increases, there is an increase in new adopters, which of course causes an increase in the adoption rate – a reinforcing feedback loop. It would seem that the adoption rate could increase endlessly, but there is also a balancing feedback loop at work. As the adoption rate increases, there are fewer adopters available to adopt, and a reduction in potential adopters will act to reduce the adoption rate. These two loops, operating together, will result in an equilibrium, where the adoption rate stabilizes at a sustainable rate, and the reinforcing feedback loop is moderated by the presence of the balancing feedback loop. It's a bit like running water into a sink that has an overflow drain. For the most part the sink will fill to the overflow drain, and the drain will drain fast enough to prevent the sink from overflowing, but increase the flow into the sink enough, and the overflow can't keep up, and the sink overflows, adding to the water on the floor indefinitely. Similarly, in a market of dramatically increased demand, propelled by panic-buying, such as the toilet paper market in the early months of COVID-19, supply can't keep up with demand, even if prices keep rising. The system is swamped and is overwhelmed until more supply can be produced. Causal loop diagrams and the reasoning that supports them helps to understand these and even more complex systems. Systems thinker have been known to stare and stare at their causal loop diagrams, wondering if there are causes they have not yet captured or relationships between causes that should be added.

The notion of causal factors, and their depiction in causal loop diagrams is intuitive and simple, but there are some subtleties and common mistakes that plague new practitioners. One is that not all causes are equal in magnitude, even though a causal loop diagram makes it appear so. College graduates earn much more over their careers than high-school graduates, and those with advanced degrees earn even more. Those attending top-tier universities also earn more than those who don't. But it's quite possible that someone with a master's degree from any university, would out earn the average undergraduate from a top-tier university. If you are already a systems thinker, you'll be wondering if attending the top tier school is causal, or if those who will ultimately be top earners It's hard to show this kind of relative causal strength on a diagram. It's also hard to show the difference between causation and correlation. Does someone with a masters degree earn more because they have a masters degree, or do people with higher earning potential tend to pursue masters degrees? The difference between correlation and causation may not make very much difference in understanding what is going on. After all, a strong correlation usually produces good predictions of system behavior, even if it is not causal. If a deeper understanding of the system is needed, however, isolating true causes among all of the other factors may be worth the effort. The little-known field of informal logic can help us here.

9.7.4 Logical Fallacies

Informal logic is the study of how logic is applied to argumentation, persuasion, and dialog. An important aspect of informal logic is the study logical fallacies, which appear to be logical, and may "sound good" but which when examined should carry no weight in support of the presented position. A well-known football star appears on television, presenting a new car model. Those who paid the star to appear and those who designed the commercial, certainly must believe that people will consider the car more desirable when it is presented by a familiar and popular personality. They are probably right, even though it makes no logical sense that the car is better because the company paid a football hero to tell you about it. That's an example of an *appeal to authority*, or *argument from respect*. We respect the *person* so we are inclined to believe him about the *product*, even though our logical minds tell us he was paid and scripted. For our purposes in systems, there are a few logical fallacies that are even more relevant.

Post Hoc

In the discussion of causation, we have noted that just because something happens after something else, it does not mean that the first event *caused* the second. This fallacy is known as post hoc *ergo propter hoc*, Latin for *after this therefore because of this*. The fact that virtually all students who attend college do so after graduating from high school does not mean that high school graduation causes them to attend college. Post hoc fallacies are everywhere. We saw one in the story of Dr. Brown's Miracle Elixir earlier, where people are tempted to conclude that because feeling better seems to only occur *after* taking Dr. Brown's, the feeling better is *because* of Dr. Brown's. In reality of course, since we instruct people to take Dr. Brown's whenever they are feeling bad, and they almost always tend toward feeling better after feeling bad, we have set people up to make this fallacious post hoc conclusion, in order to boost sales of the "product." Have you heard that genetically modified foods (GMOs) cause cancer? It can easily be shown that most people who develop cancer have eaten GMOs before the cancer occurred. The fact is of course that the vast majority of people eat GMOs (it's difficult to completely avoid them), so of course most people who develop cancer have eaten them. Even if GMO use was increasing at the same time cancer was increasing, this does not prove a causal relationship. Systems thinkers must be quite careful not to accept post hoc based theories as causal relationships when trying to understand a system.

False Cause

Another relevant logical fallacy is known as *false cause*. Say we find that most smokers also drink. Does smoking cause drinking? We also find that many drinkers smoke, but that relationship has less evidence, since there are many drinkers who don't smoke. So, we may be led to believe that smoking does indeed cause drinking. We call this a false cause argument because there is no evidence of a causal connection, only a correlation. Most cancer patients have eaten peanut butter – does peanut butter cause cancer?

Some light can be shed on these correlations by reversing them. If it is true that most cancer patients have eaten peanut butter, but most peanut-butter-eaters do not develop cancer, then it is apparent that peanut butter is not the cause of the cancer. In many false cause claims, the truth is that there is a third factor that is causing both. If smoking doesn't cause drinking, and drinking doesn't cause smoking, perhaps there is a third factor that is causing both smoking and drinking. Perhaps socializing at bars causes both smoking and drinking. If we observe that people who dress well get promoted at work, should we conclude that dressing well is the cause (or at least a partial cause) of getting promoted at work? Perhaps a third factor, like being ambitious and committed to one's job, causes both dressing well and getting promoted. Like post hoc fallacies, false cause fallacies are everywhere and systems thinkers must be careful not to assume cause when it isn't there.

In science, two important aspects are needed to begin to believe that cause is occurring. One is clear correlation, with no post hoc or false cause fallacies in the reasoning. The second is a plausible theory for how the causal relationship works. Say we are investigating whether obesity causes heart disease. First, we see if the two factors are correlated, that is, do obese people tend to develop heart disease more than otherwise similar nonobese people. If so, then we must come up with a plausible reason why obesity could physiologically cause heart disease. We might theorize that the obese person has more blood vessels and so there is more work for the heart to do, or that the extra weight carried by the person puts an extra load on the heart, or both. Now we have a plausible, though not yet proven, causal relationship in the system. We could then do further experiments by getting some obese people with heart disease to lose weight and see if their heart disease improves.

To check, we should look to see how many nonobese people develop similar heart disease and also how many obese people do not develop heart disease. Now we are on the road to an established causal relationship, and to better understanding one aspect of the human-body system.

Natural Experiments and the Scientific Method

At this point, we have worked hard to understand the system that is producing the problem and have drawn a causal loop diagram to help organize our thinking and conclusions. It may be important to take some additional time to test our understanding of the system, again apply the scientific method. If our understanding of the system is correct, we should be able to use it to make predictions about how the system will function in various circumstances. If our causal relationships are correct, if they match the system, then if we increase a factor, we can predict the results. In large-scale human activity systems of course, we probably can't test this in reality. We can't make a certain population gain or lose weight, start or stop smoking, or change the environment in which they grew up. What we often look for is what scientists call *natural experiments*, situations that occurred naturally, or accidentally, without an experimental design, but which nevertheless do what we would want an experiment to do. A natural experiment might occur, for instance, when two communities are divided by a river. The two communities are identical in environmental conditions, but the people do not intermix much due to the river. If one community changes something perhaps by passing a law, we can compare the separate but quite similar communities, and try to isolate the effects of the new law.

Steve Levitt, co-author of *Freakonomics* and an economist at the University of Chicago tells the story of working with a large retailer to try to assess the value or *return-on-investment* of newspaper insert advertising. To put the question in causal terms, does the advertising cause additional sales over the sales that would have occurred without the advertising? He discovered a natural experiment that occurred when a summer intern failed to run the inserts in Pittsburgh for about a month. Did the sales in Pittsburgh go down in that month? No, they stayed about the same. Did the company reduce advertising, saving money and adding millions to their bottom line? Also no – you can hear the full story on the Freakonomics Podcast, episode 440 (https://freakonomics.com/podcast/does-advertising-actually-work-part-1-tv-ep-440). Natural experiments are gold when it comes to trying to learn how a large system works.

9.8 System Archetypes

Once we are satisfied that we have an adequate understanding of the system producing the problem, we can begin to look for patterns in that system by looking for patterns that occur in many kinds of systems – patterns that will help us find potential solutions. These patterns are called *system archetypes* and are a wonderful shortcut to understanding many common human activity system situations. Let's start with an example. Say we have a village in a time when most residents raised livestock for a living. The village grew up around some nice pastureland with a small lake, since the land and the lake provided a handy food and water source for the animals. As the village grew, the livestock population grew with it. Since the common land and lake were free for all to use, each person reasoned that there was no reason not to use it as much as possible. If someone had 20 head of cattle, why not increase to 30, since food and water were plentiful and free? Since all the people thought this way, the number of animals using the land and the lake increased so much that the land became over-grazed and the water too polluted to use. No one person was at fault here. Each part of the village system (people, animals, land, and lake) simply operated as it could be expected to, given its goals and the opportunities available to it.

9.8.1 Tragedy of the Commons

The tragedy is that no one wanted the land and lake to become unusable; after all, everyone depended on them. We call this pattern a *tragedy of the commons,* and it occurs in many kinds of systems. A tragedy of the commons occurs whenever there is a resource available to all without cost, but which can be overused to the point of exhaustion, resulting in a detriment to all. Fishermen who over-fish an area that has no catch limits may take so many fish that not enough are left to reproduce and the area will stop producing fish for anyone. A candy dish placed on the receptionist's desk at a company is great, but since the candy is free, people take too much and the company finds they are spending too much on candy and remove it altogether. So many people drive on a road that traffic clogs the road, producing delays, and the road becomes ineffective for travel. The general solution to a tragedy of the commons is to stop making the resource available for free. Charging a toll to drive on the road reduces traffic and keeps the road effective for those who pay to use it. Requiring farmers to buy or lease the land on which their animals graze causes them to manage their own land in a sustainable way, so they can keep their farm going. Fishing and lobster licenses limit fishing to sustainable levels.

Is Internet access a potential tragedy of the commons? Not currently, for several reasons. For one, people must pay to access the Internet and most access plans have a data limit, so the resource cannot be overused at no additional cost. What can indeed be a tragedy of the commons, however, is Wi-Fi access. Regardless of the Internet speed connecting to a home, a restaurant, or a coffeeshop, too many Wi-Fi users will swamp the Wi-Fi access point and slow things down for everyone. Most Wi-Fi access points, such as those at a coffee shop or conference center don't limit the number of people connecting to them, and since no one is paying to connect to Wi-Fi, a tragedy of the commons can result.

9.8.2 The Rich Get Richer

Another very common system archetype is known as *the rich get richer*. Sometimes referred to as *success to the successful*, the archetype does not apply only to the rich in monetary terms. The idea is that in some systems, those who already possess a quantity of something, have an advantage when it comes to acquiring more compared to those who have less to start with. With money, it's easy to see how those with a lot of money have advantages when it comes to acquiring even more money. The relatively rich can spend less meeting living expenses and can use more of their money for savings, investments, to buy real estate or to become more educated and enhance their careers – all things that result in more money acquired. It is far easier for someone with a million dollars in the bank to find a way to make another million than for someone living paycheck-to-paycheck to amass the first million.

The *rich get richer* archetype applies to all kinds of "riches" or successes. Who should have the easiest time earning a master's degree? Probably someone who already has one. The person who has already earned a master's degree has many advantages over someone who has not. The path is familiar, the person knows what to expect, how to study, the work required, and even the rewarding feeling at the end of the journey. Successful football teams make more money and can thus spend more to hire the best new players, thus becoming even more successful. Some sports leagues enact salary caps and other controls to prevent a few teams from dominating the entire sport indefinitely due to this archetypal pattern. Someone learning a musical instrument will struggle with many aspects at the beginning, but someone who already plays one instrument will have a far easier time learning their second instrument. Learning a second foreign language is easier than learning the first one.

If we aim to have a fair society, with equal opportunities for many people, we must be careful to avoid systems that allow the rich to get richer at the expense of others. If a university education is too expensive, then only those with affluent parents will be able to afford it, allowing them to earn enough to send their children to expensive universities too. Universities will sometimes try to level the playing field by offering financial assistance and scholarships to those who are academically qualified, but less affluent. One might argue that the playing field is still not level, since more affluent parents live in affluent areas, with better schools and thus their children attend better public or private schools, giving them more advantages in attending universities and attaining affluence themselves. Integrating schools across socioeconomic lines is one approach, though it's understandably controversial. Redrawing county boundaries to encompass economic diversity is another way to attempt to better level the playing field. When trying to avoid the traps of a *rich get richer* archetype, the goal is not to give advantages to anyone, but to put all players on a level field where they all have equal opportunity to succeed in the game or pursuit at hand. Of course, not everyone *is* equal. Taller boys and girls will always have an advantage on the basketball court, but the idea is to compensate for unrelated differences in the system and allow the game to be fair.

Rich get richer patterns are not always bad or undesirable. There's nothing wrong with it being easier to learn a second instrument or a second foreign language more easily than someone who is working on their first. If, however, our intent is for a system to be equally available to all and to avoid special advantages given to some based on past achievement, then we must watch out for *rich get richer* situations and find ways to eliminate or compensate for the unfair elements.

Early in the history of the United States, the decision was made, and supported by government funding and regulation, to make electricity available throughout the country. The same was done for highway travel by creating the interstate highway system. By making these resources available to everyone, the playing field was made level and America continued its reputation as the land of opportunity.

An unfortunate example of the rich get richer is the pricing of groceries and other goods in lower-income neighborhoods. People in those neighborhoods may not be as mobile, and may not even own a car, so they are all but forced to shop at stores within walking distance. Some of these stores, knowing that most buyers have no alternative, have little competition in that local market and basic economics dictates that as competition is reduced or eliminated, prices will rise. Buyers in affluent areas, can easily travel several miles to shop and can choose from a variety of grocery and big-box variety discounters. So those who are already more affluent, have an advantage in becoming even more affluent by spending less for the same goods – the rich get richer. This situation could be an example of where free markets fail to do what we wish they would do which is to provide good opportunities for all people to benefit economically. There's more to this system, though, as some would claim that costs to the store owner are higher in neighborhoods due to shrinkage, crime, and real estate cost. What should happen is that new merchants should see an opportunity to move into the neighborhood, charge lower prices and create competition, but the hyper-localization may negate economies of scale: small stores serving a small number of nearby customers aren't as efficient economically and may not be able to sustain lower prices. Isn't systems thinking interesting?

We'll discuss some more system archetypes in the chapter on people systems, but for an even more in-depth treatment of this fascinating subject, I'll again recommend *Thinking in Systems – A Primer* by Donella Meadows. For now, let's continue in our problem-solving process using systems thinking. So far, we've identified a problem, found the real problem behind the problem, identified the system or systems involved, and worked hard to understand these systems, their causal relationships, and how they work. We applied the idea of system archetypes to expose any common patterns present in the system. Ideally, we have now reached a point where we feel we can intervene in the system and have a reasonable expectation of being able to predict what will happen as a result of our intervention.

9.9 Intervening in a System

While the process of considering system interventions and their effects may seem intuitive and obvious, the more common approach is to make system interventions based on a limited understanding of the resulting effects in the larger system. Common examples arise in everyday conversation. People sometimes complain about the development in their neighborhood, the building of new homes, the addition of a shopping centers, and the clearing of fields, not considering what it would mean for such development to be prevented. The land is owned by someone, and subject to city and country zoning and covenant restrictions, if any, may be used in whatever way the owner chooses. In real estate development, the concept of *highest and best use* tends to guide development toward the greatest economic benefit for the owner, whether it be a townhouse development, single family homes, or a new shopping center. An owner who buys undeveloped land will keep it undeveloped only until it is economically feasible to develop it toward a higher and better use. To legally prevent an owner from developing private land, because we like the trees growing on it, is not something that happens in our economic and governmental system. The government does buy land and turn it into parks and such, and may even force an owner to sell land to the government (at a fair price) through the principle of eminent domain, usually in order to build something for the public good, like a road. But in normal circumstances, the land and the trees are the property of the owner, who may choose to develop or not. Given this economic and governmental system, a person who wishes to prevent the development of some land has only one sure option: purchase the land and leave it undeveloped. Those who feel that some unspecified "they" should leave the trees alone are, without realizing it, imagining a completely different economic system, where the government or feudal lords own all land and make all decisions about how it will be used. To save trees under that system, one need only offer supplications to the lords and masters. It's a possible system; private ownership of real estate is not a divinely ordained right, but a choice in the design of a nation's economic system – one that has implications for economic growth, freedom, personal incentives, and sometimes, trees.

A famous example of a system intervention gone wrong comes from the old days in then Bombay, India, where cobras used to freely roam the streets. Eventually, they became a nuisance. The people complained to the government. The government, seeking to intervene in the system and to improve the cobra situation, decided on a wonderful intervention. They instituted a bounty on cobras. Anyone bringing in a dead cobra was entitled to a generous cash bounty. As the government officials were formulating this intervention to the system, we can imagine them thinking about how great this will be – a tidy incentive for people to hunt down and kill the nuisance cobras. To migrate this example to modern society in the United States, imagine that rats became rampant – a rat bounty might seem like a wonderful idea. But back to Bombay. At first the bounty seemed quite successful. Cobras were brought in, and bounties were paid. After a while, however, it began to dawn on the government officials, that the overall cobra population did not seem to be decreasing, despite the steadily increasing expenditure on bounties. What had happened is that some enterprising Indians had seen an opportunity, one created by the new bounty system. Cobras were being raised and killed to collect the bounties. After all, it's much easier to catch and kill a captive cobra than a free-roaming one. Apparently, the bounty was enough to make cobra farming profitable. Eventually the government caught on, and decided to discontinue the bounty on cobras, thinking that it would curtail the raising of cobras for "profit." Eventually, eliminating the bounty would stop the raising of cobras, but in the short term this intervention had another effect. The cobra farmers, seeing no further benefit in keeping their cobras, released them. The result is that a program intended to decrease the cobras, actually resulted in their increase. It's the kind of story that systems thinkers live for. The phenomenon of an intervention that results in an unintended consequence that is not only unintended, but which results in exactly the opposite of what was intended, has come to be known as the *cobra effect*.

The cobra effect is also an example of another systems archetype, known as a *fix that fails*, or *policy resistance*. In this archetype, an intervention is attempted, but the system pushes back, resisting the change with enough strength that the intervention must be changed or stopped. Systems, especially human activity systems, can be stubborn and resist disruption and change, even if for the better. The system's causal relationships, when fully discovered, form a highly interconnected network and a single change, unless carefully and systemically designed, may not be sufficient to modify the network or overcome the well-entrenched structure and behavior of a system. A famous example is prohibition, an attempt in the United States in the 1920s to ban all alcoholic beverages. These beverages were already too entrenched in American culture, so prohibition tended to simply drive the market for alcohol underground. Illegal goods for which there is a high demand bring the potential for big profits, organized crime, and violence, all of which flourished under prohibition. At some point, it was determined that the fix as worse than the condition it attempted to fix, and prohibition laws were repealed in most areas. Prohibition was not an example of a cobra effect, since there's no evidence that making drinking illegal actually increased drinking, but it's a great illustration of a fix that failed. Fixes that fail are usually the result of an incomplete or incorrect understanding of the system involved. With prohibition, the strength and in economic terms, inelasticity of the demand for alcohol was apparently not fully understood. If the lawmakers at the time had been systems thinkers, they would have thought about what people would use in place of alcoholic beverages. Coca-Cola was available, but didn't seem to have the same social lubrication effect that alcohol offered. Did lawmakers expect people simply to avoid bars, parties and social gatherings that previously centered around alcohol consumption? To be fair, prohibition is not extinct in the United States. Around 10% of the US population live in counties or cities that are "dry," where alcoholic sale and consumption is banned, with some fascinating exceptions (see https://www.alcoholproblemsandsolutions.org/dry-counties). Like countries that ban alcohol, dry US counties object to alcohol on a religious or moral basis. Even these areas, however, are an example of a fix that fails, since documented instances of driving under the influence of alcohol (DUI), occur at rates that exceed those in counties that allow alcohol. As in times of national prohibition, banning alcohol does not remove it, it only drives it underground, potentially increasing the distance people must go to get it as well as the distances drinkers may travel while under its influence. Aren't systems fascinating?

There are a variety of ways to intervene in a system. Meadows calls these leverage points and discusses the relative strength and impact each has on changing a system. She lists the following, in order of increasing system impact:

1. Numbers (size of flows and stocks)
2. Size of buffers
3. Intersection nodes
4. Delays (as in the Bullwhip Effect)
5. Strength of stabilizing feedback loops
6. Strength of reinforcing feedback loops (e.g. interest rate)
7. Information flows and who has access
8. Rules/Incentives/Rewards (e.g. bonuses, commissions)
9. Self-organization: the power to change system structure
10. Goals
11. Paradigms
12. Transcending paradigms

Let's explore a few of these. Changing numbers in a system can have an effect, but often not as great an effect as expected. An example is interest rates. One might expect that when home mortgage interest rates were above 10% that few homes would be sold, and that in recent years with interest rates under 3% that many more homes would be sold. To some extent this is true, but there is not a threefold difference. Auto interest rates are often under 2%, not above 15% that was seen in past economic times. Yet about the same number of people buy cars. The point is that the effect that numbers have may be overshadowed by other effects, such as the need or desire to own and home or car. Prices can make a difference, but many items will be bought, regardless of price, due to other effects. A struggling small business, for example, may think that it will increase sales by cutting prices, but may be disappointed to find that not much changes, even if prices are reduced below the level of big-box discounters. If the price changed is noticed by only a few and the store is inconveniently located, shoppers may ignore the price advantage and continue to buy from the big-box retailer. Where prices make a big difference in behavior is when purchasing alternatives are widely known and equally convenient, and where the cost of switching providers is small. Several gas stations at a busy intersection must keep prices nearly equal or business may shift quickly to the lower-cost provider.

Buffers are a fascinating, and often overlooked part of a system, but can have a big impact. A family with $10,000 in the bank but living paycheck-to-paycheck in terms of budget will have a different experience than the same family with only $500 in the bank. The larger "money buffer" allows the system to tolerate earning and spending variations. For the first family, a $1000 auto repair, while frustrating, can be managed and the money earned back slowly over time. The same auto repair will throw the second family into debt. A credit card is a kind of buffer, but an expensive one if a balance is carried beyond the current billing cycle. During the early days of the COVID pandemic, many households found that a large buffer of toilet paper made life much easier. They had what they needed and could take advantage of purchasing opportunities when they arose. A family with only a little "in stock" may suffer frequent, frantic trips to multiple stores to fill an immediate and urgent need. Large corporations that are able to maintain a buffer of qualified candidates and new hires to fill needs, is able to grow more quickly and take on new projects better than one that must only hire when there is an immediate need. Increasing buffers in a system is often a way to enable a system to tolerate problems, and operate more smoothly.

The area of hiring also shows the importance of delays in a system. From the time a need is identified to the first day of the new person's employment can be several weeks or months. The temptation is to over-hire in times of need, and then by the time the new employees start, there could be too many. This "bullwhip" effect is often seen in supply chain processes, like the one made famous by the beer game story told by Peter Senge in *The Fifth Discipline*. In brief, a single mention of a beer brand in a hit pop song caused a spike in demand for that beer. The spike was interpreted as a permanent change in demand rather than a spike, so stores, which ran out quickly, ordered more, and continued to order even more as distributors ran short and were unable to fill the orders. The factory expanded too, with the overall result that way too much of the beer was produced as a result of a lack of knowledge about how the system was responding to a new input. Delays played an important part in the story, since as both stores and the distributors were waiting for orders to be filled, they became impatient and placed even more orders.

The strength of feedback loops, both reinforcing and stabilizing is important, especially relative to the strength of other causal factors in the system. It is true that someone who exercises will tend to exercise more, implying a reinforcing loop that will lead one day to Olympic stardom, but unfortunately for the medal count, there are other causal factors that can overpower the tendency of exercise to cause more exercise. Life priorities, time availability, motivation, and even day to day feelings, can overpower the relatively weak effect of wanting to exercise again. A thermostat will indeed maintain the house at a comfortable 74° in summer, but open all the doors and windows of a Georgia home, and the system will not be able to keep up, despite the thermostat and air conditioning system's balancing feedback loop. This is also an example of an important aspect of systems known as *limits to growth*. Almost everything in a system has a limit which must be kept in mind for a full understanding of the system. The air conditioning system can only cool the home a certain amount. With unusually high temperatures outside, the system may not be able to keep up since most residential air conditioning systems can only cool the home by a maximum of about 20° (F) below the outside temperature. Pressing on the accelerator of a car increases its speed, but only up to a maximum speed, or until the engine fails. Automotive systems may even employ limiting devices to cap the vehicles speed and prevent damage to the engine.

Information flow and who has access to information is a systems phenomenon all its own. Since the late 1990s information access has been extended to many people who before had no practical means of access – the Internet has changed the way the world works in many ways. Inside a company, the degree to which various information is available can affect company productivity, security, innovation, and ability to grow. The so-called *Amazon API mandate* from about 2002, which became widely known after being leaked from the company, made it clear that all information systems in the company must make their data available to other systems in the company using computer network access methods, known as application program interfaces (APIs). Attributed to Amazon CEO Jeff Bezos himself, the mandate threatened firing as the consequence of disobeying this policy. In this one move, Bezos established a culture of open data interchange across the company, enabling a long chain of innovations based on using combinations of exiting data to offer and deliver new services and products. There are many people in many organizations who wish their CEO would place that kind of priority on information transparency and shared use within the organization. More typical is the existence of islands of information, controlled by the groups who maintain the information, who allow others access at their own whim. To these gatekeepers, it may be pointed out that they do not own the data – the organization who pays their salaries owns the data. The question is not, will someone give me access to their data, but what form of data access is the right approach for the organization?

One of my favorite kinds of interventions to study are rule changes, which include the economist's favorite tools – incentives and rewards. Changing the rules in a system can have immediate and widespread effects, assuming of course that everyone knows the new rules and that they will be enforced. As we've already noted, some rule changes can bring resistance, and result in a fix that fails, but in many cases, the rule is simply accepted as the new reality and behavior in the system begins to shift, based on the new rule. In 1978 as the US Congress enacted some new tax legislation, they included a provision buried in section 401, sub-paragraph k, that allowed employees to avoid paying taxes on compensation they would defer until retirement. This simple rule change resulted in companies developing programs that allowed employees to save pre-tax compensation in a retirement account owned by the employee, for later use in retirement, when withdrawals of this money the compensation would be taxed at the employees presumably lower tax rate. To encourage employees to use the plans, employers began to offer matching plans. From 1978 to 2021, employees have amassed a total of almost $7 trillion in 401k retirement plans, and all this resulting from a small rule change in the tax code.

When the designated hitter rule was introduced in 1973 as a temporary trial, there was concern that it would "ruin baseball." Fifty years later, there is still argument on both sides of that issue. The argument is a systems one – how does this one rule change affect the entire system of major league baseball? Team salary caps, mentioned earlier, are another rule that was designed in part to compensate for the rich-get-richer cycle in successful teams. Since teams are limited in total player salaries, even very successful teams cannot hire all of the best players and become even more dominant.

Rules and laws can be powerful system-changers, since they set up incentives. Laws come with built-in incentives to follow the laws, known as fines, courts, and prisons. Other rules guide behavior through social norms or peer pressure. But rules are not always effective. I have a paved bike trail near my house. The posted rules are that everyone – pedestrians, bikes, skaters, and kids, should stay to the right unless they are passing others. Does everyone do that? No. Do they know the rules? Uncertain. Is there any enforcement? No, but most people over time follow others and mostly stay to the right. Rules result in incentives and drive behavior. Real estate agents are meant to represent either the buyer in a transaction or the seller (in some states they are allowed to represent both), but since both the buyer's and the seller's agent are compensated only when the transaction closes, and are paid shares of a commission based on the sales price, both agents have an incentive to work for the sale. As was pointed out in Freakonomics by Dubner and Levitt, real estate agents do not have a strong incentive to maximize the selling price. While a higher selling price means a larger commission, it may also make the home harder to sell, or cause it to remain on the market longer. A bird in the hand, in this case a contract for sale, is worth two in the bush – a higher price and slightly higher commission. Is it possible that this incentive structure causes real estate agents to suggest lower selling prices to prospective sellers, rather than a selling price that maximizes profit to the seller? In Freakonomics, the authors cite studies showing that homes sell for an average of 17% more when real estate agents sell homes that they themselves own, compared to when they are selling a home for someone else. It probably goes without saying that a systems thinker must find out all the rules that are operating within a system as part of fully understanding what makes the system work the way it does.

A cousin to rules are goals. Rules drive system behavior through incentives and punishments, while goals drive behavior with the prospect of reward and avoidance of the negative consequences of not meeting the goal. The usual goal for a sales organization is sales revenue. Pursuing the maximization of revenue, could cause the sales organization to lower prices using discounts and special deals, in order to make more sales, and raise revenue. Business owners and shareholders on the other hand, tend to prefer the maximization of profit, and so may work to keep prices at a level that balances sales and profit. Narrowly or poorly selected goals can cause surprising system behavior. Years ago, I was in a client's office building for a meeting. It was summer and the air conditioning was set such that our room was getting rather chilly. In the room full of engineers, a creative solution was proposed. Since we had no access to changing the temperature setting on the thermostat (we couldn't even find it), we turned our attention to the temperature sensor in the room. Some quick thermodynamic reasoning concluded that if we could fool the temperature sensor in the room into thinking that it was already cold enough, it would stop cooling it further. We found a small Ziploc bag and put a few ice cubes in it from the lunch buffet drink tray remains. Pinning the little bag on the wall just above the sensor (cold air falls, we reasoned), we went on with our meeting and waited to see what would happen. A little while later we heard a sound that reminded us of a 747 taking off in the hallway outside the room. Investigating, we found that our solution had worked a little too well. The heating and air conditioning system, sensing that the building, or at least our meeting room, was at near-freezing temperature activated the heating system in the building, and put all the fans on high. It was about to start heating the entire building until the freezing temperature was remedied and it was going to put a maximum effort into doing just that. We quickly disposed of the Ziploc bag – the evidence of our hacking, and went back to work like nothing had happened. The heating system calmed down in a few minutes and went back to cooling the building against the summer heat. What went wrong? The goal of the heating and cooling system was too narrow – it aimed to make all rooms comfortable, no matter the effect on other rooms, so our room calling for extreme heating overruled other rooms' need for moderate cooling. Perhaps the system was not smart enough to recognize that the outdoor temperature was in the 1980s and so no room should have been near freezing inside. If one appeared to be, it would mean a system error, or maybe some hacking by a few engineers in a conference room.

Goals can change even large systems if they are put in place firmly, well-communicated, and believed by all involved in the system. In *Who Says Elephants Can't Dance* by former IBM CEO Louis Gerstner, the story is told of how IBM, a company traditionally focused on the manufacturing and selling of large computer systems, transformed itself into a services-oriented business by setting a new goal, committing to it, and then getting every aspect of the company to redirect its operation based on the new goal. To be effective, a goal must be a *real goal*; it must be known and accepted as the real goal of the organization by most of the people in the organization. Simply announcing a new goal or strategy may do little to change the system of an organization, if people view it as the "flavor of the month" and not an actual change in the organization's goal. Organizations always have goals, even if the goals are not the ones written on the plaque in the main lobby, or in the latest memo from organization leaders. The actual goals may not even be written down but may be subtly communicated through example and oral tradition. The organization is pursuing *something*, as is each division, department, and individual. If the goals are aligned throughout the organization, all the oars are pulling together in synchrony and the organization moves ahead smoothly and efficiently toward those goals. But if individuals and departments have goals that differ from the division and those differ from the organization's goals, inefficiency results in the form of local optimization. In local optimization, a part of a system operates with higher priority on its own local goals and interests, with the goals and interests of the larger system organization being of secondary concern. An instructive example of this is the shared service monopoly pattern, described in the chapter on people systems.

A paradigm is the overall concept, viewpoint, or belief system that people operate within, at a given point in time, regarding a particular subject. At one time, the prevailing paradigm was that the world was flat, and this paradigm had strong implications for travel, trade routes, exploration, and colonization. The paradigm limited scientific understanding of how natural systems in the world work. Starting from a flat-earth paradigm, it is difficult to explain the apparent movement of stars, the appearance of the horizon, gravity, and other phenomena. A paradigm is not a choice – it's more like the water within which the fish swims, unaware that water is even a thing. Companies, nations, and individuals have their own paradigms. Paradigms are so powerful because they are unrecognized and therefore unexamined and unquestioned. Fleas put in a covered jar, it is said (though the science may be in doubt), will soon learn they can jump only so high without hitting the lid, and continue jumping only that high after the lid is removed. A paradigm is akin to the box in the expression, "thinking out of the box."

To challenge the paradigm within which a system operates, one must first identify the paradigm and bring it to light, so that it can be examined. Once the paradigm is exposed, it is easy to ask questions like: What if the paradigm were not true? How would things be in a world where that paradigm were not true? Questions like these could have led watchmakers to explore non-mechanical watch movements (even the term *movement* for the insides of a watch is steeped in the paradigm of watches being mechanical), or camera makers to explore cameras that don't use film to capture images. A family struggling to decide between living in the city or living in the rural mountains, could think beyond the paradigm of having just one home, and consider two less-expensive homes instead of the traditional single home at the maximum level the family can afford.

COVID-19 has caused many companies to challenge the paradigm that employees are most productive when working in an office in a company building, and are shifting to allowing work from home, not only during COVID times, but even permanently. Some corporations are even considering reducing their office real estate holdings – an additional benefit to the work-from-home paradigm-busting. Exploring this paradigm shift a bit more, we might wonder whether there is evidence that working from home is as productive as working from a company-provided office. In trying to quantify this system, we first note that most jobs do not have easily measured productivity metrics. So which way should we phrase the hypothesis we are testing? Consider the difference between the claim that productivity is no worse with employees working at home than it is with employees working in offices, and the claim the remote employees are just as productive. With the former claim, we can consider evidence like the lack of studies, and the lack of anecdotal claims that productivity is lower when employees work from home. With the latter claim, we would need to produce conclusive studies that show productivity parity between remote and in-office employees, and this is harder to do – not because the claim is in more doubt, but because it's hard to do studies on something that is hard to quantify. Subjective claims that remote employees are equally or more productive, when made by the employees, may be viewed as biased, since many employees prefer to work at home. A broad claim that employees are not less productive at home, however, may be easier to support by showing that there is no evidence to its contrary and that there would indeed be evidence if it were true. The era of

COVID-19, which forced many to work from home temporarily may have shifted the burden of proof from those that try to claim that productivity is the same or better from home (hard to show evidence) to those that claim that productivity is no worse at home (easier to show evidence). If companies cannot tell the difference in productivity when people work from home, then this supports the idea that working from home does not harm productivity. In statistical terms, the null hypothesis is that working from home is just as productive as from the office – there is no significant difference. Convincing evidence must then be produced that forces us to reject the null hypothesis. If there is no such evidence, then the null hypothesis cannot be rejected.

As we wrap up our discussion of systems thinking and how it can be used to solve hard problems, let's consider a brief case study that illustrates several concepts we've covered in this chapter. There is probably no one who wouldn't agree that highly addictive substances are a big problem in the world today. Let's consider heroin, one of the most dangerous controlled substances due to its highly addictive nature and the danger of fatal overdosing. Following the problem-solving process outlined in this chapter, we first aim to formulate the problem. What is the problem with heroin in the world today? Is the problem that too many people use heroin, despite its illegality? Is the problem that there is other crime that comes with heroin usage, such as theft by addicts, and violence among those who supply, distribute, and sell? Is the problem that heroin is illegal? Here we must be careful we don't slip into considering a solution presented in problem clothing. Making heroin legal might be someone's idea of a solution to something, or perhaps just someone's agenda based on an ideology, such as the idea that all substances should be legal, but it would not be useful to state the problem as the fact that heroin is illegal. The real problem is probably twofold: the problem for the addict and the problem for society. Heroin addicts damage their own health, sometimes fatally, and may turn to crime. In society, the system that provides heroin brings with it additional violent and criminal behavior. To this we could add the problem of pain and suffering of those close to the addicts, and economic loss of productive citizens, both addicts and criminal drug dealers. The next step is to understand the system within which the problem exists.

The heroin-crime system is used as an example in the MIT publication, *Building a System Dynamics Model* (https://ocw.mit.edu/courses/sloan-school-of-management/15-988-system-dynamics-self-study-fall-1998-spring-1999/readings/building.pdf) in order to demonstrate the fundamentals of system dynamics. For our purposes here, we'll note the important causal relationships, as shown in Figure 9.1 from the MIT paper. Like any free market, the heroin market is governed by price, supply, and demand. Increases in one of these three factors will cause an effect in one or both of the other two. Increasing heroin demand will either result in increased supply, increased price, or both. With no other factors

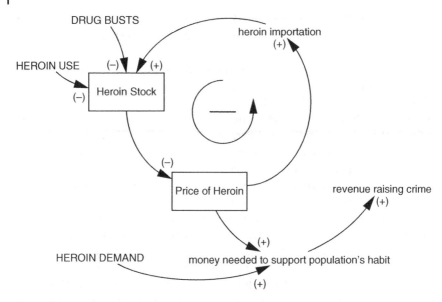

Figure 9.1 Causal loop diagram of the heroin crime system. Source: From Building a System Dynamics Model, MIT (1997), used with permission from MIT.

involved, it would seem that an increasing demand, with a constant supply will cause prices to increase indefinitely, but there are two limits that moderate that. For one, there is a limit to what addicts can pay, or the amount they can steal in order to pay. There's also some limit to what addicts are willing to pay. Addiction is strong, but addicts do sometimes quit using even a highly addictive substance for a variety of reasons. That the addictive substance becomes more unaffordable could be a factor. Smokers do quit due to cost, but one study shows that only 3% quit for that reason (https://pubmed.ncbi.nlm.nih.gov/22644233). Heroin addicts unfortunately sometimes die, thereby reducing demand, which would tend to exert downward pressure on prices or ultimately reduce supply. The demand for heroin is fairly inelastic, meaning that even as prices rise, demand doesn't change much – most addicts will remain addicted and continue to do whatever they can to pay for the next fix.

Heroin is not an ordinary market though, since its possession, use, and sale are all highly illegal and severely punished in many countries. Illegal markets must operate beyond the law, and thus cannot and do not rely on laws for regulation, paving the way for gangs or organized crime to become the "governing" influence. As demand for heroin increases, causing the price of heroin to increase as well, there is increased incentive for others to enter the business. Competition among suppliers, distributors, and dealers is not regulated in the normal way, but may involve violence, coercion, bribery, and the like, and at the same time provides larger incentives for market dominance. Eliminating the competition may be something that is carried out quite literally.

Next in our process we consider whether there are any archetypes or common patterns in this system. We can certainly see a *rich get richer* archetype at work; as distributors of heroin become stronger, they have advantages in becoming stronger still. One can easily imagine an enterprising individual or small group, attempting to enter a market, perhaps buying from suppliers other than those that dominate that market. The strong, entrenched distributor can exert pressure in several ways, such a tipping police to the presence of the new distributor, threatening dealers or customers with retaliation if they buy from the new distributor, or even direct violence or the threat of violence against the distributor. The illegality, and therefore nonregulation of the market allows for an unfair playing field among would-be competitors. We can also see a *limits to growth* archetype, since there is a practical limit to how many people will actually become addicted to heroin. Distributors and dealers can try to gain new "customers," but they take on new risk of exposure and arrest. While legal businesses may grow by attracting new customers, developing efficiencies of scale and competing on price with new market entrants, in a brazenly illegal market, new entrants can be easily stopped by strongly entrenched players.

We can also see the archetype of escalation in the heroin-crime system. Imagine a market area that is nearby two rival gangs who each control the market for heroin in their respective territories. Competition for dealers or buyers in the overlapping, disputed areas can easily turn to violence between individuals, then escalate to small groups, and then to full gang war, as each side steps up its efforts to control the new market. Since legal limitations and regulations are not honored in such a system, dominance of the new market area may be a matter of how far each side is willing to go to control that market.

At this point, we may plausibly think we understand the system well enough to consider some system interventions that may solve or reduce the problems we have identified. One of the first natural interventions would be for the police to try to find and arrest drug dealers and distributors. Say the police can find a dealer, make an arrest, and seize the dealer's stock of heroin. It's probably a small amount of heroin compared to all the heroin in the market area, but from a systems perspective, what effect does removing a dealer and some heroin from the system have? Removing heroin decreases the available supply of heroin, at least temporarily. It is unlikely that demand will decrease – heroin addicts supplied by the now-incarcerated dealer will simply find another dealer, or more likely another dealer will find them. When supply decreases, but demand stays the same, prices will tend to increase. The increase in street price may act as an incentive for more people to get into the heroin business, replacing that dealer and perhaps seeking to expand their distribution of heroin, which is now even more profitable than before. The intervention of drug busts and seizing stocks of heroin may actually help the heroin trade to expand. It's a cobra effect – achieving the opposite of what we want. As we said at the outset, this is a hard problem. No one is suggesting we stop enforcing drug laws, but it's useful to think through the possible effects on the entire system that seemingly helpful interventions may have. Let's take it even one step further.

Suppose in a giant drug bust operation we could eliminate all of the distributors, dealers, and heroin stock from an entire city. What would happen the next day? Would heroin addicts simply stop using heroin and switch to drinking beer? More likely, new distributors and dealers would move into the city, perhaps from nearby cities. If the distribution in the city had been controlled by a large gang or syndicate, then that organization might simply be able to reallocate new resources to meet the need in the city. To be fair, a reduced supply and thus reduced availability may result in fewer new potential addicts being introduced to heroin since the neighborhood dealer isn't there to provide that first free fix. This temporary reduction in new heroin users is of course this is a good thing, but the point is that reducing supply of a substance like heroin will not, by itself, eliminate the problem.

Let's consider a different intervention. A story from radio station WBEZ in Chicago (https://interactive.wbez.org/puertoricochicagopipeline) and told in depth on the podcast program *This American Life* (https://www.thisamericanlife .org/554/transcript) illustrated a different, though perhaps unintentional intervention to the heroin-crime system. Puerto Rican authorities apparently sent at least some drug addicts to Chicago for rehab and treatment. The reason it became a story is that it seems that the addicts did not get the promised rehab programs and ended up on the streets of Chicago. Their tickets from Puerto Rico to Chicago were one-way. By removing addicts from Puerto Rico, the authorities effectively reduced the demand for heroin in Puerto Rico by some amount. Assuming that addicts could be removed faster than new addicts could be cultivated, there would be a net decrease in demand for heroin. A decrease in demand with no change in supply would ultimately cause prices to fall, and even more importantly, would reduce the incentive for new entrants to join the heroin trade. To go further, imagine a city, or an island, where all of the addicts have been removed overnight. The heroin trade would collapse, distributors and dealer might turn to other activities. We could also expect that enterprising dealers would offer some samples of heroin at reduced prices to try to get new people addicted, but it's unlikely they could produce enough new customers to replace the ones lost. Heroin activity in that area would decrease, perhaps permanently.

Another kind of intervention would be to change the rules of the system. Previously illegal drugs have been legalized in some states and countries, with the predictable effects of increasing tax revenue, and decreasing crime, violence, and the incarceration of drug users. Naturally legalizing a previously illegal substance makes it available to more people and could predictably result in additional addictions. But with legalization, the trade of a drug may become more like the handgun trade, with various avenues for legal distribution, sale, and use, but still a black market in operation as well. The systems thinker can use the tools described here to reason about systemic effects of these changes. Reasoning about the heroin-crime system and possible interventions does not guarantee a solution, let alone a practical or feasible one, but it does illuminate the complexity of the system and allow useful thought experiments about potential interventions and a more complete estimation of their resulting effects.

9.10 Testing Implementing Intervention Incrementally

After carefully following the systems thinking problem solving process outlined in this chapter, it may be tempting to believe that the intervention chosen is the best one and is sure to succeed. With this abundant confidence, it is even more tempting to implement the intervention broadly, for all the employees, groups, cities, or companies involved. It is easier and simpler to go "all in" on an intervention, but it's risky. If the intervention does not have the intended result, a large mess may result. If a law is passed affecting the entire country, the entire country is affected by any unintended consequences. Perhaps there is a better way to roll out a new intervention.

We can learn from how large web sites, those with millions of users implement changes. They don't make the change active for all their millions of users all at once. Instead, they implement a split approach by implementing the change for say, 1% of users or even 0.1% of users, by randomly directing this small subset of users to the new web page, while the majority of users still experience the old page. This way, they can see if the new page works for a small set of users. If there's a problem, they'll receive dozens of error reports instead of thousands, and if some kind of fix is necessary, there are far fewer users affected.

In society one way of implementing incremental interventions is to encourage a city or state to adopt the new law or method and see what happens. The state of Colorado, for example, was the first to legalize recreational use of marijuana, so it served as a test case and exemplar of what might happen if the United States where to make the same change across the country.

Implementing a change incrementally, with fewer people affected, reduces the load on those performing the implementation as well as the volume of feedback received in the early stages. Feedback received early can be used in later increments, improving acceptance of those who receive the changes later.

9.11 Systems Thinking and the World

This chapter has introduced the fascinating field of systems thinking, an approach to problem solving in systems. Systems thinking can be applied to any kind of system, but is most often applied to human activity systems, those containing both technology and people. In a way, systems thinking is a field of fields. To be a systems thinker often means learning about a variety of other fields in order to understand the operations of systems – it's perfect for lifelong learners. Fields such as psychology, sociology, anthropology, economics, behavioral economics, probability, statistics, design of experiments, business, marketing, biology, and the other sciences all shed light on aspects of a large, complex, and diverse system.

In the Chapter 10, we explore a particular kind of human activity system – the people system. People systems, which consist only of people, have a unique set of dynamics, causes, and behaviors that explain a surprisingly large scope of what we experience, and are puzzled by, in life.

10

People Systems: A New Way to Understand the World

10.1 Reviewing Types of Systems

In the Chapter 9, we have described how in the paradigm of systems thinking, there are three types of systems:

- Engineered Systems are those designed and built by people, including aircraft, vehicles, satellites, smartphones, computers, and factories.
- Human Activity Systems are those that contain both humans and engineered systems, but are motivated primarily by human action, including companies, governments, families, economies, and armies. These systems may incorporate technological systems, but the overall system operates based on human will and action.
- Biological or Natural Systems are those which are biological in nature, and are observed to operate in a systemic way to achieve apparent purposes. These include animals, human beings, trees, forests, schools of fish, ant colonies, and oceans.

Naturally, systems *engineering* is concerned primarily with *engineered systems*. As technology advances, systems engineers may also be concerned with how to build more intelligent capabilities into engineered systems, making them better performing, more capable and more autonomous. Systems thinking, is concerned with the broader field of *human activity systems*, and the many problems that occur when trying to optimize or change them. Many – perhaps all – problems in human society are *systems* problems, and can be analyzed using systems thinking.

This book describes how to make all kinds of systems more intelligent through the use of good design, understanding user needs, and adding artificial intelligence technology. We might consider how the system of a restaurant be made more intelligent in the way it interacts with its customers, or how the combination of introducing new technology and changing how humans act make the system of an organization perform better and produce a superior customer experience.

Engineering Intelligent Systems: Systems Engineering and Design with Artificial Intelligence, Visual Modeling, and Systems Thinking, First Edition. Barclay R. Brown.

Most people have experienced less than excellent service from a large company and its people. By applying systems thinking and intelligent systems principles to the human activity system that is the company, we discover that simply replacing people with computers or robots may not yield better customer experiences or a more successful business enterprise. What we need to do is transform the business into a more intelligent system. For instance, Stackers, a restaurant in California, eliminates waiters in favor of tabletop touchscreens and friendly hosts, allowing diners to order their food, drinks, and refills without waiting for a server. The restaurant needs fewer staff, and diners who accept this new way of interacting with the restaurant system can experience a new level of efficiency, as well as a reduction in errors, increased security, better privacy, and even a more consistently pleasant experience. For Stackers, it's not just about introducing a touch screen at the table, it's about rethinking the entire restaurant as a newer, more intelligent system to serve the customer.

10.2 People Systems

In this chapter, we consider a new type of system – the *people system*. A people system, simply put, is a system that consists only of people. A people system is a kind of human activity system – one without any technology.

A people system is a biological system because it contains only biological elements (people), but it's a special kind of biological system because – let's face it – people are special. At least on planet earth, human beings have a very special status in that we appear to be the only species with our level of intelligence, consciousness, language, and creativity, which has allowed us to become the single dominant species on the planet. Since only human beings are reading this book, we can feel free to focus here on what is most relevant to us – the systems that involve only us – people systems.

It is certainly true that animals form systems, as they interact with their own species, other species, and human beings, but it is systems of people that run the world. People have unique and complex forms of communication that other biological species, on earth, do not have, or at least of which we are unaware. Regardless of what animals may be doing behind our backs, people, and their interactions with each other, form a very special kind of system. People systems are responsible for nearly everything we know, experience, and love in our lives. We live in a world of systems, people systems.

People systems are a kind of human activity system, since they are systems that are motivated by human activity. The difference between a people system and other human activity systems is that a people system contains only people, while a human activity system includes people, but can also contain technological systems, devices, vehicles, buildings, satellites, spacecraft, weapons, and all forms of tools, including of course computers and artificial intelligence. A large corporation, when considered as a system, contains many technological systems, but without the people, would simply stop cold. It is also valid to claim that the corporation could not function without its technology, but were all the technology destroyed, the people could still rebuild the corporation, while in the reverse situation, the technology could not simply replace the people – though that would make for a fascinating science fiction story.

Even though human activity systems contain people, people systems are still a very special form of human activity system. When the system contains only people, all communications are person-to-person, and all interconnections are of the complex human kind. A very large part of what drives people systems is human emotion, whether the one-word headline emotions like love, envy, anger, or grief, or the more subtle emotions that require phrases like fear of failure, motivation, fear of success, the will to power, the drive to excellence, resignation, discouragement, hopelessness, or suspicion. Engineered systems, even those using artificial intelligence, use calculation to make decisions, but in people systems decisions are made by humans, in a human-like way. Levels and layers of logic, emotion, and motivation, mix together to produce decisions that drive people systems. Decisions to take a new job, sign a deal, start a business, marry a person, move to a new country or have another child, along with countless smaller decisions, drive the people systems of relationships, families, garden clubs, churches, cities, companies, and nations. One can say without overstatement that the study of people systems is the study of human life.

Taking the systems approach to the study of human life can bring new insight and understanding. The more common way of analyzing human activity is to consider either the *individual*, with his or her psychology, behavior, and motivation, as a nonhuman machine or device, or to consider a large group of people, such as those in a company or country, as a kind of machine or device. These dual views are seen in both science and everyday conversation. A story may help illustrate. In my very first employment experience after earning my bachelor's degree, I had approached a large computer company. There were several interviews, and I was hopeful of receiving an offer letter, but what I received was a rejection letter. Actually, I received two rejection letters from the same company. I felt, quite reasonably, that I had been rejected by the company, and was tempted to move on. Then I received an offer letter from the same company. How can a company both reject and accept me in the same week? The apparent contradiction comes from the assumption that it was the company that rejected me. Companies do not reject job applications. People in companies reject job applicants. In this particular case, what had happened was that my resume had been sent to a number of offices and departments within the company, each with its own recruiting staff. Some staff in those offices had not needed someone like me so they sent me rejection letters. One manager in one office, in fact the manager and office whom I approached initially, recommended that the recruiting staff in that office send me an offer, and I worked at that company for many years. The key to understanding this kind of apparently contradictory behavior is to think of the hiring function in a company (and perhaps even the entire company) as a people system. Naturally, the human resources people use computer systems, but for the most part, the system is all driven by humans and their incentives, preferences, and decisions. Considering the hiring function of a company to be a people system makes it much easier to understand the behavior we observe.

A people systems approach is one that treats people as individuals, not as members of a group or organization. As we've noted in Chapters 6 and 7, we can describe systems at one of many levels of abstraction. When it comes to people, our organs form systems, such as the circulatory system or the nervous system, and our cells are also systems. Our families are systems and so are our companies, economies, and countries. For the purpose of understanding people systems, however, we make the choice to simplify all of these levels of abstraction to level where the individual person is the fundamental system part. People systems are made up of people, and we ignore all other levels of abstraction, both below and above the level of the individual person. From a broad systems perspective, people systems are simple – they are just made up of individual people. At the same time, everything we've learned about systems applies to people systems. People systems can be designed using the principles of systems engineering. Like engineered systems, people systems can exhibit some degree of intelligence. People systems can be illuminated by the application of systems thinking methods such as the application of system archetypes, causal loop analysis, and the use of leverage points. As we begin to build an understanding of people systems in this chapter, we'll apply some of these systems ideas.

10.3 People Systems and Psychology

People systems as an approach has some common ground with disciplines such as psychology and anthropology that study people and behavior but is a unique perspective and approach. As my first master's degree is in psychology, I have a great affinity for that subject. Psychology, for the most part, deals with the individual and the individual's relationship to other individuals, but not with the overall characteristics of systems composed of individuals. Sociology and anthropology tend to treat groups of individuals as a unit, without considering the individuals. In a system, as we have seen, we focus on the behavior of the system and the way that behavior is produced by the parts of the system. In people systems, this means we focus on the behavior of people as a system, and examine how that behavior is produced as a result of the behavior of the individuals in that system. In people systems, only people have behavior; groups, families, companies, and nations do not.

A people system is fundamentally a system, with all the characteristics of any kind of system, including a function or purpose, parts within the system that interact and a specific arrangement of those parts. In a people system, the parts of the system are people and their interactions are only with other people, so in a way, it's quite a simple kind of system. There is only one kind of part, and one level of abstraction since we don't decompose the people into parts. From a systems perspective, it is valid to do that but examining the biological parts of people doesn't lead to any increased understanding of their behavior. A people system is a system, composed of people. In people systems, we also don't use a higher level of abstraction, that is, a system of people systems, since the part of that higher level system would be people systems, not people. A fundamental idea in people systems is that it is people, as individuals who make decisions and exhibit the behavior that moves the system. A people system does not have some kind of mind of its own, or some decision-making ability. It is always only the people.

Ray Bradbury's science fiction short story, *There Will Come Soft Rains*, published in 1950 and revised in 1999, tells the story of a fully automated home that survives a nuclear bomb, unlike its occupants who are all killed. The house is automated enough, that most of its functions, including cooking and serving meals, cleaning and even attempting to entertain the children, are carried out on schedule in the complete absence of any of the people. It's a picture of a future where systems will operate without us. We are not there yet. Most of our systems are completely dependent on people, and it is the people who are important in nearly all functioning systems. Without the people, the automated systems stop or eventually breakdown, but without the systems, people carry on, either without the automated systems or by building new ones. In our world, it's the people that count.

Perhaps the most important concept in understanding people systems is incentives. People are primarily motivated to act, or refrain from acting, based on incentives or disincentives, or their perceptions of these. They eat, sleep, work, and play based on what they perceive to be the incentives related to each activity and its outcome. In the Chapter 9, we looked at how changing the incentives in a system can be a powerful leverage point to alter the system. The systems in which we live provide many and varied incentives. We eat to satisfy our hunger, but also because we want to continue to live. We go to work so that we can provide the best life we can for ourselves and our families. We make purchases of homes, cars, clothes, and electronic gadgets that we feel will provide comfort, enjoyment of fulfillment. These are obvious incentives; there is little mystery to how they work and how they incentivize and motivate people. Let's consider something more subtle.

In generations past, it was normal for young adults graduating from colleges and universities to seek employment, usually right away, in the field in which they were educated. The incentive of developing a career in that field, and living independently, having a family, and so on, was powerfully present in the culture of that time. The career-oriented path was not required but was clearly the norm. Prior to attending college, those young people had grown up observing their parents and other adults having successful, happy, and apparently fulfilling lives following this path. Wanting the same outcomes, they pursued the same path.

In the early part of the twenty-first century, things changed. The terrorist attacks of 9/11, coupled with the real estate boom and bust, may have made the traditional career process appear less attractive to some young people with the result that the lifelong-career pattern became less compelling. Long-term career success was no longer the strong incentive it once was, leading some new college graduates to "take some time off," move back home with their parents, take jobs outside their field of education, and generally eschew the traditional career path, at least in their immediate post-graduate years. But a deeper analysis of this new-graduate people system is more subtle, since in really all the ways we can measure objectively, the incentives and rewards of a career-building process had not changed. It was only the perceptions of the cost vs. the benefits of the career process that had changed. Perhaps needless to say, the perception of incentives is as powerful as the actual incentives.

Incentives are related to time, both in cost and benefit. Investing the time and effort to earn an advanced degree now has large costs, both financial as well as the opportunity costs of missing out on other activities and pursuits. The benefits to that advanced degree come later, often much later, in the form of new opportunities, career growth, and income. How a person values *now* vs. *later* can be a significant driver of behavior. Purchasing a new car, and financing it over time, has an automatic and immediate incentive – driving the shiny new car and the apparent prestige and status that may be assumed to come with that. The triple negative financial impact of the depreciation cost from the rapid loss of value in the car, the interest paid on the loan, and the opportunity cost of not investing the money instead are only incentives in the long run.

For many years, economists tended to think that they had incentives all figured out, and the people operated according to what they called utility theory, a rational weighing of the costs and benefits of alternatives. The field of behavioral economics, with prime mover Daniel Kahneman, and his field-launching book, *Thinking, Fast and Slow*, introduced the idea that people have important and consistent patterns in their thinking and behavior, but these patterns do not follow the neat mathematical rules that economists and engineers would assume. It turns out that people are not that much like machines. A few examples will illustrate. One of Kahneman's many experiments was to give half of a group of people a gift of a coffee mug, and the other half a gift of a candy bar. Then, he offers people in both groups the option to trade what they have for the other item. One would expect that the willingness to trade would not depend on which item was given initially, but it does. Eventually called the endowment effect, what Kahneman and others confirmed over many kinds of experiments, is that people tend to value something they *have* about twice as much as the same thing when they do not yet possess it. The pet shop owner is wise to suggest that a family "just take the puppy home for tonight – you can always give him back tomorrow" knowing that the family will bond with the cute little animal and consider it "theirs," triggering an endowment effect.

10.4 Endowment Effect

In people systems, people "own" all sorts of things, not just physical items, but titles, power, areas of responsibility, positions of influence, control over other people, or the ability to approve or disapprove decisions and proposals. People often speak of "owning" the hiring process or a new product strategy in an organization. Due to the endowment effect, people do not give these things up easily, because they value them about twice as much as something equivalent that they do not possesses. When trying to make organizations changes in a company, for instance, leaders must realize that for people to give up a position happily, or relinquish an area they "own" they must believe they are gaining something twice as valuable (some researches claim it's closer to 2.4 times) as what they feel they are losing. Organizational change experts advise leaders to carefully consider what each person involved in the change is giving up and work to help them handle that part of the transition. People will protect what they have more – two times more – than pursue something they want. In the same way, people stay in jobs unless they believe that a new job will be twice as good as their current position. Career-wise, it's counterproductive. Pushing back against the endowment effect's mental influence, and seeking a new job that may only be 25% or 50% better than the current one is a more reliable path to a successful career than waiting for a position to come along that is 200% better. At least some of what is referred to as resistance to change may actually be driven by the endowment effect.

The endowment effect also shows up in family systems. Friends and relatives may look at a family situation and wonder why the people involved stay. Why doesn't she divorce him? Why doesn't he break up with her? Why doesn't the 30-year-old son move out on his own? There can be many aspects to why a family system is working the way it is, but one is the endowment effect. As reflected in the idiom, *better the devil you know, than the devil you don't*, people often prefer to stay with what's familiar, even if it's far from what they really want, rather than take the risk of looking for something (or someone) new. The endowment effect would predict that someone would have to perceive the possibility of a new husband or wife, as twice as good as what they currently have, to be willing to give up their current situation no matter how negative or destructive it may be.

10.5 Anchoring

The developing field of behavioral economics has revealed many ways in which human beings behave that are not logical, but which are remarkably consistent and predictable. The concept of anchoring, also described by Kahneman, shows the tendency of numeric guesses, say the number of beans in a jar or the price of a home, to be highly influenced by numbers that the person heard just before making the estimate, even if those numbers are completely unrelated to what's being predicted. Dan Ariely, in *Predictably Irrational*, shows how most people cheat, but only a little, when they are in a situation when they believe the cheating will not be detected. Other studies show how people are strongly influenced by what they believe others are doing, even more than they are by the idea of doing what's right. Kahneman and Sunstein's development of what they call *libertarian paternalism*, shows that most people will accept a given default choice instead of acting to change it. When asked to check a box if they are willing to donate organs at the time of their death, few will check the box. When asked to check a box if they are *unwilling* to be an organ donor, again few will check the box. To the engineer or economist, this makes no sense, but it is how humans operate. When encouraged to enroll in a savings plan for retirement at work, only about half will do so, but when automatically enrolled, with the ability to opt out, very few will opt out of the savings plan. Thaler and Sunstein's book, *Nudge*, introduced this kind of *choice architecture* over a decade ago, and have updated their findings in *Nudge: The Final Edition*.

So far, we have looked at two important aspects of understanding people systems: *incentives*, which guide people in rational ways, and *cognitive biases*, the curious and irrational but reliable behavioral patterns described by the field of behavioral economics. Next, we look at something even deeper than behavioral characteristics – values. In Chapter 7 we've described how systems of all kinds have both structure and behavior, and how these two are interrelated. In engineered systems, the structure, often called system architecture, is designed to support the needed behavior. In natural or biological systems, structure and behavior evolve together, optimized over many generations. The structure of a system can be seen through a physical or a functional (sometimes called *logical*) perspective. In people systems, the physical architecture of a human being, the organ systems and cells are not usually of much interest, except as they affect human behavior (and they do). People systems treat each person as a physically, single, indivisible unit. What is of interest is what we might call the functional architecture of a human being. Though they may not have used this term, psychologists and philosophers have made many attempts at describing the functional architecture of human beings and these mental models can be helpful in understanding how human beings operate in people systems, or may at least give us concepts and language to reason about human behavior.

10.6 Functional Architecture of a Person

A common model is to think of a human being as having two parts: mind and body. Originated by philosopher, mathematician, and scientist Rene Descartes in the seventeenth century, this model has come to be known as Cartesian dualism (Descartes, 2013). The model has become so common that most people who have not studied philosophy accept it without another thought. In this view, a person has a physical body and also a nonphysical mind. That persons have physical bodies is not often disputed, but the invention of a nonphysical mind brings many difficult questions and challenges, some of which Descartes himself proposed, and which philosophers have been arguing for centuries. For instance, there is the problem of *interaction.* If the mind is not physical, how does it interact with and even control the body? We have no other examples in the world of something nonphysical affecting something physical. Of course, other nonphysical entities can be posited, including a variety of gods, spirits, and demons, along with presumed nonphysical forces that act on the physical world, like psychic phenomena, collective consciousness, and a wide variety of "energies." It is particularly confusing to speak of nonphysical energies, such as those of a so-called psychic or faith healer, since science also uses this term to refer to detectable, measurable energies like magnetism, electricity, and gravity. Along with the idea of a mind, these nonphysical entities and forces become articles of faith, usually beyond the reach of scientific or systems investigation. To embrace Cartesian dualism, one must have some way to explain the interaction of a nonphysical mind with a physical body.

Whether humans have an actual mind or not is a fascinating philosophical debate, but either way, we can use the idea of a mind metaphorically or colloquially to refer to the collection of conscious and subconscious thoughts possessed by a person. The idea of human beings having a mind and a body can be extended a bit to include emotions. Emotions could be considered as special kinds of thoughts, but perhaps it's best to give them their own category. This functional architecture of a human being then consists of mind, body, and emotions. Whether this three-way division matches the actual underlying physical structures is less important that whether it enables us to better understand how people operate in people systems.

At an everyday level, it is apparent that humans have thoughts about things and also feelings and emotions about things, and these seem to be somewhat independent of each other. The body too seems separate; humans experience body sensations as something quite different from thoughts and emotions. In extreme cases, these boundaries may blur. Chronic pain can eventually cause personality changes – no one can think very clearly when ill. As another illustration of this human architecture in use, the field of psychotherapy has shifted in recent decades from a fairly strict separation of mind, body, and emotions, to a more integrated approach. In the past, proponents of behaviorism focused almost solely on what a person did, how they behaved, and much less on what the person thought or felt. Depth psychologists, including psychoanalysts and "how did that make you feel?" counselors, focused on thoughts and feelings, including those potentially hidden in the subconscious. In recent years, psychology has for the most part shifted to an integrated approach known as cognitive behavioral therapy. As its name implies, this approach considers both behavior and thought to be important and interrelated. When patients behave in certain ways, they have thoughts and feelings that come before, during, and after the behavior. Bringing the light of critical thinking to those thoughts can change the thoughts and result in changed behavior. Cognitive behavioral therapy is also practiced by psychiatrists, who may use drugs, such as anti-depressants to help bring patients to a state where they can deal successfully with their thoughts, thus addressing all three parts of the human architecture: body, mind, and emotions.

The three-part architecture is not the only one that has been used to help understand people. Another popular approach involves categorizing people into types, usually based on a concept of personality. The popular Myers–Briggs personality test assigns one of sixteen personality types using combinations of four dynamics: extrovert/introvert, intuitive/sensing, thinking/feeling, and judging/perceiving (Myers, 1962). After taking the test, a person might come out an ENFP – extroverted, intuitive, feeling, and perceiving. Other personality type or behavioral type systems are in wide use as well, and include categories such as analytical, amiable, driver, and expressive, or using the colors of red, green, blue, and yellow as described in the wonderful *Surrounded by Idiots*, by Thomas Erikson. It's not that any of these type schemes are *true* in some universal sense, as if we could perform exploratory surgery or a deep brain scan and conclude that yes, this person is indeed an ENTJ and a red. Typing frameworks give us a tool for thinking, in a systematic way about how people interact, and what may be behind their interactions. Even if a person is not, in some inherent way a blue, the fact that we have observed them operating in the way Erikson describes as blue, could sensibly lead us to believe that the person may function that way in the future. If we can better understand how people, and various types of people behave, we can better understand people systems. Of course, people always reserve the right to behave differently than they have in the past, which makes people systems even more interesting.

Applying schemes of types to people can extend to interactions between people, and to groups. The psychological theory of *transactional analysis*, developed in the 1950s by Eric Berne, applies types to interactions (Berne, 1968). A person might be having a parent–child interaction with a subordinate one day, a parent–parent interaction with a fellow manager or a crossed interaction in which one person is acting in a parent–child pattern, but the other is responding in an adult–adult mode. Transactional analysis is based in psychoanalytic theory, which focuses on subconscious thoughts and believes as key drivers of behavior. Types have also been applied to groups of people using terms like groupthink, mob psychology, or the wisdom of crowds. Each concept shed some light on how a people system is likely to operate in a particular situation. Having an awareness of multiple frameworks can support a more complete understanding and even prediction of both individual and group human behavior – the driving forces in people systems.

10.7 Example: The Problem of Pollution

Let's apply what we have covered so far to a difficult problem in people systems – the environment, specifically the pollution of planet earth. Pollution comes in many forms – litter on the streets, plastics in the oceans, light in the night skies, noise coming through our windows, space junk in orbit, and antibiotics in our drinking water, to name a few. Everyone one agrees that pollution is a problem, and it would seem that the solution is simple – just get people to stop polluting. Assuming that a simple solution will be effective reflects a lack of understanding about people systems. A systems thinker, especially a people systems thinker, will first seek to understand the system before trying to solve the problem. While it may appear that corporations pollute, they do not. People pollute. It is people who make the decisions in corporations that result in pollution. Why do they make those decisions? Is it because they want to pollute, either based on some perverse personal agenda, or because the corporation has published a set of value statements that advocates pollution? Likely not – I doubt anyone has an intentional goal of polluting. When the executives in a manufacturing company somewhere in the world decide to dump waste into the nearby river, their goal is not to pollute, their goal is to manufacture their goods as efficiently as possible. Dumping waste in the river is seen as the most effective way. We might argue that the executives should not make decisions this way, but should consider other factors, like the health of the river and the life forms (including humans) that depend on it, but the executive's job is not to cater to those people, but rather to his or her superiors, to the corporation's board of directors, and they in turn to the owners and shareholders of the company. The people system that controls the corporation is aimed at a goal of long-term profitability. Of course, the next step in our thinking is that we should simply make pollution illegal, and fine the corporation if we catch them dumping waste into the river. Most parts of the world do exactly that. The executives then have a choice – pollute and pay the fines, ship the waste to some other country for disposal or pay for systems to purify the waste before discharge. Assuming the fines are large enough to make paying them the most expensive option of the three, it would seem that the problem is solved. But wait – who pays for the waste disposal? It would seem that the corporation does. A bit of systems thinking, however, makes it apparent that it is the corporation's customers who pay for disposal through increased prices. If the corporation does not raise prices, then the profits of the company suffer, and the shareholders will demand that prices be increased as long as the increased prices do not significantly reduce sales. If it is not possible to increase prices enough and maintain competitiveness in the market, then customers may shun this manufacturer and buy from others, perhaps those in areas that have less stringent, and therefore, less costly pollution regulations. By enacting stricter anti-pollution laws in one country, we may have reduced the competitiveness of the manufacturers in that country compared with some other countries. There are two possible answers at this point – either get the other country to enact similar pollution laws to level the playing field or apply tariffs to imported goods from that country to seemingly make products from both countries competitive. A closer look reveals that in many cases, countries supply each other with raw materials and parts, so a tariff on imported goods results in increased costs to the importer and thus results in more price increases and lack of competitiveness. If we apply tariffs only to finished goods, the supplying country's manufacturers may simply set up assembly and finishing operations in the original country, buying raw materials from their home country, or from whatever source is most cost effective.

An alternative for the manager who makes the decision to dump the waste into the river, since it is the most cost-effective way, is to make the decision on moral grounds rather than economic. In doing so, the manager fails to operate in the best interests of the owners and stakeholders and may find his or her job eliminated, or at least the decision-making power removed. The owners may decide that they will back the moral, anti-pollution stance, and risk become less competitive, or may use their fact that they don't pollute as a selling feature of their higher-priced products.

In this example so far, the people systems are driven by economics and markets. Markets may be mostly free but are always subject to various kinds and levels of government-imposed standards, regulations, tariffs, or taxes. Let's turn to a different part of the global system that causes pollution – the ultimate consumers.

Can't we solve pollution by just getting consumers to stop buying from supplies who pollute, forcing those companies to stop polluting to sell their goods? To some extent this can work – if widespread public perception is that a certain company is a major polluter then it could influence sales. For this to be effective, the system overall must be able to provide complete, correct, and unbiased information on all providers in the market, to all consumers involved. It's hard to see how this would be practical or economical. Companies could counter any pollution accusations with advertising campaigns of their own, at considerable expense, but perhaps worth it to maintain sales. It would become a war of information. Setting that aside, consider what motivates a consumer. Consumers are obligated to make the best decisions for themselves and their families, buying the most cost-effective goods to make the most of their income. The more affluent may have more choices but follow mostly the same logic. The rich may make consumer choices without regard for cost effectiveness, but these relatively few symbolic choices are unlikely to change market preferences overall. It is for this reason that campaigns like "Buy American" which encourages Americans to buy goods made in America never seem to make a large impact. For one thing, many "American" products contain many, perhaps all, parts made outside America. Tracking down the "pollution-pedigree" of a product is a daunting, even impossible task for consumers, so they likely buy what appears to give the most value, performance, or image, for the best possible price.

Let's turn our attention to an aspect of pollution that has perhaps more promise – recycling. Recalling that our first task is to understand the system, and remembering that the system is *always working*, we need to consider what causes recycling. If we were drawing causal loop diagrams, we might be tempted to show that the more people recycle, the more material is recycled, but this is an obvious, *rocks are hard* relationship, not a cause. What we mean by more material being recycled is that more people are recycling – it's a definition, not a causal factor, so we haven't said anything. As an aside, this is a common problem faced by new students of systems thinking – stating definitional relationships as causes and thinking that they have documented important causal relationships. Showing that more murders in a city causes more violent crime is not a cause – murder *is* violent crime. Returning to recycling, we might be tempted to think that the capacity of recycling centers is what causes the amount of recycling, implying that if we increased the size or availability of recycling centers, we would have more recycling. If this were true, we would expect currently overflowing recycling centers, lines of people waiting to drop off recyclables, etc. We probably need to keep looking to see what causes more or less recycling. Is it public attitudes – are some people somehow against recycling and thus do not recycle on principle? Perhaps recycling is, for most people, a matter of convenience. Most people *want* to recycle, and see the wisdom of it, but will only recycle if it is sufficiently convenient. I once lived in a town where every resident had town-provided garbage collection. With the regular brown rolling trash bin, the town delivered a blue rolling recycle bin, with instructions for what could go into it – paper, glass, cardboard, plastic bottles, etc. Recycling when living in that town could not have been more convenient. In another town where I lived, the state collected a small additional fee when products were purchased that came in recyclable containers. The empty containers could then be turned in for a refund of the fee later. Some people paid the fee and then did not bother to turn in the containers, placing them in the garbage instead. As you might expect, this provided an opportunity for others to make a little money by going through trash containers, extracting recyclable containers and turning them in for the small deposit. Those who didn't want their trash picked through, went to locking up their trash cans, defeating the system. Here we see incentives in action and clear causal relationships. Convenience can be an important factor in driving behavior, and financial incentives, even if small, can be important as well. Whether these factors are important, and which ones drive behavior best, can be determined by conducting trials and experiments, or through natural experiments – studying recycling rates in different areas that have different conveniences and incentives in place. As in all systems thinking, the goal is to understand the system that is producing the problem, before attempting interventions.

When considering the systems that are causing pollution and environmental problems, we might be tempted to simply assume that the problem is that people aren't altruistic enough, and thus the solution is to keep telling people to do the right thing and that somehow this will make them operate more altruistically, even if it costs them additional time and money. There are two problems with this overly simplistic thinking. One is that only a relatively few people will consider altruism as their top goal, ahead of economic efficiency and personal benefit. The other is that many people will consider it an obligation to behave in a way that is most beneficial for their companies, countries, families, and themselves. Environmental concerns are a tragedy of the commons, and unless economic incentives are in place that either incentivize people to act in a more positive environmental way, or at least make it an economically neutral choice, it is likely that most people will still act in a way that benefits themselves and their families.

Here's an example of a way incentives could be provided. Say each resident in a town were given a large recycling collection container, but only a small garbage collection container, with low or no monthly charge for collection. There would be an incentive to sort waste into recyclables vs. trash so as to avoid exceeding the trash volume requirements. Additional trash containers could be obtained at additional collection cost for those who insist on producing more trash. The incentive would be for people to try to minimize trash and maximize recycling.

The additional recyclable material could offset the cost of collection and result in a lower cost service for residents with environmental benefits built in. Systems thinkers will also note how a system of this kind would also provide an incentive for consumers to choose products that come in recyclable containers, so they don't have to use their precious garbage space allowance to discard nonrecyclable containers. Consumers might be more careful to crush non-recyclables, benefiting themselves and making the landfill more efficient. Of course, this kind of program would only work in an area with town-provided garbage and recycling collection. If garbage collection is a free and open market, people could opt for a private garbage collection service, if available at a lower cost, and thus avoid recycling. Even in a free market, landfills, which are normally managed by the city or county, could raise fees to increase the incentives for more recycling. There are in fact some towns and countries where some of these ideas have been implemented, and these areas could be studied to see the results.

The point here is that systems thinking, especially when we take the perspective of people systems, opens many new possibilities for creative solutions to hard problems. If the problem boils down to people and their decisions, working through the incentives, financial, and otherwise, will bring a fuller understanding of the system, and usually result in creative solution possibilities.

People systems often do not stand alone, isolated from other kinds of systems. Continuing with the theme of garbage, recycling, and pollution, we might consider whether there is a technological solution to some of these issues. The Freakonomics podcast (freakonomics.com), hosted by *Freakonomics* author Stephen Dubner, could easily become the systems thinker's favorite listen, due to its insightful and logical examination of interesting issues of all kinds, and the financial and behavioral economics approaches they apply. In episode 17 from 2011, the podcast examined the problem of trash. Pair their examination of the subject with episode 554 of *This American Life* (www.thisamericanlife.org) a less logic and science-based podcast, but an equally interesting listen, and you'll come across a possible solution to trash – burn it. I won't spoil the story, but incinerators could be a part of the solution, if they are efficient enough, economically viable and don't produce too much pollution themselves. What if garbage could be processed on site, perhaps with a small incinerator that could help heat the home, or perhaps by some other process? Some homes have their own septic tanks and underground drainfields, allowing on site waste processing instead of pumping waste to a large communal waste processing system. Could something similar be done for garbage?

Another excellent aid to systems thinking is science fiction. How does the Starship Enterprise handle trash? I would assume they have a way to vaporize it, or to reuse the molecules as raw material for their replicators. Curiously the Imperial ships in Star Wars (Episode 4: A New Hope) had large-scale garbage smashers and then dumped their garbage into space – another reason to prefer Star Trek to Star Wars. Science fiction invites us to ask, *what if?* What if we could build a device that would vaporize garbage, or transform it into something else? What if we could burn garbage but transform the polluting by products into something else. What if such a device already exists, but is somehow not economically feasible, or people aren't aware of it? Could the people systems around such technologies be modified in make the technology workable?

10.8 Speech Acts

Returning to our main discussion of architectures and frameworks that help us understand people systems, let's consider the topic of *speech acts*. The topic because is little known, but with even a cursory understanding, reveals a great deal about how people systems work when they communication between people is an important part of the system – and it almost always is. We've probably all experienced a meeting where there was a great deal of discussion about work topics, many alternatives were explored, and action steps mentioned. Nevertheless, at the next meeting a month later nothing had been done. How could this be? Are the people all lazy? It's more likely that the right speech acts did not occur. Have you ever had a lengthy conversation and left it wondering, what did that person really want anyway? – only to find later that the person had left the conversation expecting you to take some action? Or have you ever wondered how some people seem to be able to make things happen, while others appear mired in bureaucracy or derailed by processes and procedures? Speech acts may explain all of these mysteries.

Mitchell Green (2021) gives a useful introduction to the general field of speech acts in the Stanford Encyclopedia of Philosophy explaining how the topic is one of deep investigation by linguists and philosophers. Here, we will reference the more colloquial aspects of speech acts – the conversational constructs we use, or forget to use, in everyday conversations and meetings. Psychologist Joel Friedman in *The Grammar of Committed Action: Speaking That Brings Forth Being* (mentalhelp.net/blogs/the-grammar-of-committed-action-speaking-that-brings-forth-being) explores the origin of speech acts and the role of Fernando Flores. Flores focused on the everyday aspect of speech acts and described how the ideas became a foundational aspect of several major self-development programs including est, LifeSpring, and the Landmark Forum. For our purposes here, we will introduce the six main speech acts and show how they help understand what happens in people systems when people talk. Speech acts can be thought of as moves in a game, or plays in an athletic competition, where an act is performed, and then someone else may perform an act in response. Speech acts can be referred to as committed speaking, since in many cases they commit the speaker to a position, or an action. Here are the six speech acts described by Flores:

1. **Request.** A request speech act is one in which a person makes a request of another person do something. Examples are: Will you please take me to the circus on Friday? Please submit the report by noon tomorrow; Will you go with me to the store today? Notice that a request must be specific enough to be able to be accepted or rejected and needs to be recognized as a request in order to be successful. The responses to a request can be for the receiver of the request to accept the request and make a promise, to decline the request, to make a counteroffer, or to ignore the request altogether.

2. **Declaration.** A declaration is a statement that needs no support or basis, but which creates a new reality by its being said. To be effective the person making the declaration must have the social authority to make the declaration to those who are hearing it. Declarations may require the declarer and others to change their behavior based on the newly declared reality. When U.S. President John Kennedy said, "We will go to the moon in this decade" it was not a statement based on plans, or even on the current capabilities of the United States space program, nor was it a prediction. He declared it, and others took action to make it real. A declaration can be about a goal, a condition, a decision or about anything which relies on the intent and will of people to fulfill. When a state is hit by a hurricane and the governor *declares* a state of emergency, he or she is not describing something, but creating a new reality, one in which outside aid can be given, supplies sent, and agencies mobilized.

3. **Promise.** A promise can be made about anything to anyone, in response to a request, or on its own, unprompted. It is unilateral – the promiser is committing to a specific course of action, often producing some result within a specified time. A promise made without a specified or understood time is still a promise, but it may lack practical value since no one can count on the promiser acting within any particular time. A promise is not a guarantee, but is an expressed intention and commitment. No response is needed to a promise, but those to whom the promise is made enter into an accountability relationship with the one making the promise, allowing them to inquire about the status of the fulfillment of the promise and ultimately hold the person to account for the outcome. Of course, promises are not always fulfilled, and when a promise is unfulfilled, the promiser can acknowledge the non-fulfillment and cancel the broken promise, or make a new promise, allowing for more time, more resources, or a different approach. In some cultures and families, it is taught that "one should always keep one's promises" making promise-keeping a moral matter and generally causing people to make only promises that are

sure to be kept. Others say that promises are useful tools to harness and direct intentionality, and are more powerful speech acts than the statement of a goal, which may in reality be only a descriptive assertion of possession: *I have a goal*, is no more powerful than *I have a cat*. To say it another way, a goal does not move people until the people make promises to act in fulfillment of the goal.

4. **Offer.** An offer is similar to a promise in that it is a statement of intent by the one making the offer, but offers are conditional upon the acceptance of the person being made the offer. Offers, like requests, can be accepted, rejected, or ignored, or a counteroffer can be made. Examples include: How about if we go to the theater on Tuesday? I'll cook if you will wash the dishes; We are offering you the position of Vice-President of Operations; I will prepare the presentation by next Monday if the team agrees. Contract law is based on the idea of offers and acceptance. If a person offers to purchase a product at a given price, and that offer is accepted, a contract is created, whether written or verbal. If instead the response is a counteroffer, then the roles switch, and the counteroffer is now an offer being made to the original offeror, and it must can be accepted, declined, ignored, or another counteroffer made. In contract law, when a counteroffer is made the original offer is made void and can no longer be accepted later, unless offered again.

5. **Assertion.** An assertion is a statement made by a person, claiming that the statement is true and that there is evidence for the statement being true. The statement is made based on evidence and the person making the assertion should be willing to produce the evidence upon which the assertion is based. An assertion is a statement that can be tested using the methods of science and experiment or the methods of logic and argument. Examples include: The COVID vaccine is effective for the prevention of the disease; The best time to plant roses is the spring; Jocelyn has a PhD in mathematics. Assertions are the bread and butter of science, medicine, philosophy, and politics. Assertions are made and then evaluated based on evidence, science, or logic. Assertions can either be true, false, or as-yet unproven. Scientists tell us that nothing is ever proved true in science – assertions or hypotheses are either proved false or are shown to be more likely true than false, given the evidence collected thus far. The point is that an assertion can be tested by evidence.

6. **Assessment.** An assessment is a value judgment, carrying only the weight of the credibility or expertise of the one making the judgment. Examples include: *The Hunt for Red October* is an excellent movie; Lexus is a more reliable brand of car than Jaguar; Jim should get a big raise this year; Recumbent bicycles are more comfortable than upright bicycles; If I get that additional degree, it will help my career. Assessments cannot be true or false – they are opinions, conclusions, or viewpoints owned by the person making them. Assessments may come with explanation, reasoning, or even evidence, but are still matters

of judgment. I can assert, based on evidence, that about 30 000 people are killed in traffic accidents each year in the United States, but the idea that self-driving cars will cut that in half, is an assessment. Since assessments are neither true nor false, a statement like, *you will never make it to Carnegie Hall*, no matter who is making the statement, is still only an assessment and not a fact.

This is not an exhaustive list of speech acts; many more can be identified. For our purposes here, we want to examine how the knowledge of speech acts can help us understand how people systems work when people are talking or writing to each other, which is pretty much all the time. In the examples given earlier, there are several possible causes for the difficulties when seen in the light of speech acts. One difficulty is when a speech act is not clearly understood to be that particular speech act. For example, someone makes a request, but those hearing it do not understand it to be a request. In this case, it is unlikely that the request will be fulfilled. Since there is no specific language that identifies requests and the word *request* may not even be included, it can be unclear whether a request is being made. For example, if in a meeting if someone says, "We really need to get that marketing campaign sketched out and then figure out the pricing," he or she may feel like they have asked the marketing and accounting departments to work on these tasks, but the marketing and accounting people may feel that they agree with the assertion that "we" need to get these things done, but feel that they have not yet been asked to do anything. Some knowledge of speech acts could prompt the marketing people to ask if a request is being made, or to make an offer in order to clarify the communication and move things forward. In work cultures that begin to understand speech acts, people will sometimes make their requests explicit and unambiguous, by prefacing them with, "I have a request ..." but this is not required and cannot be expected in every situation.

The efficiency of people systems can suffer from cases of mistaking one speech act for another. If a marketing analyst says, "It would sure be helpful if we had more analysis of this market," the speaker may be making an offer, or not. The statement could be taken as an assessment – a value judgment, with no action following. If others agree with the sentiment, someone might respond with, "Are you making an offer?" to which the marketing analyst would likely respond affirmatively and perhaps with enthusiasm for the task that is now commissioned. Complaints are an interesting kind of statement. A complaint may be an assessment, but it could also be a request or an offer: "I sure wish the projects files were better organized," could be a request for someone to organize the files, an offer for the speaker to organize them, or just am assessment. Clarifying which is intended in each situation will lead to a more effective people system. Offers can go to waste, and even result in discouragement of the offeror, when they are not recognized as offers and accepted, rejected, or counteroffered.

In a work environment, offers are perhaps the most underrated of the speech acts. Requests routinely flow in waves from higher levels of management to lower levels and promises or at least the request to make a promise, are elicited by the imposition of deadlines and targets. Offers, however can come from anywhere – anyone can make an offer to anyone else. The offer obligates only the one making the offer, and only if it is accepted. One can literally offer one's way into all kinds of new projects, new responsibilities, and new relationships. Making an offer to solve a problem or exploit a new opportunity, when made to the person who as able to accept the offer, is a powerful intervention in a people system, and a great career-developer.

Speech acts, and confusion over them reach far beyond business conversations and meetings. Disagreements, arguments, and even religious wars have been fought over the difference between assertions and declarations. When a religion refers to a historical religious text, such as when Christians refer to the Bible, should they interpret what is written as a set of requests, declarations, assertions, or assessments. Does the fact that a book of the Bible documents what a certain people did and the way the God responded, mean that God is requesting all people to act accordingly? Does the fact that the early Christian church prohibited women from speaking or teaching men, mean that churches today must not ordain women ministers? When someone uses the phrase, "God is love," is it meant as an assertion, for which there is claimed evidence? If so, the statement may be evaluated and tested using the scientific method, leading to a conclusion regarding its truthfulness or falsity. Or, is the statement meant as a declaration, which cannot be true or false, but is a free creation of the author or person declaring it. To move outside religion, what about someone who announces, "we live in a time of great opportunity"? Do we take this as an assertion backed by economic statistics and other evidence, or do we take it as a declaration and perhaps an invitation to behave consistent with the declared sentiment? The difference between assertions and declarations can do much to quiet the seeming conflict between science and the religious, mystical, or spiritual. A mystic may make a statement that sounds like an assertion, leading the scientist to try to prove or disprove it, but perhaps the statement is best understood as a declaration, meant to inspire, open the minds of listeners, and create a new world by its utterance.

People systems rely completely on communication in language, spoken and written, as the main means of coordinating people to operate as a system, either effectively or poorly. Familiarity with speech acts can reveal much about why communication achieves the intended result or fails. On a personal level, making clear what speech act one is performing when one says or writes something can bring a clarity, effectiveness, and economy to the communication.

10.8.1 People System Archetypes

In the Chapter 9, we introduced system archetypes as one of the main tools systems thinkers use to understand all kinds of systems. Here, we'll look at a few more archetypes, ones that are particularly important in people systems. *Accidental adversaries* is an archetype described by Meadows in *Thinking in Systems*, and refers to a people system in which two people start out as allies, partners, friends, or even lovers. At some point, one person does something that the partner interprets as negative toward the relationship. It may be interpreted as undermining, thoughtless, or even actively hostile. The act may be small, and even unnoticed by the partner who perpetrated it, but it results in the partner developing some negative feelings, perhaps subtle, like a slight suspicion, or a mild annoyance. Having this newly negative attitude, the person acts just a little differently toward the partner, perhaps carefully, perhaps not as open as before, perhaps less generous in how the partner is heard in conversation. There may even be a negative action in return, which is interpreted by the partner, not as a transactional retaliation, but as confirmation of the worsening condition of the relationship. With each interaction, the relationship comes to be more and more adversarial, until finally, both partners consider the other an enemy, even though neither intended things to go that way, and neither can even identify the reasons for the shift. In most cases, each feels that it was the other who did things to damage the relationship and send it on the seemingly irreversible path it's on. Neither wanted it to be this way, and yet it has become exactly this way. People systems can contain many accidentally adversarial relationships between people in different functions of an organization, between marketing people and salespeople, engineers and manufacturing people, or corporate executives and business unit managers. Remember, in people systems, there is no "group," so the problem is not between the marketing and sales groups – it's between particular marketing people and particular salespeople.

There is a paradox here. As we have noted, the system is more powerful than any individual. Replacing an individual in a people system (or any system) may result in the new person becoming part of the existing system and taking on the roles, attitudes, and behavior patterns of his or her predecessor. That's why changing out a person in an organization may not solve a systems problem. An example comes to mind from my work in engineering. In large engineering organizations, there is normally an IT (information technology) group that provides the computers, networks, and software that the engineers use. Engineers, many of whom also understand IT, networks, and software may complain about the IT function and its practices, arbitrary decisions, delays, unnecessary (in the engineers' view) caution regarding security. Engineers feel that IT just doesn't fully understand them and their needs. Sometimes the engineers take it on themselves to create software and then feel that IT doesn't respect their work. IT for their part believe they are just trying to provide a secure, fair environment for everyone. All this can result in some tension between engineering and IT. A possible solution is to have some engineers transfer to, or at least take a turn working in IT, assuming they have the needed skills. When someone from the engineering group moves over to an IT job, the engineers may feel that things will get much better instantly, since now there is "one of us" in that IT group. Before long, however, the transferred engineer begins to take on the attitudes, behaviors, and approaches of the IT people. He or she finds out that there are reasons they do what they do, and have the attitudes they have, and the engineer turned IT person starts to fit into that IT system and is soon, to the frustration of the engineers, more like an IT person than an engineer. One side of the paradox is that the system is more powerful than any individual person.

The other side of the paradox is that in people systems there is no group or body making decisions or having attitudes. It is individual people who make decisions, have attitudes, write emails, hold meetings, and set schedules. We may speak of delays from the IT group, but the delays are from people, not from some machine called IT. From the perspective of people systems, there are only people. IT is just a category of people, based on a department code in each person's human resources record. When understood this way, we don't speak of IT making decisions, or even of a particular person making a decision on behalf of IT, but of each person making a decision, based in part on the incentives experienced by that person at that time. By carefully analyzing the incentives, rewards, rules, and relationships of each person, we can usually understand why that person made the decision in the way it was made. Our hypothetical engineer who transferred to the IT department now reports to new people, whom he or she naturally wants to please in order to get along, be successful (by the standards of their new managers) and be promoted so it should be no surprise that that engineer's behavior begins to change to that of an IT person, to the disappointment perhaps of the engineering team the person (literally and figuratively) left behind.

In trying to understand how people operate in people systems, we are faced with no lesser a question than the psychological and philosophical favorite, why do people do what they do? We've looked at some of the frameworks used by philosophers, psychologists, and economists to try to answer that question, but there is much we have not covered. Part of the fun and challenge of being a systems thinker, and of working with people systems in particular, is the variety of scientific, mathematical, and humanistic fields that all try in one way or another to help explain human behavior. A people systems approach can help integrate and use all these insights together to explain a people system. Let's explore this approach with another example or two, and in doing so, develop a new systems archetype.

10.8.1.1 Demand Slowing

Imagine entering a sandwich shop, the kind where you line up and describe your dream sandwich and they make it right in front of you – bread, filling, toppings, sauces. You grab some chips, pay, and then take it away or sit down and enjoy consuming it on the spot. The question is, how fast do they make your sandwich? It seems that if you arrive in the middle of the lunch rush hour, and there are ten people in line ahead of you, the sandwich makers are working at full speed. After all, they have a lot of sandwiches to make, and they know people don't want to wait too long – lunch hours are precious time. On the other hand, if you walk into the same sandwich shop at say 3 pm, you might be the only customer in the place. The sandwich maker must stop cleaning, restocking, or baking bread in order to make your sandwich. Since you are the only customer in the shop, however, the sandwich maker can and often does take much more time to make the sandwich. They aren't in a rush, since there are no other customers waiting. From your perspective as the customer, this doesn't seem fair – you are required to wait longer for your sandwich to be made than if you had arrived at lunchtime, with a long line of customers behind you. So why is the people system slower when there is less demand? If we examine the incentives, motivations, and thoughts of the sandwich maker the reason is clear. With a long line, there is urgency to get to the next customer, so things move quickly. When there are no more customers, the prospect of returning to cleaning the floors may be less appealing than making sandwiches so why not take a bit of extra time making the sandwich? The assumption is that when there is less demand, there is no need to provide fast service; when there is high demand, service must be faster to meet the demand. We call this system archetype *demand slowing* and it can be seen in many kinds of situations. In a way it's a special case of the famous Parkinson's law, which states that any job will expand to fill the time allotted, where the time allotted for the task varies depending on the current demand. When there is low demand, the service-provider feels that there is more time for each job, so each job takes longer. Where demand is high, urgency is felt, and less time is allotted to each job. To the recipient of the service, it doesn't seem fair – why should my sandwich take longer when there are no customers behind me, but less time when I'm the only customer left in the line (even when there are the same number of service providers (sandwich makers) available)? When the people system is analyzed from the perspective of the sandwich company, the answer is clear. To overcome the tendency of demand slowing, the company could introduce standards for how long it should take to make the sandwich, regardless of the current demand or the size of the customer queue.

Demand slowing is everywhere and seems to be a kind of unconscious, natural human tendency. If I have six-yard chores to do, I hurry along to get them done, but if there is only a single chore, I take my time and that chore takes longer than if it were one of the six. The only difference between my yard and the sandwich shop is that no one is waiting on me to finish that lawn chore, so it's only my own schedule that is affected.

10.8.1.2 Customer Service

Let's apply our new knowledge of people systems to the issue or poor customer service from large corporations. It's well documented that customer service from large consumer-service firms, such as telephone, cellular, and cable companies is the worst among all kinds of companies. To understand why, think about the people systems involved. These companies have thousands or even millions of customers, so they must employ many, many customer service representatives. They have a hard time employing enough customer service representatives to meet the needs, resulting in large numbers of new and less-trained representatives. Much of the time, things work out fine – people call in with simple requests that any customer service representative can handle successfully. In the face of any kind of special request or unusual problem, the difficulty begins. The customer service representative, who is likely managed on call completion rates, is eager to complete the call. The customer is more concerned about a correct resolution than a quickly ended call. It is very likely that mistakes or shortcuts taken by the customer service representative won't be detected or at the very least won't be traced back to the representative. Even if they are, it may be better for the representative to complete the call quickly, even if not correctly, so that he or she can move on to complete several easier calls. Who hasn't had a customer service person make a promise that is never fulfilled?

Struggling to find the right solution for a difficult customer situation may not be rewarded in the world of the customer service representative. The lack of a case successful resolution incentive may result in the well-known approach of asking the customer to try something, then closing the case, and asking the customer to call back and open a new case, likely with another representative to continue their quest for the right resolution. It takes higher skill and knowledge levels for customer service representatives to resolve more difficult situations, and the high turnover rates of these jobs may not offer the time to gain that knowledge. If customers cannot detect that the customer service representative is providing a quick guess of a solution, or none at all, and can't come back to that representative for follow up, the representative is actually better off "completing" all calls quickly, regardless of the results for the customer. It could explain why it often takes multiple calls to resolve issues with these kinds of companies – the calls are repeated attempts to find a customer service representative who is able and willing to address the issue.

10.9   Seeking Quality

Among the fields that can help inform our understanding of people systems, including the customer service function at a large utility company, is the field of quality management, especially values-based quality management as taught by the Quality Management Institute (https://www.qualitymanagementinstitute .com). Larry Kennedy, CEO of the Quality Management Institute quotes the Gallup organization's poll results on employee engagement, which show that fully 64% of all workers are disengaged – either actively disengaged (13%), or just not engaged (51%) in their work, and describes how this level of disengagement can't help but affect the way employees treat customers. Disengagement leads to turnover, which leads to lower average experience and training levels among workers, which also affect customer service negatively. Engagement, says Kennedy, is a stronger predictor of quality in the products and services produced by a company, than years of experience or technical skills. At one level, the solution is to simply hire engaged workers, or at least, be willing to let disengaged, especially actively disengaged workers, move on quickly to their next jobs. At a deeper level, Kennedy and his company train organizations to recognize, teach, and promote values like a zero-defects attitude, vocational certainty, and personal authenticity. Organizations that adopt and live these values from executive levels and down through the organization, will naturally attract workers with the same values, and just as naturally make the most disengaged workers uncomfortable enough to leave on their own, resulting in a reinforcing feedback look that solidifies a quality culture and perpetuates it.

In the field of quality management, the meaning of quality is meeting the customer's requirements. Designing a human activity system (like a restaurant) to meet those needs, even when the customer does not know his or her own needs, or cannot articulate them, requires an intelligent systems approach. An intelligent system with people, machines, and technology all performing their most appropriate role, will result in the customer's best experience. Traditional quality management approaches focus on defect rates and statistical measures of processes, but it's the values the people hold that are the ultimate drivers of quality products and results. The idea of personal values is a useful shorthand for the complex system in each person that drives behavior. A person's background, culture, upbringing, experiences, attitudes, personality, and preferences all contribute to the values from which the person operates. People may not even be aware of the values that drive their behavior – values can live in the unconscious.

Values can trump every other internal and external influence on a person. At any moment, a person can choose to operate from a value that may be completely new, compared to those from his or her background. But that's rare – most people are who they are, and who they are is seen in their values. Of course, people can also act contrary to their values, for good or for bad, at any time. Values can change over time, but they are not changed easily or quickly and cannot be taught directly, since they are not based in information, but in the internalization, over time, of concepts, principles, ideologies, and interpretations of the world. There is much about values in the Christian Bible for example, but simply reading it will not magically transform someone into someone with the values taught and illustrated. Values are what a person will hold onto, even in the face of dire circumstances and adversity, or while enjoying great fortune and success. It has been said that wealth does not change a person's values; it reveals and amplifies them, meaning that the values someone has when they are not wealthy, will be, for the most part, the values they have after gaining great wealth. While values are vitally important to the productivity and effectiveness of employees, they are very difficult to assess during the interview and hiring process. It's pretty useless to ask a job seeker about his or her values directly; even sincere answers can be distorted by a person's limited awareness of their own values. Values are only seen in a person's behavior in a variety of situations, over time.

While values cannot be trained into people and cannot be assessed quickly or directly, they can be cultivated. Creating a quality culture in an organization will over time cause people with those same values to stay, and others to leave. Company culture is not a matter of catchy vision statements or grandly stated strategies. To take a rather odd example, it is as if an organization developed a culture of physical fitness, with sports and exercise made available and encouraged, and most people embracing and adopting a fitness lifestyle. Those who don't care about fitness won't be attracted to the company in the first place, and those already there will become increasingly uncomfortable and will gradually find their way to an exit. In the same way, a quality culture supports the hiring, development, and retention of people committed to quality. A quality culture is a people system, powered by the values of the people.

While people systems are composed of only people, those people can be affected by subtle and not-so-subtle influences from sources that are not human. Years ago, I was in a meeting at a research facility in a company that makes heating and air conditioning systems. On the wall of the conference room was a device, that displayed a number that appeared to vary between about 500 and 900. Being engineers, we began to speculate about what this device might be reading, until someone informed us that it showed the carbon dioxide (CO_2) level in the room. Along with temperature and humidity, CO_2 level is an aspect of indoor air quality. High CO_2 levels, above 2000 parts per million (ppm) are known to cause sleepiness and headaches, and levels above 40 000 ppm can cause coma and death. We became intrigued about whether the variations we were seeing would have any effect on us. I started to notice that when the levels crept up into the 800+ range, the meeting began to sort of slow down, with people becoming less creative, less interested, less attentive, and less vocal. There might have been an occasional yawn, but no one was actually becoming sleepy, which would have been more noticeable. The effect was subtle. The room was on the outside periphery of the research complex, so it was easy to open the door and let in some fresh air or step out for a break or a walk. Bringing outside air into a room is the only practical way to reduce CO_2 concentration – outside air is typically 350–450 ppm of CO_2. I remember at one point, the meeting had slowed to a crawl, and with the CO_2 meter indicating about 750, someone suggested going for a walk. We opened the door, to set out on our walk, and opened a window or two for good measure. When we returned a few minutes later, the CO_2 level had dropped to about 450 from the influx of fresh air, and we were all revitalized. Was it the few minutes in nature, or was it simply the CO_2 levels in the room that affected our moods and intellectual productivity?

I became intrigued with the subject of CO_2 and productivity and found that there were few if any studies that examined the effect of moderate CO_2 concentrations on worker attitude or productivity. I bought myself a portable CO_2 meter, and began to take it to meetings I attended, often placing it in the middle of the conference room table. Sure enough, when it began to rise about 750 or so, the meeting pace slowed, with creativity and interest in the subject dropping as well. It was as if people just didn't care much anymore about the work. If we had the good fortune to be in a room with windows that could open (a rarity in US offices and conference rooms) I would quietly open them and watch the effect. In a few minutes, the CO_2 level dropped a few hundred points and people were suddenly more interested, creative, and verbal. If the windows didn't open, even opening the conference room door and exchanging air with the hallway was of some help. Of course, CO_2 will rise most in a small room with a tightly packed group of large mammals (the people) breathing up the oxygen and exhaling CO_2. The effects were subtle – no one is falling asleep or even noticing that the mood of the meeting had changed. If someone became uninterested in the subject, both the person and the others in the meeting would conclude that, in fact, the person wasn't interested – but the disinterest could be physiological, not psychological.

In the years since I began my very informal study of CO_2 and behavioral effects, I have come to wonder if the industrialized world, especially in the United States, but also other steel- and glass-based cities around the world, is losing a potentially unfathomable amount of productivity due to moderately high CO_2 levels in work areas. Many new buildings in the United States and elsewhere are sealed; windows do not open and fresh air exchange with the outside may be limited or nonexistent. In cold weather, even windows that can be opened tend to remain closed. Closed office spaces, especially where there are many people in a small area, naturally generate higher CO_2 levels. Ironically, the very occasions where we want people to be at the sharpest and most creative, in meetings and workshops, are held in rooms likely to have the worst CO_2 concentrations. Do people hate meetings because they are boring and lifeless, or is it the CO_2? Most people relish the idea of an outdoor meeting. Is it because of the inspiration of being in nature or the wonderfully low CO_2 levels in outdoor air? Does taking a break in a meeting reinvigorate the participants because they needed a mental break, or because the CO_2 levels were getting high and the room needs to be opened to exchange the air and let people find some new air to breathe in the hallway or outside?

As systems thinkers, we are naturally led to ask why this CO_2 foggy brain phenomenon, if valid, isn't studied more. Part of the answer is degree. Indoor air quality regulations studies tend to require that CO_2 levels be kept below 5000 ppm, since it is established that negative human health effects result from levels above about 7000 ppm. Researchers tend to ignore potentially far more subtle effects of what are, to them, CO_2 levels that are well down into the "safe" range. It is similar to when physicians and public health specialists study overweight issues in a society. They would tend to ignore people who are less seriously overweight. The medical definition of overweight for a 6′ tall man is a weight over 205 lb, but to meet US Army standards, that 6′ tall man can weigh no more than 190 lb, if under age 20. In that range between what the doctors consider overweight and what the US Army considers fit for duty, is a range of 15 lb. Does the presence or absence of that 15 lb, or some part of it make a difference in the health, appearance, self-image, and general fitness of a person? Many people would think that it does. While our 6′ tall man may not be overweight by medical standards, he may feel some significant negative effects from that extra 8, 12, or 15 lb. (My personal theory on the ideal weight for a person is the weight the person would be if they were exercising hard for over 30 minutes, at least four times per week, and eating sensibly, for at least six months.) Since medical standards count the 6′ tall, 200 lb man as a healthy weight, little research is done into the effects of the extra weight below the healthy threshold. It's the same for CO_2. Research focuses on CO_2 levels much higher than most people experience in their homes and offices, so the effects of CO_2 levels between say 700 and 2000 receive little attention. The question of whether moderately high CO_2 levels are stealing our attention, creativity, and productivity may be something that deserves more research.

10.10 Job Hunting as a People System

Earlier we mentioned the hiring process as an example of a people system and noted how it's a mistake to assume that an entire large corporation has rejected a job seeker, simply because one rejection letter was received. It's an interest of mine, as I've been teaching job hunting skills for a few decades, as a sideline. I call my approach job stalking, since it's more akin to the way a hunter goes after prey with a bow and arrow, than the way a fisherman goes after fish with a net. The specific methods that work best in job hunting have changed little, even with the advance of technology. Here's a summary of the job stalking approach, along with comments on how this approach is based in a people systems view of the hiring process.

10.10.1 Who Are You?

Ultimately, if you are seeking a job, you are seeking to become a part of the people system which is the hiring company or organization – but what kind of part? In this first step, you explore yourself, professionally, trying to understand yourself the way a hiring manager would. By looking closely at your skills, professional background, and other qualifications, you form a description of yourself that is as close as possible to the way a professional recruiter would describe you to another professional recruiter. This description generally won't be the specific job title you hold (or held) in any particular company, nor will it be a generic category of worker. *Electrical engineer* is far too broad and general, but *Director of Mission Systems Avionics* is too specific to a company and doesn't reveal who you are across the field of job seekers. Realize that terms may have changed since you last tried to describe yourself. In past years, you might have described yourself as a statistician, with a focus on manufacturing defect rates and process improvement. Today you might revise that to *data scientist, with a focus on manufacturing quality control and assurance*. The words you use should match the words that hiring managers would use in a job description – it's *who you are* in today's job description terms, so you may have to update your language. When you complete this step, you will know how a potential hiring manager will view you when you arrive on his or her desk in the form of a resume.

10.10.2 What Do You Want to Do?

For some people who have had a smooth career progression in a single field and specialty, understanding who you are in the job market leads directly and specifically to what you want to do. For others with a more diverse set of professional experiences, it may take some thinking to determine what you want to do. Like the *who are you* question, this one should be answered in the terms of the current job market. You may have held positions known as *business analyst* or *applied statistician* before, but perhaps you now see that you want to be what is currently called a data scientist. Your target job should be based on your existing skills and abilities, and the positions you have held in the past, but also on your current interests and career goals. It can be a step beyond your past, but it is probably impractical to make a giant leap beyond your current career trajectory. Thinking in people systems terms, ask yourself how the person making the hiring decision, would describe the person they want to hire. You want to be able to see a clear alignment between what the hiring manager wants, what you are, and what you want.

Cal Newport's brilliant *So Good They Can't Ignore You*, describes it as *adjacency*. The place of best opportunity for a job seeker who wants to expand and grow a career, is to look in an adjacent area, not a far-flung one. In my career, I worked in software engineering, and then made an adjacent jump to systems engineering, the bridge being model based systems engineering, which had its origins in object-oriented software methods. Then later, as I became interested in artificial intelligence, I worked my way into roles where systems engineering was trying to use artificial intelligence methods, and then later into a pure artificial intelligence research position. Each step was a small one. Another way to think about this is that if you want to change your career, consider changing either your job type or your industry, but not both at the same time. If you're an accountant in the automotive industry and would love to work in finance in the entertainment industry, that could be a leap too far. First, either move to finance in the automotive industry, or move to accounting in the entertainment industry, and then later make the step to finance in entertainment. Hiring managers will respect someone making a one-dimensional jump, and even see potential benefit: can you help bring the detailed rigor of accounting in the automotive industry to the entertainment company? But not many may believe you have the background to make a two-dimensional jump, without demonstrated expertise in both the new profession and the new industry.

This step can involve a lot of research. You can make good use of online job listing sites – not as a way to get a job, since applying to jobs online is often a way to be screened out by human resources, or worse, a computer algorithm, but as a research tool. Be searching and reading job descriptions, you can get a good, general view of what companies are looking for, what skills they list and the terms they use to describe them. You can discover companies you had never heard of, and then research those companies. You can find job titles in a company, then search those titles on LinkedIn, finding people in those jobs and reading about their backgrounds. You might learn that *principal investigator* is a title used in academic and industry research and that most people in that kind of role have a PhD in their field. That's useful to know. You might find that systems engineers are usually engineers with an undergraduate degree in another kind of engineering, and that some have a master's degree in systems engineering. The job stalking process I'm describing here hasn't changed at all since I began to develop it in the early 1980s, but the tools available to carry it out are so much better than the pre-Internet era had to offer.

10.10.3   For Whom?

Now that you know what you want to do, and you have described it in current job market terms, it's time to figure out for whom you want to do that work. Most professional skillsets can be applied in many jobs, in different industries, and in various types of companies. Since it is up to the job seeker to go after the opportunities that may exist, targets must be chosen. While it might seem that the best approach is to "stay open" and try to get any suitable job, and this attitude can be fueled by increasing desperation or financial pressure on the part of the job seeker, it is also counterproductive when the people systems viewpoint is considered. A hiring manager would rather hire someone who is closely matched to the specific position being hired, rather than someone who seems to be a general sort of match for a variety of positions. A hiring manager looking for a materials engineer to work on composites for aviation systems, would much prefer someone with that precise interest, and qualifications to match, over a general materials engineer who seems to be looking for any job in materials engineering in any industry.

Answering the *for whom?* question includes considering the industry and type of company within which you wish to work, as well as the required work location (or perhaps the lack of one in the case of remote work). The people system of a large corporation differs from that of a medium sized company of a few hundred people, or that of a small company, of a few dozen or less. Large corporations have the likely (but not guaranteed) advantage of job stability, in that a large company is not likely to simply go out of business, and they tend not to fire people quickly or capriciously. Large companies do experience downsizing and layoffs, but probably less that the average failure rate of small companies. Companies both large and small can be acquired by still larger companies, and this is not necessarily a negative impact to most of the employees. All that said, in the many-jobs style of career progress prevalent in recent decades, the idea of staying with one company, no matter how large, may not be the biggest advantage of joining a large company. Biased by my own experience, I view large companies, assuming they are even moderately successful, as offer lots of opportunities. If there are thousands of people in the company doing your particular kind of work, there are likely many opportunities to move both across and up in the organization. Getting a new job inside your current company is much like job hunting externally, but is much easier, since you are already "in the family." On the lighter side, if there is someone you don't like or don't gel with in your part of a large company, there is enough movement that eventually either you or your nemesis will be reorganized or promoted into a different group before long.

Small companies and startups offer their own advantages. Being in on the ground floor of building something new can be exhilarating. Jobs may be wider in scope and may provide more opportunities to learn and grow. In a small company, it's easy to see how your work contributes directly to the success of the company. Size of an organization is more than just a number; it may mean a different kind of people system. I tend to think of joining a small company as more like marrying into a family, while working in a large company is more like joining a local softball team. You may like your teammates, and play hard for the team, but you are free to leave, or choose another team in the league. In a small company, you are married to them – leaving can feel more like a divorce. Some people like this close connection – others prefer more freedom.

Once you focus on exactly what you want to do and for whom, based on who you are, you can refine your resume to show clearly how you are perfect match for your chosen job. The resume, which is not a person and thus not part of the people system, is best seen as a part of the communication between you and the hiring manager. Resumes are not some kind of literal history of your career – they are more like a movie about a person's life. The two-hour movie doesn't show all fifty years of life. Highlights are chosen to best exemplify the person's life according to the director's intent for the movie. In the same way, you can craft your resume to make the best possible case for your new job target – in speech act terms, you are making an offer. Resumes are more flexible than some job seekers assume. There are no laws about what you can and cannot put on a resume. Of course, you must be truthful in what you state, but nothing says you must include all jobs you've ever had (though be ready to explain any long gaps), or that you have to reveal your age by showing your first job after college graduation, or your graduation year. Resume job descriptions can't possibly describe everything you did in a job, so emphasize the aspects of past jobs that most qualify you for your target job. If the job title you had doesn't relate well to the job you seek, rephrase it, or state your official job title and then add more descriptive terms. Official job titles are often generic and undescriptive. Perhaps you were a Senior Software Engineer by title, but a more accurate and descriptive label might be Senior Machine Learning and Image Recognition Software Engineer, assuming that matches your job target. Or if your job target now is to move more into reinforcement learning, you might state the job as Senior AI Software Engineer, and then emphasize the reinforcement learning knowledge you have.

10.10.4 Pick a Few

In this step, you refine your thinking in the *for whom* step, and choose specific company targets, making a list of about 10 to start. You'll find these companies by reputation, but also through your knowledge and active participation in the people system that is the industry. If you are targeting aerospace firms, you should be reading Aviation Week and going to AIAA conferences. There, you are finding out who is doing interesting work and what companies seem to have the culture and philosophy you like. Your targets will be specific like *radar avionics design engineer for Boeing*, or *accounting manager for MediaMonger*, a high-tech social media startup. List your 10 targets and then go on to the next step.

10.10.5 Go Straight to the Hiring Manager

In this step, you will make a direct approach to people who can hire you for the job you seek. Doing this is not the way most job seekers proceed. First, it must be noted, personnel, human resources people, and internal recruiters do not hire anyone, except of course, to work in the human resources department. They do not make the hiring decisions, but they do make decisions to screen people out. A core principle of negotiation, which I first learned from Roger Dawson, is that you should never negotiate with someone who can only say no, but who can't say yes. Human resources people can say no, and screen you out by simply not passing you on to anyone who can hire you. They are people who can only say no to you, so there's no point in approaching them first. It is much better to find and approach the person who can hire you for the job you seek. That's how the hiring system works – people hire people, generally to work in their own part of the company. The hiring manager for an engineering job is probably the engineering manager or director to whom the new engineer would report, or a level or two above that. Job postings do not generally reveal the name of the hiring manager, so most job seekers, simply apply for the job, end up on a human resources person's desk, and get screened out, since they are only one resume in a large pile. Technologies like online job listing sites make it easy to run your resume across many human resources desks and be screened out at record rates. The better way is to go straight to the person who can make the hiring decision. Technologies like social media, LinkedIn, and web searching can help you locate the person most likely to be the hiring manager for each job on your target list.

I have referred to this job hunting approach as job stalking, and this is where the stalking part comes in. Once you identify the person who can hire you, you study the person, read about him or her, and read what he or she has written. You try to come up with ways you might connect with the person, how you might make the approach and what would get the conversation going. A great start is to formulate a question that shows your knowledge of the industry and of the company, and which is also sincere – it's something you really want to know. In my very first job hunt after university graduation, my go-to question for hiring managers was, do you look for engineers or for business people in your computer sales groups? I was an engineer looking to go into computer sales, and so wanted to know how they approach this question, which was a matter highly debated in the industry at that time, so it always brought interesting answers. As the conversation continued, eventually the hiring manager I was speaking with would inquire about why I was calling and asking. That was my opportunity to explain that I had just come out of the university and was looking for my next opportunity and was looking to see if the company was a good fit. The response was often an invitation to send in a resume to the manage with whom I'm speaking. In making this approach it doesn't matter whether company has posted job openings for what you want to do or not. A great paradox in the people system of hiring is that posted openings are often jobs that are not really available, since they are already filled unofficially, or may be generic and not a specific position, while many available jobs are not posted, since they are still being formalized, or because they exist only as a general hiring goal. A growing or even just a stable company is always looking for good people and can often create a job when the right person comes along. Even if there isn't a position for you at the time you make the direct approach, you have made a good connection in the company, who may pass you on to others, or think of you when a possible position does come up, say when someone announces they are leaving, or when a group needs to be expanded.

In the pre-Internet days when this process was developed, there was little choice in how to go straight to a hiring manager. You found out their name by talking with others at the company, and then simply placed a call. My favorite opening was something like, "Hi Jim, this is Barclay Brown from Atlanta and I wonder if you can help me today. I'm trying to find out if you look for engineers or business people in your computer sales groups." I practiced until I could get that out in one breath, not leaving Jim any room to interrupt or think I was calling to sell something. You may be thinking – can I really call a hiring manager by his or her first name in a situation like this. The answer depends on the conventions in your industry and region. If I would call Jim by his first name *after* I'm hired, then I can call him by his first name *before* I'm hired. This live telephone call approach still works, but technology offers other ways that can also work. You can use email or an introductory message with a connection request on LinkedIn, though it's best to keep it very short and try to interest the person in talking about the subject you introduce. You don't want to come across as, "I'm Barclay and I'm looking for a job" – that will usually get you redirected to human resources.

Many people think that networking is the key to job hunting success. Networking of the kind people tend to mean, which involves random connections that your close friends and associates may know, is just that – random. There are always many more people I don't know than people I and my friends know. If professional success is about whom you know, then simply get to know more people – you're not stuck with whom you know *now*. I may use my professional connections to find out more about a company, or maybe to get a name, but expecting someone to "get you into" a company is usually ineffective. I've sometimes been approached by a stranger, or an associate of an associate, for help in getting the person into a company I work for. The person is in effect asking me to go to a hiring manager

I know in my company and say, "Hey Emily, this guy just messaged me on the Internet. I have never met them and I have no knowledge of their work or their background, though their resume looks like it could be a fit for a position here. Do you want to talk to them?" It's a long shot at best. People who have not worked with you can't credibly recommend you for a job – doing so could actually harm their own credibility. As a job seeker, you even risk it backfiring, if your friend is not well regarded in the company, his or her recommendation could even be a negative. I've always preferred to make my own approaches. That said, I do follow up on opportunities that come my way. I was once at a social gathering and while chatting with a friend about my current job search, she said that I'd be perfect for this new fast-growing company in town, and she would put me in touch with a manager she knew there. I thought, long shot. But I followed up and was hired by the company in a great job almost immediately. Blue birds like that can come your way, but you can't count on them as part of your job stalking strategy.

10.10.6 Follow Through

As you approach your first ten targets, be sure to follow through with each one after your initial conversation. Of course, send the resume, with a tailored cover letter recapping the conversation and asking for an interview. If the hiring manager asked that you send a resume or application through the normal human resources channels, do that, but follow up with the manger as well. If there's no immediate position for you in the manager's sphere of influence, but you feel the interaction is positive, follow up regularly afterward, perhaps approximately monthly, not with "do you have a job for me yet?" but in a way that develops the connection. Send an interesting article in an industry publication that relates to your mutual interests. Ask if the manager is coming to the next industry conference and if he or she is speaking so you can be there to listen.

I once stalked my way into a 17-year career at IBM with this approach. IBM had just acquired Rational Software and I wanted to join IBM and work in the Rational Software area. I had no connections and knew no one in there, but I hunted around and found that Murray Cantor was one of the leading thinkers in the company in the area of systems engineering and the application of Rational Software methods. I read articles he wrote, and eventually wrote to him, with a question or two about his approaches and his work. I didn't hide that I was looking for a new opportunity, but the focus was on his work and my interest in it. Of course, I was not at the top of his professional priority list, so sometimes I would get a reply and sometimes not, which was no problem for me. I realized we would both be at the next big conference, and of course he was speaking. I sent him a note to see if we might meet at the conference for a brief conversation. He was willing and asked if I could send a resume. Of course I had sent him my resume several times, but again, I didn't expect to be at the top of his priority list, so no problem there. But there was a problem now. We were both already at the conference, and I wanted him to have a hard copy of my resume, since this was in the days before smartphones. It took some work, but I found a way in the hotel to get my resume printed (this was also before there were handy printers in hotel lobbies) and then I asked the hotel to deliver it to his room, which they did. When I met up with him later in the conference, he had the resume in his pocket and his first comment was, "your timing may be perfect – we just offered a position to someone like you, and he turned us down." The process I had begun many months earlier paid off and I spent 17 fulfilling and successful years with the company.

10.10.7 Broaden Your View

You've chosen ten targets and approached them and are maintaining contact with the people you've met, but you still haven't found a job. The first thing to do, is simply choose 10 more targets and go after them. But after a few rounds of this, if you can't seem to get traction with your targets, or you run out of targets, you may need to broaden your view a bit. Maybe all your targets so far have been the largest companies in your industry. Perhaps go for some of the smaller companies in the same business area. Or if you were focusing on only one type of your position based on your skills and interest, perhaps broaden that a bit. What optimizes the entire process is that targets are broadened only after you work your top targets.

10.10.8 Step Two

After broadening your view, go back to step two, what do you want to do, and cycle through the process again. Observant readers may have noticed that the steps, when abbreviated, rhyme a bit, making the sequence easier to remember: who you, what do, for whom, pick few, straight to, follow through, broaden view, step two.

Your career is a people system, unless perhaps you are a solo wood carver living on an otherwise deserted island, and even then, you have to sell and ship your product. What you call your career is a set of connections you have with people – people who hire you, buy from you, supply you with raw materials, read your books, and use your work. Making this system as functional and enjoyable as possible is an important component to a happy and fulfilling life, and each job hunting opportunity is a chance to improve that system. Sometimes you seek out those opportunities and sometimes they find you. I mentioned how I got the job at IBM earlier. What I didn't mention was that the only reason I was looking for a job was that I had just been fired from another job. That's right – I said fired. Not let go, not downsized – fired. The reason given was the look on my face. I was fired for the look on my face. This little company had an IT department that had a lot of problems. I was brought in as a project manager to help straighten things out, but it turned out no one wanted my input or my help on how to do things better. They continued in their ways and my disappointment and disapproval apparently showed on my face in enough situations that I was fired for it. Months later the entire department and its management were also fired. Being fired was the occasion for the beginning of an amazing career in IBM, which focused on me teaching state-of-the-art methods for software systems development to companies all over the world. Ironic. Poetic. So don't worry about being fired – it's simply a signal that you are in the wrong job. Use the chance to go find the right job.

Before you consider broadening your view, make sure you are going for the jobs you really want. In many cases, I've talked to people who are looking for a job that's very similar to the one they just left, but who would really like to be doing something else. They often assume that they can't get that other job, because they lack the education, the experience, or some other reason. My very first question when beginning to coach someone on job hunting is, "what is your dream job"? I don't mean a fantasy job, like being a rock star or an astronaut, unless you have the background to step into one of those jobs – I mean a realistic, feasible job that you would love and that you know you can do well. Start there; see if there's a way to go after those jobs and see if you can get one. Sometimes I talk with someone who is thinking about going after another university degree, enabling them to get a better or different kind of job. I'm a huge fan of formal education – I think it opens doors, broadens awareness, increases confidence, adds credibility, and generally results in much higher earnings over a career. It may be however, that you can get the job you are after without the degree, or at least well before you complete the new degree. I believe it was Richard Bolles, author of *What Color Is Your Parachute* – a classic on job hunting, who said that the best jobs do not go to the most qualified people; they go to the people who know how to get the best jobs.

At the risk of making this chapter too much about me and my career, I'll use myself as an example again. Somewhat late in my career I decided to focus on artificial intelligence as a new career direction. I began by thinking I should get a third master's degree, or maybe even a second PhD in order to qualify me in that field. Doing that would have taken years and would have been a great adventure in itself, and I may still do it, but then I took my own advice and tried to see if I could get my job in AI without adding another degree. I took some online courses (I'm a huge fan of Andrew Ng and his deeplearning.ai programs, offered through Coursera.org) and even tried to take a few courses at Stanford, which I found to be beyond my abilities at the time. I used adjacency, as described earlier, looking to see how I could leverage my background in systems engineering and aerospace, and my growing knowledge of the relationship between AI and systems engineering, and see if I could find a job that would leverage that knowledge. I would try to move from a systems engineering job, where I was doing some AI, to an AI job where my knowledge of systems engineering would be an asset. As soon as I formed this idea, I began to take action on it, trolling around in the job postings of my company to see if anyone might be looking for AI and systems experience in the same job. To make the long story shorter, I found a listing in a different division of my same company, which led me to a group that ended up hiring me in exactly the kind of position I wanted. All it took was understanding who I was professionally, and what kind of moves I could make, with little bit of courage to give it a try.

10.11 Shared Service Monopolies

It may be unwise to attempt to solve a problem that few even recognize as a problem, but let's give it a try anyway. In large organizations, there are often complaints that this or that group doesn't provide good, responsive service, or is slow, inefficient, and puts up too many roadblocks to getting the job done. It could be an in-house recruiting function, shipping or marketing services, or an information technology (IT) department. These internal service providers are separated out and made into their own departments in order to unify the function and become more efficient due to economies of scale, or the department may have been created to bring a service in house that was once provided by an outside source, a consulting company or other provider. Leaders of the company reasoned that they could save money by creating an inhouse public relations department, for instance, instead of continuing to use an expensive outside PR firm.

Things start out well, but eventually the inhouse PR department's service declines, and others in the organization are tempted to hire outside PR help, a move which is forcefully resisted by the internal PR group. It seems that the internal service provider becomes more focused on its own identity as a department, perhaps aiming to grow its size and influence, and pays less attention to providing excellent service to the other parts of the organization.

If we look at the situation from a people systems perspective, it become clear what may be happening. When the company went outside for the needed service, there were many possible providers, each of whom had to compete for the firm's business. When the service was brought inhouse, the provider became a monopoly. Departments throughout the organization were required to use the inhouse service; after all it should save the company money. Monopolies, however, have a very specific way of behaving, and it's not a way that's great for their customers. Monopolies are monopolies because their customers have no choice, therefore they don't need to work to please their customers – their customers are forced to use them. Monopolies become inefficient, because they have no reason to be efficient for customers. Consider how your local cable company, almost always a monopoly in a single area, treats customers. Also notice how their behavior changes when a new competing provider, say a fiber optic service, comes to town and customers have a choice.

The internal group is no different when it becomes what we'll term a *shared service monopoly*. The service they provide is used by multiple groups, so there is no common management between the internal customer groups and the shared service group, and there is no choice – the shared service monopoly is the only game in town and must be used. A group that becomes a shared service monopoly will begin to exhibit the same characteristics as any monopoly, including less emphasis on customers and more emphasis on lowering costs to improve their bottom line, and if possible expand to offering more monopolistic services. Customers may perceive this as cutting corners or a reduction in service quality, but what matters to the shared service monopoly is lower costs, not better customer service. Better customer service levels won't help them – they are already a monopoly. From a people systems perspective, we see what's happening, but what kind of intervention might we try?

How about, if instead of creating a single service provider group inside an organization, we create two? Using IT as an example, say we create IT group A, and IT group B, both with similar capabilities and resources initially. Internal customers, when they need some software created or a new server provisioned, must choose either A or B. How would they choose? Perhaps they provide an RFQ (Request for Quote) to both groups, who respond with a quote for how many hours or dollars they will spend to fulfill the request. Cost isn't everything though; the two groups will also have to provide a time within which they will complete the project. As time goes on, say group B becomes more motivated to do better or faster work, and thus wins more internal jobs. As they succeed, they'll need more people to handle the workload, and with the increased internal "revenue" and momentum, may be able to entice some of the better performers in A to join them. B grows a bit, as A shrinks. Like any market competitor, A will respond by lowering prices, offering better or faster service, and working harder to win new business. If they start to prosper, they might be able to steal people back from B. The ultimate winner is the rest of the organization, who now have two internal groups competing for their business, each causing improvement in the other, just as competitive firms in a free market do. The small additional overhead of the selection process that each customer department must undergo, and the time A and B must spend bidding and quoting jobs, will be easily overcome by the increased productivity and efficiency of the micro-competitive market established inside the company. If A and B should fall into a complacent duopoly, say by diving types of work between the two, a move which would be which would be illegal collusion in the open market, then a disrupter, C might be allowed to enter the game, challenging both A and B to compete or risk losing resources. It's admittedly an unusual solution to a subtle problem, but it illustrates how thinking in people systems can explain what may otherwise remain mysterious.

In this chapter, we've introduced the idea of people systems, systems that are composed only of people, and how these systems are everywhere. Our lives consist of our participation in multiple people systems including our relationships, families, community, company, industry, bowling league, church, country, and many more. Successful participation in, or judicious withdrawal from these systems determines our happiness and fulfillment in life. It may not be an exaggeration to say that people systems are the most important things in life. While there may be other ways to think about how we participate in life with other people, using the concept of people systems can be illuminating. With a combination of insights into how people behave, based on their incentives, personality, and values, coupled with insights into how systems operate in general, much of the mystery of why the world operates as it does is unveiled and brought to light. Understanding how a key system like the job hunting and employment people system works, can make career progress easier, more productive, and more enjoyable. Life is a people system, waiting for you to optimize it.

References

Berne, E. (1968). *Games People Play: The psychology of Human Relationships*, vol. 2768. London: Penguin.

Descartes, R. (2013). *Meditations on First Philosophy*. Broadview Press.

Friedman, J. (n.d.). *The Grammar of Committed Action: Speaking That Brings Forth Being*. Retrieved from https://www.mentalhelp.net/blogs/the-grammar-of-committed-action-speaking-that-brings-forth-being (July 30, 2022).

Green, M. (2021). Speech Acts. *The Stanford Encyclopedia of Philosophy* (Fall 2021 Edition). Edward N. Zalta (ed.), retrieved from https://plato.stanford.edu/archives/fall2021/entries/speech-acts (July 30, 2022).

Myers, I.B. (1962). The Myers-Briggs Type Indicator: Manual (1962).

Index

*Engineering Intelligent Systems: Systems Engineering and Design with Artificial Intelligence,
Visual Modeling, and Systems Thinking,* First Edition. Barclay R. Brown.
© 2023 John Wiley & Sons, Inc. Published 2023 by John Wiley & Sons, Inc.

Printed and bound by CPI Group (UK) Ltd, Croydon, CR0 4YY

27/10/2024

14580472-0004